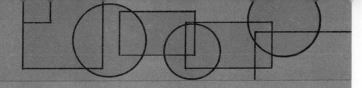

A. W. GOODMAN
J. S. RATTI

University of South Florida

finite
mathematics
with
applications

THE MACMILLAN COMPANY, NEW YORK
COLLIER-MACMILLAN LIMITED, LONDON

The Macmillan Company
866 Third Avenue, New York, New York 10022

Collier-Macmillan Canada, Ltd., Toronto, Ontario

Library of Congress catalog card number: 72-119146

PRINTING   78910   YEAR   3456789

preface

The concept of a course in finite mathematics is now well established. The guiding principle is to present to the reader a slice of mathematics that is interesting, meaningful, and useful and that at the same time does not involve the calculus.

With these limitations the content of a text for the course is almost uniquely determined, and we arrive at the usual topics:

Group I—Theory
    A. Logic
    B. Sets
    C. Combinatorial
        analysis
    D. Probability
    E. Vectors and matrices

Group II—Applications
    A. Linear Programming
    B. Game Theory
    C. Applications to
        social sciences
    D. Graph theory

In this text these nine topics are arranged into twelve chapters. Further, we include three appendices and an introductory chapter which cover supporting material. The table of contents gives the full details of the organization.

It is impossible for the authors to present a universal course outline because of the natural variations in the local calendars, previous preparation of the students, and objectives of the course. However, the following comments may prove helpful.

The text has 74 exercises (including the 6 in the appendices). Consequently the entire book can be covered by a class meeting three times a week, either for two semesters or for three quarters. If the time allotted is significantly less (one semester or two quarters), then some subset of the material must be selected.

The two chapters on logic and sets are not absolutely essential for the remainder of the book, and the authors feel that these two chapters can safely (but not wisely) be omitted. These two topics (logic and sets) are interesting in their own right and serve well to train the student to think clearly. If logic and sets are to be omitted, the teacher should seriously consider covering Sections 6 and 7 of Chapter 2 ("The number of elements in a set" and "Tree diagrams") since these items form a natural introduction to the combinatorial analysis in Chapters 3 and 4. If there is time to cover only one of the first two chapters, then the authors recommend that the chapter on logic be omitted.

It is not intended that the teacher cover the three appendices. This material is included merely for the comfort and convenience of the reader. However, the teacher should feel free to incorporate these appendices into his lectures if he feels there is a need.

The following table may be useful as a guide to those who are planning a course based on this text.

Time allotted	Chapters to be omitted	Number of exercises to be covered
Two quarters Three credits	1 and 2	50
	or 9, 10, 11, and 12	53
One semester Four credits	1 and 2	50
	or 9, 10, and 11	58
One semester Three credits	1, 2, 11, and 12	43
	or 7 through 12	41
One quarter Four credits	6 through 12	35
	or 1, 2, and 9 through 12	35

We follow the usual custom of placing a star (★) on those sections which may be omitted without loss of continuity. However, a competent teacher will certainly make his own selection. We also use a system of stars to mark those problems which may be omitted without loss of continuity or which are relatively difficult. The following diagram gives the relative dependence of the various chapters.

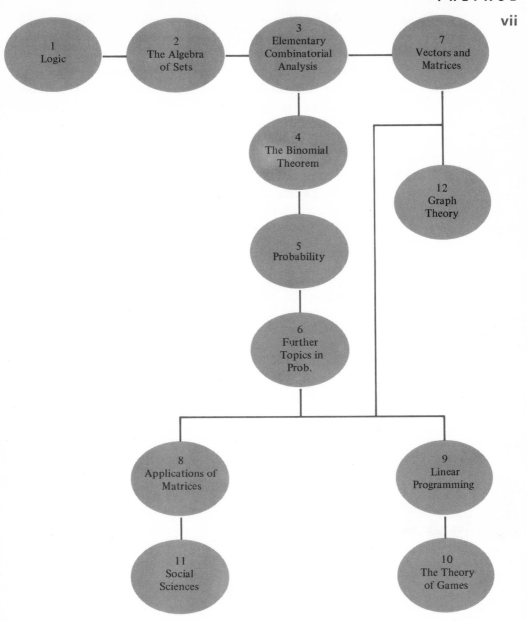

Custom also dictates that answers be supplied only to the odd-numbered problems. Here we respect this well-established practice, but we do not follow it slavishly. Answers to all the odd-numbered problems are indeed supplied, but wherever we feel that there are good reasons, we also give the answer to an even-numbered problem. As a result, answers are supplied in the text for approximately two-thirds of the problems, and answers for the complementary set are available (as usual) in a supplementary pamphlet.

We are grateful for the help of many of our friends and colleagues, and we take pleasure in acknowledging our indebtedness.

Many suggestions for improvements were supplied by Professor M. P. Fobes (College of Wooster), Professor William Ziemer (Indiana University), Professor Klaus Witz (University of Illinois), and Professor Bernard Kolman (Drexel University). The two chapters on probability were carefully examined by Professor Bernard Lindgren (University of Minnesota) and the chapter on applications to the social sciences was reviewed by Professor Gordon Bower (Stanford University), Professor Douglas Nelson (University of South Florida), and Professor G. H. Mellish (University of South Florida). With the able assistance of these kind critics, we managed to remove many inaccuracies and to clean up some (we even hope all) of the obscure writing.

We are also indebted to our very capable student Miss Cynthia Strong (now Mrs. Norman Mansour) who worked every problem in the text and corrected our many slips.

Finally, our thanks also go to Mrs. Janelle Fortson for her cheerfulness and accuracy in typing the entire manuscript, together with numerous revisions.

Tampa, Florida

A. W. G.
J. S. R.

contents

2

the algebra of sets   39

3

elementary combinatorial analysis   79

4

the binomial theorem and
related topics   95

5
probability   123

6
further topics in probability   171

7
vectors and matrices   207

appendix

1

functions and function notation   **400**

appendix

2

inequalities   **412**

appendix

3

mathematical induction   **424**

CONTENTS

finite mathematics with applications

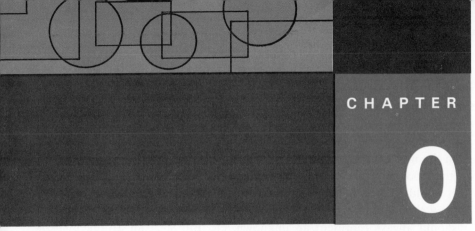

preliminaries

The student who is about to start a solid course in finite mathematics should be familiar with a large portion of elementary mathematics.

We resist the temptation to make a brief list of these elementary items for review. First, such a list is usually dull and boring, second it should be unnecessary, and third the list (together with review problems) is never brief.

Our objective here is to touch on just a few selected points that may have been slighted, omitted, or misunderstood.

1. mathematical symbols

The reader is already familiar with the more common mathematical symbols such as $+$, $-$, $=$, etc. To appreciate the advantages offered by good symbolism, one might compare the statement "For every pair of real numbers x and y

$$(1) \qquad x + y = y + x"$$

with the nonsymbolic form: "For every pair of real numbers, the sum is independent of the order in which the addition of the two numbers is performed."

The energetic reader (who still doubts the advantages of symbols) should try to restate, without using symbols, the problem "Solve the equation $x^2 - 7x + 10 = 0$."

Certainly the use of symbols shortens the labor of writing and speeds the thinking process. One might deduce from this that the more symbols we have available, the better off we are. To some extent this is true, but we cannot introduce 50 new symbols all at once. The new symbols must be introduced gradually, and with each new symbol the student must be given ample time to practice using it, and to master it, before being confronted with the next one. For the reader's convenience, a list of the symbols used in this book is given just before the index.

It would be nice if each symbol represented a unique element or concept. Thus A should always represent area, C should always represent a constant, x should always represent an unknown, etc. But such an arrangement is impossible, because there are many more concepts in mathematics than there are symbols. Some symbols must bear the burden of being employed quite often with a variety of meanings. Here the best rule we can make is this: In any particular problem or proof, do not use one symbol with two different meanings.

2.   subscripts and superscripts

Suppose that we have a problem involving the areas of four triangles (the exact nature of the problem is unimportant). It is natural to use the letter A to represent the area of the first triangle and to use a for the area of the second triangle. For the third area we might use α (Greek letter alpha) because it corresponds to the English a. But now we are stuck for a suitable choice of a symbol for the area of the fourth triangle. The way out is quite simple. We return to the letter A and put little numbers called *subscripts* just below the letter, thus: A_1, A_2, A_3, A_4, and use these to represent the areas of the four triangles. These symbols are read A sub-one, A sub-two, etc. If we are in a hurry we may say A-one, A-two, etc. Clearly the device of adding subscripts greatly enlarges the number of symbols available for our use.

We can also use *superscripts*. Thus we might write $A^{(1)}$, $A^{(2)}$, $A^{(3)}$, and $A^{(4)}$ to denote the altitudes of the triangles. Here we want to avoid A_1, because presumably in the problem at hand it has already been assigned the meaning of area. The symbols $A^{(1)}$, $A^{(2)}$, . . . are read A upper-one, A upper-two, etc. The superscripts are enclosed in parentheses to distinguish them from powers. Thus $A^{(2)} \neq A^2$, because the latter is AA or A squared, whereas $A^{(2)}$ is merely a symbol; in this example it is the symbol for the altitude of the second triangle.

A subscript or superscript is also called an *index*. The indices may be represented by a letter such as k. As a matter of shorthand, we can indicate the four altitudes $A^{(1)}$, $A^{(2)}$, $A^{(3)}$, $A^{(4)}$ by merely writing $A^{(k)}$ ($k = 1,2,3,4$).

3.   the three dots notation

In Appendix 3 on mathematical induction we shall prove the beautiful formula

(2) $$1 + 3 + 5 + \cdots + (2n - 1) = n^2.$$

But to follow the proof, one must *first understand* what the formula means. Many students do not understand the three dots ($\cdots$) which appear in equation (2). These dots mean that the terms proceed in the manner indicated by the first few already given. For example, if $n = 6$, then $2n - 1 = 11$, and equation (2) states that

$$1 + 3 + 5 + 7 + 9 + 11 = 6^2.$$

In this case the three dots represent $7 + 9$. But if $n = 10$, then equation (2) states that

$$1 + 3 + 5 + 7 + 9 + 11 + 13 + 15 + 17 + 19 = 10^2.$$

and in this case the three dots represent

$$7 + 9 + 11 + 13 + 15 + 17.$$

The three dots do not always mean that terms are *missing*. If $n = 3$, then $1 + 3 + 5 + \cdots + (2n - 1)$ *means* $1 + 3 + 5$. If $n = 2$, this same expression *means* $1 + 3$. If $n = 1$, the left-hand side of equation (2) has only *one* term, namely 1.

Why do we use the three dots? In the first place it is shorter. Thus if $2n - 1 = 99$ in equation (2), we would waste considerable time in writing all of the terms. If $2n - 1 = 9,999,999$ in equation (2), it would take about 58 days to write all of the terms of the sum, writing at the rate of one number per second and not stopping to eat or sleep.

Further, the three dots notation is a necessity, because our assertion is that equation (2) is true for any n. It would be impossible to express this idea properly if we did not have some notation such as the three dots.

The three dots notation is not restricted to sums or numbers. It may be used to indicate a sequence

$$x_1, x_2, x_3, \ldots, x_n,$$

where the notation merely means that we have n elements in some particular order and that the kth element in the sequence is x_k.

3

The three dots can also be used for products. For example, either

$$P = 1 \cdot 2 \cdot 3 \cdot \cdots \cdot n$$

or

$$P = 1 \times 2 \times 3 \times \cdots \times n$$

means that P is the product of all the integers from 1 to n inclusive. The symbol $n!$ (read "n factorial") is always attached to this product (for further details see Chapter 3).

4.   what is finite mathematics?

Roughly speaking, finite mathematics consists of those parts of mathematics that are concerned with finite sets. Perhaps the reader is now comfortably waiting for the definition of a finite set, but here he may be disappointed. Of course the idea of a finite set is certainly easy, and we all have a strong intuitive understanding of this concept. But the precise definition is not simple, and we prefer to avoid it because it is never really needed in this book. For the reader who absolutely insists we offer

> **Definition 1. Finite Set.**   A set is said to be finite if it cannot be put into one-to-one correspondence with a proper subset of itself. A set that is not finite is said to be infinite.

We shall have much more to say about finite sets in Chapter 2, and, indeed, finite sets form the underlying theme of the book. However, we shall often use infinite sets in our examples and problems. The most important of these infinite sets are

(3) $N = \{1,2,3,4,5,\ldots\},$

the set of all natural numbers;

(4) $Z = \{0,\pm1,\pm2,\pm3,\ldots\},$

the set of all integers; and

(5) $P = \{2,3,5,7,11,\ldots\},$

the set of all primes (see Section 5).

It may seem improper to introduce sets with *infinitely* many elements in a book called Finite Mathematics. However, it is very convenient to use familiar objects for our examples, and everyone is familiar with the set Z of all integers and many of its subsets such as N, P, D (the set of all odd integers), and E (the set of all even integers). The reader or teacher who is disturbed by the infinite set Z may change it to a finite set by considering only those integers that are less than 10^{10} and greater than -10^{10}.

The reader should note carefully the use of the three dots (. . .) in equations (3), (4), and (5). Here the situation is slightly different from that encountered in equation (2). In equation (2), the three dots represented a *finite* set of numbers to be included in the sum. In equations (3), (4), and (5), the same three dots represent an infinite set of numbers (and of course no addition is intended). In equation (2) we can list the last term of the sum, whereas a listing of the last element is not possible for the sets N, Z, and P.

5. prime numbers

A number n is said to be *composite* if $n = ab$, where a and b are both integers each different from 1. Thus $10 = 2 \cdot 5$ and $57 = 3 \cdot 19$ are both composite. An integer n, greater than 1, that is not composite is called a *prime*. The set P of all primes is indicated in equation (5) by listing the first five elements of the set.

The number 1 is also not a composite, and hence we might expect it to be included in the list of primes. However, the number 1 is exceptional, and there are good reasons for *not* calling it a prime. By general agreement it is excluded from P, and hence 2 is the smallest prime number.

6. the absolute value notation

It is convenient to have a notation which tells us to "throw away" the negative sign when x is negative. The symbol of this is $|x|$ (read "absolute value of x" or "numerical value of x"). For example $|3| = 3$, while $|-3| = 3$. Unfortunately, if we just look at the symbol x there is no way of telling whether it represents a positive number or a negative one. Therefore in a formal definition the phrase "throw away the negative sign" cannot be used. If we recall that $-(-3) = +3$, then the way out of our difficulty is easy.

Definition 2. If x is any number, then $|x|$ is defined as follows:

$$|x| = x, \qquad \text{if } x \geq 0 \text{ (if } x \text{ is positive or zero),}$$
$$|x| = -x, \qquad \text{if } x < 0 \text{ (if } x \text{ is negative).}$$

According to this definition, $|13| = 13$, $|0| = 0$, and $|-7| = -(-7) = 7$. On the other hand, there is no way to simplify expressions such as

$$|a - b|, \qquad |a - b + c - d|, \qquad \text{and} \qquad \left| \frac{x + y - z}{-x + y - z} \right|$$

without further information about the variables involved. Between these extremes we may cite

$$q = |\sqrt{23} - \sqrt{57} + \sqrt{159} - \sqrt{97}|.$$

Of course q is positive, but it will take some effort to determine whether

$$q = \sqrt{23} - \sqrt{57} + \sqrt{159} - \sqrt{97}$$

or

$$q = -\sqrt{23} + \sqrt{57} - \sqrt{159} + \sqrt{97}.$$

7. the proof is completed

It is convenient to have a mark to signal the end of a proof. Thus if the reader has trouble understanding the proof, he can at least locate the place where the proof is completed and then reread the proof until it does become clear. In the past this place was often indicated by the letters Q.E.D., which abbreviate "Quod Erat Demonstrandum," the Latin phrase for "which was to be demonstrated." In recent times it has become the custom to use the symbol ▮ with exactly the same meaning. In this book we will use ▮.

The symbol ▮ is called the Halmos symbol, after Professor Paul Halmos, the man who first introduced it.

8. more minor items

There are still other preliminaries not yet touched. Among these are (1) functions and function notation, (2) inequalities, and (3) mathematical induction. These are a little too complex to be dismissed with a brief paragraph, and yet the reader should be somewhat familiar with these three topics. Our compromise is to cover them in some detail in three appendices. These sections at the end of the book are not to be regarded as a proper part of the finite mathematics text but are included only for review.

We urge the reader not to hesitate longer on the periphery but to take the plunge into Chapter 1.

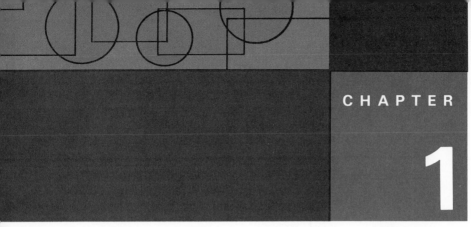

CHAPTER

1

logic

The use of symbols such as x, y, a, b, ... to represent numbers and the symbols $+$, $\cdot$, $=$, ... to represent operations and relations among numbers marks the beginning of elementary algebra. The reader is certainly aware of the many advantages of algebra: Problems that are otherwise difficult to solve become easy when algebra is used.

The reader may be surprised and perhaps pleased to learn that there is also an algebra of logic and a related algebra of sets. We develop the algebra of logic in this chapter and defer the treatment of sets to Chapter 2.

It is customary to use letters such as p, q, r, s, ... to represent statements and other mysterious looking symbols such as $\wedge$, $\vee$, $\sim$, $\longrightarrow$, and $\longleftrightarrow$ to connect the statements. The meaning and use of these connectives will be discussed in the following sections.

1. statements

A *statement* may be defined as a declarative sentence. By a declarative sentence we shall understand any collection of words concerning which it is meaningful to say that its content is true or false. Therefore a *statement* will always mean a declarative sentence which is either true or false, but not both simultaneously.

EXAMPLE 1. Which of the following are statements?

(a) Money always brings happiness.
(b) $7 + 11 = 711$.

7

(c) Mathematics is useful.

(d) New York is a good place to live.

(e) A genius is antisocial.

(f) *Carmen* is the name of an opera.

(g) The exception proves the rule.

(h) George is very smart.

(i) A socialist is usually more interesting than a vegetarian.

(j) Mary Poppins is supercalifragilisticexpialidocious.

(k) Each of the above is a statement.

Solution. Clearly (d), (i), and (j) have the form of a statement. However, (j) is meaningless so it is impossible to decide whether it is true or false. Sentences (d) and (i) are meaningful but cannot be regarded as statements (as we use the term) because there is certainly disagreement about whether they are true or not. The remaining sentences are acceptable as statements. Clearly (a), (b), (e), and (g) are false, while (c) and (f) are true. Whether (h) is true or false depends upon the particular person, named George, under consideration. Finally, (k) is a statement, although in this case it is false.

A statement is said to be *compound* if it has other statements as its components; otherwise it is called *simple*. All of the statements in Example 1 are simple statements. The statement "This is not a good course" is a condensation of the more elaborate "It is not true that this is a good course" and is called the *negation* of the statement "This is a good course." It is customary to call the negation of a simple statement a compound statement.

EXAMPLE 2. Which of the following statements are compound and which are simple?

(a) The sun is shining.

(b) I am not playing tennis.

(c) Ten is a prime number.

(d) The sun is shining and I am playing tennis.

(e) I am playing tennis or ten is a prime number.

(f) The sun is not shining.

(g) If ten is a prime number, then the sun is shining.

(h) The sun is shining if and only if ten is a prime number.

Solution. (a) and (c) are simple statements, while (b), (d), (e), (f), (g), and (h) are compound statements.

We note that in Example 2 the words "and," "or," "not," "if . . . , then," and "if and only if" were used to form compound statements from the simple ones. The words "and," "or," "not," "if . . . , then," and "if and only if" are called *logical connectives*.

EXAMPLE 3. Which of the following statements are simple and which are compound?

> **(a)** Tom is either at the riot or in the library studying.
> **(b)** John and Mary went to school.
> **(c)** Al and Betty make a fine pair.

Solution. (a) is obviously a compound statement made up of two simple statements, "Tom is at the riot" and "Tom is in the library studying," by using the logical connective "or."

  (b) is also a compound statement formed by combining the simple statements "John went to school" and "Mary went to school" with the help of the logical connective "and."

  (c) is not a compound statement, although the word "and" is used in the statement. This is a simple statement. On close examination of this statement, it is apparent that the word "and" is not used as a logical connective, for the two parts, "Al makes a fine pair" and "Betty makes a fine pair," are not individually meaningful statements. Thus the reader is cautioned against jumping to conclusions.

Exercise 1

1. Classify the following into nonstatements, simple statements, and compound statements:
    (a) Stop smoking.
    (b) Roses are red and violets are blue.
    (c) Beware of thieves.
    (d) Jeane is blonde but she is under twenty-one.
    (e) All cats are black.
    (f) I have to study hard if I want to get an A in this course.
    (g) John can't do his homework.
    (h) Do you want to bet?
    (i) If the sun is shining, then it is raining and conversely.
2. Use at least three of the following simple statements to formulate compound statements involving the logical connectives:
    (a) I have money.
    (b) I want to buy stocks.
    (c) My shoes need shining.
    (d) I would like to be a senator.
    (e) My old man is stingy.
3. For each compound statement in problem 1, indicate the logical connective.

2. symbols for the three simplest connectives

It is customary and convenient to use letters such as p, q, r, s, t, ... to represent statements. Thus we may let p denote the statement "It is raining," and we may let q denote the statement "The martini is dry."

Suppose we wish to make the compound statement "It is raining and the martini is dry." We may symbolize this compound statement by $p \land q$. The symbol $\land$, which may be read "and," is our first connective.

Let us consider the compound statement "It is raining or the martini is dry." This statement has two interpretations:

(1) It is either raining or the martini is dry, but not both.
(2) It is either raining or the martini is dry, or possibly both.

In the first interpretation we used "or" in the *exclusive* sense, and in the second interpretation we used it in the *inclusive* sense. The theory of logic can be built using either one of the two interpretations for the word "or." But once we make a decision, we should be consistent. We shall use "or" in the inclusive sense unless specifically mentioned otherwise. This inclusive "or" is our second connective and is symbolized by $\lor$. Thus our compound statement "It is raining or the martini is dry" is written $p \lor q$ (read "p or q"), meaning "It is either raining or the martini is dry or both."

Our third connective is used for the negation of a statement. It is symbolized as $\sim$. Thus the compound statement $\sim p$ means "It is not raining."

To summarize, we have the symbols

$$\land \text{ denoting the connective "and,"}$$
$$\lor \text{ denoting the connective "or," and}$$
$$\sim \text{ denoting the connective "not."}$$

Our symbols can now be used to write briefly some rather complicated statements. For example, $\sim p \lor q$ symbolizes[1] the compound statement "It is not raining or the martini is dry."

EXAMPLE 1. Let p be "I study" and let q be "I pass." Give a verbal translation of the following:

(a) $p \land q$. (d) $\sim(p \lor \sim q)$.
(b) $\sim p \lor q$. (e) $\sim(\sim p \lor \sim q)$.
(c) $\sim p \land \sim q$. (f) $(p \land \sim q) \lor (\sim p \land \sim q)$.

---

[1]Whenever the symbol $\sim$ (negation) is used without parentheses, it is understood that it applies only to the adjacent statement. If we wanted the negation of the compound statement $p \lor q$, we would write $\sim(p \lor q)$.

Solution (a) I study and I pass.

(b) I do not study or I pass.

(c) I do not study and I do not pass.

(d) It is not true that I study or I do not pass.

(e) The statement "I do not study or I do not pass" is not true.

(f) I study and I do not pass, or I do not study and I do not pass (simplification: I fail).

Exercise 2

1. Let p be the statement "John likes Mary," and let q be the statement "Mary likes John." Match each verbal statement in column I with its symbolic form given in column II.

I	II
(a) John and Mary like each other.	(1) $\sim p \wedge \sim q$
(b) John likes Mary but Mary dislikes John.	(2) $p \wedge q$
(c) John and Mary dislike each other.	(3) $p \vee \sim q$
(d) It is not true that John and Mary dislike each other.	(4) $\sim p \vee \sim q$
(e) It is not true that John and Mary like each other.	(5) $p \wedge \sim q$
(f) Either John likes Mary or Mary dislikes John.	(6) $p \vee q$

2. Let p be the statement "This answer is right." Give the symbolic form for the following statements:

(a) This answer is right and wrong.

(b) This answer is either right or wrong and is not both right and wrong.

(c) This answer is wrong.

(d) It is not true that this answer is wrong.

3. Let p be the statement "Hubert is poor," and let q be the statement "Muriel is rich." Write in symbolic form the following compound statements:

(a) Hubert and Muriel are both poor.

(b) Neither Hubert nor Muriel is poor.

(c) It is not true that Hubert and Muriel are rich.

(d) Hubert is rich but Muriel is poor.

(e) It is not true that Hubert is rich or Muriel is poor.

4. Let p be "The wind is blowing," and let q be "It is snowing." Give a simple verbal translation of the following:

(a) $\sim(p \wedge q)$.　　(e) $\sim(p \vee q)$.

(b) $\sim(\sim p)$.　　(f) $\sim(\sim p \wedge \sim q)$.

(c) $\sim p \vee \sim q$.　　★(g) $\sim(\sim\sim p \wedge \sim q) \vee \sim p$.

(d) $\sim p \wedge \sim q$.

5. Write the negation of the following statements:
  (a)   Some cats are pets.
  (b)   All cats are pets.
  (c)   No cat is not a pet.
  (d)   All cats are not pets.
  (e)   It is not true that some pets are cats.
  (f)   Some cats are not pets.

6. Give the negation of each of the following statements:
  (a)   The girl is either blonde or under twenty-one.
  (b)   The girl is not blonde or she is not under twenty-one.
  (c)   The girl is blonde and she is under twenty-one.
  (d)   The girl is not blonde and she is under twenty-one.
  (e)   The girl is not blonde and she is not under twenty-one.

7. Let p be the statement "The girl is not blonde," and let q be the statement "The girl is under twenty-one." Write in symbolic form the negation of each of the statements in Problem 6.

3.   truth tables

When we discussed statements in the previous sections, we emphasized the fact that each statement must be either true or false but not both simultaneously. We say that a statement has truth value T if it is true and has truth value F if it is false. Our aim is to determine the truth value of a compound statement if the truth values of its components are known. Suppose we wish to determine the truth value for the compound statement $p \wedge q$, called the *conjunction* of p and q. We know that the statement p could have the truth value T or F and so could the statement q. There are four possible pairs of truth values for these statements, namely

  (1) p is true, q is true;
  (2) p is true, q is false;
  (3) p is false, q is true; and
  (4) p is false, q is false.

We wish to know in each of the above four cases whether or not the statement $p \wedge q$ is true. It seems reasonable that the only case in which p "and" q ($p \wedge q$) is true is if p is true and q is true. For example, the statement "Jack and Jill went up the hill" is true if Jack went up the hill and Jill went up the hill, and the above compound statement would be false if either Jack or Jill or both went swimming. Thus we assign the truth value T to the statement $p \wedge q$ if p has truth value T and q has truth value T. In the other three cases the statement $p \wedge q$ is assigned the truth value F.

TABLE 1

	p	q	$p \wedge q$
Case 1	T	T	T
Case 2	T	F	F
Case 3	F	T	F
Case 4	F	F	F

These truth values are arranged systematically in Table 1. This table is called *the truth table for $p \wedge q$.*

The truth table gives all the information about the connective $\wedge$. Although we arrived at Table 1 by appealing to the usual meaning of the word "and" in the English language, we want to emphasize that Table 1 is a definition, the definition of the truth values for the connective $\wedge$, and as such is independent of the persuasive arguments used to obtain it.

Next we consider the compound statement $p \vee q$, called the *disjunction of p and q.* The compound statement "$p \vee q$ is true" means either p is true or q is true. Here, as asserted before, we use the word "or" in the inclusive sense; i.e., $p \vee q$ will be assigned the truth value T if p has truth value T and q has truth value T. Further, if one statement is true and the other one false, then the disjunction is true; while if both statements are false, then the disjunction is false. The defining truth table for $p \vee q$, the disjunction of p and q, is given in Table 2.

TABLE 2

p	q	$p \vee q$
T	T	T
T	F	T
F	T	T
F	F	F

If a statement p is true, the symbol $\sim p$, called the *negation of p*, means p is false. Thus if p is false, then $\sim p$ is true. The defining truth table for the connective $\sim$ (read "not" or "negation") is given in Table 3.

TABLE 3

p	$\sim p$
T	F
F	T

With the help of Tables 1, 2, and 3, we can construct the truth table of each compound statement having $\wedge$, $\vee$, and $\sim$ as its connectives.

EXAMPLE 1. Construct the truth table for the compound statement $p \wedge \sim q$.

Solution. Notice that this compound statement is the conjunction of two statements p and $\sim q$. We begin the construction of its truth table by writing in the first two columns the four possible pairs of truth values for the statements p and q. For consistency we repeat the pattern already established in Tables 1 and 2. In the third column we write the corresponding truth values for $\sim q$ from Table 3. For example, in the second case q is false, and hence in this case $\sim q$ is true. We then use columns 1 and 3 to determine the corresponding truth values for $p \wedge \sim q$. These are listed in column four with the help of Table 1. Columns 1, 2, and 4 of Table 4 constitute the truth table for $p \wedge \sim q$.

TABLE 4

p	q	$\sim q$	$p \wedge \sim q$
T	T	F	F
T	F	T	T
F	T	F	F
F	F	T	F

EXAMPLE 2. Construct the truth table for the compound statement $(p \vee \sim q) \wedge (q \vee \sim p)$.

Solution. The reader should check each entry in Table 5. Columns 1, 2, and 7 of this table constitute the truth table for the given statement.

TABLE 5

p	q	$\sim p$	$\sim q$	$(p \vee \sim q)$	$(q \vee \sim p)$	$(p \vee \sim q) \wedge (q \vee \sim p)$
T	T	F	F	T	T	T
T	F	F	T	T	F	F
F	T	T	F	F	T	F
F	F	T	T	T	T	T

EXAMPLE 3. Construct the truth table for the compound statement $(p \wedge \sim q) \vee r$.

Solution. Since this compound statement involves three components, p, q, and r, we shall need $2^3 = 8$ rows in the truth table to include all possible cases. Table 6 contains the truth table for the compound statement in question. The reader should check each entry.

TABLE 6

p	q	r	$\sim q$	$(p \wedge \sim q)$	$(p \wedge \sim q) \vee r$
T	T	T	F	F	T
T	T	F	F	F	F
T	F	T	T	T	T
T	F	F	T	T	T
F	T	T	F	F	T
F	T	F	F	F	F
F	F	T	T	F	T
F	F	F	T	F	F

Exercise 3

In problems 1 through 10, construct the truth tables for the given compound statements.

1. $\sim p \vee \sim q$.

2. $\sim(p \vee q)$.

3. $\sim(p \wedge q)$.

4. $\sim\sim p$.

5. $(\sim p \wedge q) \vee (p \wedge \sim q)$.

6. $(\sim p \vee \sim q) \wedge (p \vee q)$.

7. $(p \wedge q) \vee r$.

8. $(p \vee \sim r) \wedge (q \vee \sim r)$.

9. $\sim(p \vee \sim r) \wedge (q \vee \sim p)$.

10. $[\sim p \wedge (\sim q \wedge \sim r)] \vee (p \wedge \sim q)$.

11. Prove that the truth tables for $(p \vee q) \vee r$ and $p \vee (q \vee r)$ are the same.

12. Repeat problem 11 for the pairs:

    (a) $p \wedge q$ and $q \wedge p$.

    (b) $p \vee q$ and $q \vee p$.

    (c) $(p \wedge q) \wedge r$ and $p \wedge (q \wedge r)$.

13. Examine carefully the arrangement in Table 6 and prove that this table includes all possible cases for the truth values of p, q and r.

14. Suppose that a compound statement involves four simple statements, $p, q, r,$ and s. How many rows would be necessary for a truth table that includes all possible cases? How many rows are necessary if the statement involves n simple statements?

15. Find compound statements S, T, and U, involving p and q, which have the truth values indicated in the following table:

p	q	S	T	U
T	T	F	F	F
T	F	F	F	F
F	T	F	T	T
F	F	T	F	T

16. Construct the truth table for $([p \wedge q \wedge r] \vee [q \vee s]) \wedge (p \wedge s)$.

4. two more connectives

In everyday life we frequently make statements of the form "If I get a lot of money, then I am going to invest the money in stocks"; "If I do my homework regularly, then I will get an A in this course"; etc. Such statements are not direct assertions but are conditional statements. The first part is the condition or hypothesis, and the second part is the conclusion. In general if p and q are two statements and we form a compound statement "if p, then q," this new statement is called a *conditional* and is denoted symbolically by $p \longrightarrow q$ (read "if p, then q"). Thus $\longrightarrow$ is our new connective. The truth table for the conditional, on which all logicians agree, is given in Table 7.

TABLE 7

p	q	$p \longrightarrow q$
T	T	T
T	F	F
F	T	T
F	F	T

We remind the reader that the precise definition of the connective $\longrightarrow$, given by Table 7, is by agreement. Some of the entries in the table may not appeal to our intuition. If p is true and q is true, then the statement $p \longrightarrow q$ is certainly true. If p is true and q is false, then it is reasonable that $p \longrightarrow q$ is false. Thus the first two rows of Table 7 can easily be filled in. We now consider in detail the last two rows of Table 7.

Let p be the statement "Lincoln was the first president," and let q be the statement "Xeno is smart." The statement p is certainly false. Not knowing too much about Xeno, the statement q may be true or it may be false. If, indeed, Xeno is smart, then $p \longrightarrow q$ is covered by the third line in Table 7, and if Xeno is not smart, $p \longrightarrow q$ is covered by the fourth line. In either case we learn that "If Lincoln was the first president, then Xeno is smart" is a true statement. This conclusion may at first seem unacceptable to either the casual or the critical reader, and he may well ask why one makes the definition embodied in the last column of Table 7. Would some other selection of T's and F's do better? There are a variety of arguments that can be given as a motivation for Table 7. We present one of these here and indicate a second argument in problem 10 of the next exercise.

Let us test all reasonable definitions of $p \longrightarrow q$. As already mentioned, the first two rows of Table 7 are quite satisfactory, so we concentrate on the last two rows. Given that we do not want to leave them blank, there are four different ways of completing the truth table for $p \longrightarrow q$, and these are given in Table 8.

16

TABLE 8

p	q	Definitions of $p \longrightarrow q$			
		1	2	3	4
T	T	T	T	T	T
T	F	F	F	F	F
F	T	F	F	T	T
F	F	F	T	F	T

We test each of these definitions with the compound statement $p \longrightarrow p \vee q$. Now a little thought shows that the statement "if p, then p or q" should be true in every case. In Table 9 we give the truth table for $p \longrightarrow p \vee q$, using each of the four definitions for $p \longrightarrow q$ from Table 8. We find that only definition 4 gives T for $p \longrightarrow p \vee q$ in all cases. Consequently, if we wish $p \longrightarrow p \vee q$ to be true in all cases, then we are driven to the definition adopted in Table 7.

TABLE 9

p	q	$p \vee q$	$p \longrightarrow p \vee q$			
			1	2	3	4
T	T	T	T	T	T	T
T	F	T	T	T	T	T
F	T	T	F	F	T	T
F	F	F	F	T	F	T

As a memory aid, we note from Table 7 that the conditional $p \longrightarrow q$ is false in only one case, namely when p is true and q is false.

Our next connective is the *biconditional* denoted by $p \longleftrightarrow q$ (read "p if and only if q"). The biconditional is actually the double conditional: $p \longrightarrow q$ and $q \longrightarrow p$. The defining truth table for the biconditional is given in Table 10. The reader should prove that the last column gives the truth values for $(p \longrightarrow q) \wedge (q \longrightarrow p)$.

TABLE 10

p	q	$p \longleftrightarrow q$
T	T	T
T	F	F
F	T	F
F	F	T

We can now use the truth tables for the five basic connectives—conjunction, disjunction, negation, conditional, and biconditional—to construct the truth tables for more complex compound statements.

EXAMPLE 1. Construct the truth table for the compound statement $(p \wedge \sim q) \longrightarrow (q \vee p)$.

Solution. As before, we write in the first two columns the four possible pairs of truth values for the statements p and q. In the third column we write the corresponding truth values for $\sim q$. In the fourth column we write the corresponding truth values for $(p \wedge \sim q)$ using Table 1, and in the fifth column we write the truth values for $(q \vee p)$ using Table 2. Finally, in the sixth column we write the truth values for $(p \wedge \sim q) \longrightarrow (q \vee p)$ by using Table 7 together with the entries already made in columns 4 and 5. Thus columns 1, 2, and 6 of Table 11 constitute the truth table for the compound statement $(p \wedge \sim q) \longrightarrow (q \vee p)$.

TABLE 11

p	q	$\sim q$	$p \wedge \sim q$	$q \vee p$	$(p \wedge \sim q) \longrightarrow (q \vee p)$
T	T	F	F	T	T
T	F	T	T	T	T
F	T	F	F	T	T
F	F	T	F	F	T

Observe that the last column in Table 11 has only T's. In other words the given statement is true in every case. When this occurs the statement is called a *tautology*. Such statements will be studied in detail in Section 5.

Exercise 4

1. Construct the truth tables for the following statements:
   (a) $[\sim(p \longrightarrow q)] \vee [\sim(q \wedge p)]$.
   (b) $\sim[p \longrightarrow (p \wedge q)] \longleftrightarrow (\sim p \vee \sim q)$.
   (c) $(p \longrightarrow q) \longleftrightarrow (q \longrightarrow \sim p)$.
   (d) $[(p \longrightarrow \sim q) \wedge (q \longleftrightarrow \sim p)] \longrightarrow \sim(\sim p \vee \sim q)$.
   (e) $\sim[(\sim p \longleftrightarrow q) \longrightarrow \{(p \vee \sim q) \wedge q\}]$.

2. The truth table for a certain statement (x) is given below:

p	q	(x)
T	T	T
T	F	T
F	T	F
F	F	F

Which of the following compound statements could be statement (x)?

(a) $(\sim p \vee q) \vee (\sim p \wedge \sim q)$.

(b) $(p \wedge q) \vee (p \wedge \sim q)$.

(c) $(\sim p \vee q) \vee (\sim p \vee \sim q)$.

(d) $(p \wedge q) \wedge (p \wedge \sim q)$.

(e) $(p \vee q) \vee (p \vee \sim q)$.

(f) $(p \wedge \sim q) \longrightarrow (\sim p \vee \sim q)$.

3. Let the connective $\square$ be defined by the truth table

p	q	$p \square q$
T	T	F
T	F	F
F	T	T
F	F	F

Find the truth values for

(a) $(\sim p \vee q) \square \sim q$.

(b) $(p \longrightarrow q) \square q$.

(c) $\sim[(p \longrightarrow \sim q) \longleftrightarrow (p \square q)]$.

(d) $p \longrightarrow [\sim p \square (\sim q \wedge p)]$.

4. Let the connective $\odot$ be defined by the truth table

p	q	$p \odot q$
T	T	F
T	F	F
F	T	F
F	F	T

Find the truth values of the statements (a), (b), (c), and (d) of problem 3 when $\square$ is replaced by $\odot$.

5. Do problem 3, replacing the connective $\square$ by $*$, where $*$ has the following truth table:

p	q	$p * q$
T	T	F
T	F	T
F	T	F
F	F	F

6. Let $p \veebar q$ denote the compound statement p or q but not both (we are using "or" in the exclusive sense). Construct the truth table for $p \veebar q$.

7. Find statements involving p and q using only the connectives $\lor$, $\land$, $\sim$, $\longrightarrow$, and $\longleftrightarrow$ which have the same truth values as

    (a) $p \,\square\, q$.   (b) $p \odot q$.

    (c) $p * q$.   (d) $p \veebar q$.

8. Find the truth tables for the following statements:

    (a) $(p \longrightarrow q) \lor r$.   (b) $(p \longrightarrow q) \longrightarrow r$.

    (c) $(p \lor q) \longrightarrow (p \land r)$.   (d) $(p \longleftrightarrow r) \land \sim q$.

    (e) $[p \longleftrightarrow (r \longrightarrow \sim q)] \lor [(\sim p \longrightarrow r) \land q]$.

9. Check each entry in Table 9.

10. Referring to Table 8, which gives the various possible definitions for $p \longrightarrow q$, prove the following:

    (a) With Definition 1, $p \longrightarrow q$ and $p \land q$ have the same truth table.

    (b) With Definition 2, $p \longrightarrow q$ and $p \longleftrightarrow q$ have the same truth table.

    (c) With Definition 3, $p \longrightarrow q$ and q have the same truth table.

Consequently if we wish the statement $p \longrightarrow q$ to be essentially different from the statements $p \land q$, $p \longleftrightarrow q$, and q, then we must adopt Definition 4.

5. tautology and contradiction

A compound statement is called a *tautology* (or *logically true*) if it is true for all possible truth values of its components. A compound statement is called a *contradiction* (or *logically false*) if it is false for all possible truth values of its components.

Consider Table 12. We observe that both T and F are present in the third column and also in the fourth column. Hence the statements $p \lor q$ and $(p \lor q) \longrightarrow p$ are neither tautologies nor contradictions. Since all the entries in the fifth column of Table 12 are F's, the statement $(p \land q) \land \sim q$ is a contradiction. Since all the entries in the sixth column of Table 12 are T's, the statement $p \longrightarrow (p \lor q)$ is a tautology.

TABLE 12

p	q	$p \lor q$	$(p \lor q) \longrightarrow p$	$(p \land q) \land \sim q$	$p \longrightarrow (p \lor q)$
T	T	T	T	F	T
T	F	T	T	F	T
F	T	T	F	F	T
F	F	F	T	F	T

Exercise 5

In problems 1 through 13, determine if the given statement is a tautology, a contradiction, or neither.

1. $p \vee \sim p$.
2. $p \wedge \sim p$.
3. $p \longrightarrow \sim p$.
4. $(p \vee \sim p) \longrightarrow q$.
5. $p \longleftrightarrow (p \wedge q)$.
6. $\sim(p \wedge q) \longleftrightarrow \sim p \vee \sim q$.
7. $(p \longrightarrow q) \longrightarrow \sim p \vee q$.
8. $(p \vee \sim q) \longrightarrow (p \longrightarrow \sim q)$.
9. $\sim q \longrightarrow \sim(q \wedge r)$.
10. $[p \wedge (p \longrightarrow q)] \longrightarrow q$.
11. $\sim[(p \wedge \sim p) \longrightarrow q]$.
12. $q \longrightarrow (p \wedge \sim p)$.
13. $[(p \longrightarrow q) \wedge (q \longrightarrow r)] \longrightarrow (p \longrightarrow r)$.

6. implication and equivalence

In the previous sections we were concerned with the formulation of new statements from a set of given statements with the help of the connectives. We now consider the relation between two given statements. A statement r is said to *imply* a statement s if s is true whenever r is true. We write this relation as $r \Longrightarrow s$ (read "r implies s") to distinguish it from the statement $r \longrightarrow s$ (read "if r, then s"). We emphasize that the implication ($\Longrightarrow$) is not a connective[1] but a relation between the statements r and s.

Although the relation of implication is closely related to the conditional, we must be careful not to confuse the two. The conditional is a way of getting a new statement from two other statements, whereas the implication is a relation between the two statements. In fact we say $r \Longrightarrow s$ if and only if the conditional $r \longrightarrow s$ is a tautology (is true in every case). From Table 12 we see that the conditional $p \longrightarrow (p \vee q)$ is a tautology, and therefore it makes sense to write $p \Longrightarrow (p \vee q)$. But $p \vee q$ does not imply p, since the conditional $(p \vee q) \longrightarrow p$ is not a tautology [the statement $(p \vee q) \longrightarrow p$ is not true in every case].

For example, "If I am cold, then I am cold or hungry" is true regardless of the truth or falsity of the component statements and hence is an implication. Whereas "If I am cold or hungry, then I am cold" is a statement, which is not true in every case.

Our next relation is that of *equivalence* between two statements. We say that the two statements r and s are *equivalent* if $r \Longrightarrow s$ and $s \Longrightarrow r$. When this occurs we write $r \Longleftrightarrow s$ (read "r is equivalent to s"). It is easy to see that r and s are equivalent if and only if the biconditional $r \longleftrightarrow s$ is a tautology.

EXAMPLE 1. Are the statements $\sim(p \vee q)$ and $\sim p \wedge \sim q$ equivalent?

Solution. The truth table for the biconditional $\sim(p \vee q) \longleftrightarrow \sim p \wedge \sim q$ is given in Table 13.

---

[1]Some of the confusion between the statement "if r, then s" and the implication "r implies s" stems from the fact that most mathematicians use the two phrases interchangeably. Thus in writing a theorem, the form "If ... , then ... " is quite common, but a theorem is really an implication.

TABLE 13

p	q	$\sim(p \vee q)$	$\sim p \wedge \sim q$	$\sim(p \vee q) \longleftrightarrow \sim p \wedge \sim q$
T	T	F	F	T
T	F	F	F	T
F	T	F	F	T
F	F	T	T	T

Since all the entries in the last column of Table 13 are T's, the biconditional $\sim(p \vee q) \longleftrightarrow \sim p \wedge \sim q$ is a tautology. Hence the given statements are equivalent.

Sometimes the symbol $r \Longleftrightarrow s$ for the equivalence of r and s is replaced by the more familiar symbol $r \equiv s$. We shall use these symbols interchangeably. We could also define $r \equiv s$ if and only if r and s have identical truth tables.

EXAMPLE 2.   Prove that $p \vee (q \wedge r) \equiv (p \vee q) \wedge (p \vee r)$.

Solution.   The required truth table is given in Table 14. Since the truth tables for $p \vee (q \wedge r)$ and $(p \vee q) \wedge (p \vee r)$ are identical, the statements are equivalent.

TABLE 14

p	q	r	$q \wedge r$	$p \vee (q \wedge r)$	$p \vee q$	$p \vee r$	$(p \vee q) \wedge (p \vee r)$
T	T	T	T	T	T	T	T
T	T	F	F	T	T	T	T
T	F	T	F	T	T	T	T
T	F	F	F	T	T	T	T
F	T	T	T	T	T	T	T
F	T	F	F	F	T	F	F
F	F	T	F	F	F	T	F
F	F	F	F	F	F	F	F

Exercise 6

In problems 1 through 7, two statements are given. In each case determine whether the two statements are equivalent, or if one of the statements implies the other.

1. $p \longrightarrow q$; $\sim p \wedge q$.
2. $\sim(p \wedge q)$; $\sim p \vee \sim q$.
3. $p \longleftrightarrow q$; $(p \longrightarrow q) \wedge (\sim p \longrightarrow \sim q)$.
4. $p \longrightarrow q$; $\sim q \longrightarrow \sim p$.
5. $\sim q \longrightarrow p$; $\sim p \longrightarrow \sim q$.
6. $p \wedge \sim q$; $\sim p \vee \sim q$.
7. $\sim(p \longrightarrow \sim q)$; $p \wedge q$.

8. Which pairs of the following statements are equivalent?

    (a) If I lose, then I dislike the game.

    (b) It is not true that "I lose and I like the game."

    (c) If I like the game, then I win.

    (d) I win or I dislike the game.

9. Construct compound statements equivalent to the following, using only the connectives $\sim$ and $\vee$:

    (a) $p \wedge q$.   **(b)** $p \longrightarrow q$.

    (c) $p \longleftrightarrow q$.   **(d)** $p \longrightarrow (p \wedge q)$.

10. Among the following statements which pairs are equivalent?

    (a) $p \wedge q$.   **(b)** $p \longrightarrow \sim q$.

    (c) $\sim p \vee \sim q$.   **(d)** $\sim(\sim p \vee \sim q)$.

    (e) $\sim p \vee q$.   **(f)** $q \longrightarrow \sim p$.

11. Consider the following table with the given truth values for the statements (a), (b), and (c):

p	q	(a)	(b)	(c)
T	T	T	T	F
T	F	F	T	F
F	T	F	F	F
F	F	T	T	F

    What implications exist between any two of the statements (a), (b), and (c) in the table?

12. Prove that if $r \Longrightarrow s$ and $s \Longrightarrow r$, then $r \longleftrightarrow s$ is a tautology.

13. Prove that if $r \Longrightarrow s$ and $s \Longrightarrow t$, then $r \Longrightarrow t$.

7. laws of the algebra of statements

Perhaps the reader has already met the commutative, associative, and distributive laws in algebra. There are similar laws for statements. The important ones are stated below as theorems. For a proof, it is sufficient to construct the truth table to show that the stated equivalence is indeed true. This easy task is left for the student.

In the following theorems p, q, and r are arbitrary statements.

Theorem 1. The Commutative Laws.

(1) $p \wedge q \equiv q \wedge p$.   (2) $p \vee q \equiv q \vee p$.

Theorem 2. The Associative Laws.

(3) $(p \wedge q) \wedge r \equiv p \wedge (q \wedge r)$.   (4) $(p \vee q) \vee r \equiv p \vee (q \vee r)$.

Theorem 3. The Distributive Laws.

(5) $p \lor (q \land r) \equiv (p \lor q) \land (p \lor r)$.

(6) $p \land (q \lor r) \equiv (p \land q) \lor (p \land r)$.

Theorem 4. The Idempotent Laws.

(7) $p \lor p \equiv p$.   (8) $p \land p \equiv p$.

Theorem 5. The Identity Laws. If t is any tautology and c is any contradiction, then

(9) $p \lor c \equiv p$.   (10) $p \land c \equiv c$.

(11) $p \lor t \equiv t$.   (12) $p \land t \equiv p$.

Theorem 6. The Complement Laws. If t is any tautology and c is any contradiction, then

(13) $p \land \sim p \equiv c$.   (14) $p \lor \sim p \equiv t$.

(15) $\sim t \equiv c$.   (16) $\sim c \equiv t$.

(17) $\sim\sim p \equiv p$.

Theorem 7. DeMorgan's Laws.

(18) $\sim(p \land q) \equiv \sim p \lor \sim q$ (19) $\sim(p \lor q) \equiv \sim p \land \sim q$.

EXAMPLE 1. Simplify the statement $p \lor (\sim p \land q)$ by using the laws of the algebra of statements.

Solution.

Statement	Reason
$p \lor (\sim p \land q) \equiv (p \lor \sim p) \land (p \lor q)$	Distributive Law
$\equiv t \land (p \lor q)$	Complement Law
$\equiv p \lor q$	Identity Law.

The reader should check the equivalence of $p \lor (\sim p \land q)$ and $p \lor q$ by constructing the truth tables.

Exercise 7

1. Simplify the following statements by using the laws of the algebra of statements:

(a) $\sim p \vee (\sim p \wedge q)$.

(b) $p \vee (p \wedge q)$.

(c) $\sim(p \wedge q) \vee (p \wedge \sim q)$.

(d) $\sim(p \vee q) \vee (\sim p \wedge q)$.

2. Prove Theorems 2 and 3 by constructing truth tables.

3. Prove that

(a) $[(p \wedge q) \vee (p \wedge \sim r)] \wedge r \equiv (p \wedge q) \wedge r$.

(b) $(r \wedge p) \vee [(r \wedge q) \vee (r \wedge \sim q)] \equiv r$.

4. In Theorems 1, 2, and 3 replace $\vee$ by $+$ and $\wedge$ by $\times$. If p, q, and r represent real numbers and $\equiv$ is replaced by $=$, which of the statements obtained are always true?

8. conditional statements and variations

In Section 7 we observed that there is an analogy between the way the symbols $\wedge$ and $\vee$ operate on statements and the way the symbols $+$ and $\times$ operate on numbers. For example, $p \wedge q$ is equivalent to $q \wedge p$, and $p \vee q$ is equivalent to $q \vee p$. However $p \longrightarrow q$ is *not* equivalent to $q \longrightarrow p$ (see Table 15). Thus the connective $\longrightarrow$ lacks the commutative law. In the next exercise we ask the students to show that the associative law for $\longrightarrow$ does not hold.

Let us now consider the conditionals formed from the statements p and q and their negations. The statement $q \longrightarrow p$ is called the *converse* of $p \longrightarrow q$. The statement $\sim q \longrightarrow \sim p$ is called the *contrapositive* of $p \longrightarrow q$. Finally, $\sim p \longrightarrow \sim q$ is called the *inverse* of $p \longrightarrow q$. Let us look at the truth table of these statements (Table 15).

TABLE 15

p	q	Conditional $p \rightarrow q$	Converse $q \rightarrow p$	Contrapositive $\sim q \rightarrow \sim p$	Inverse $\sim p \rightarrow \sim q$
T	T	T	T	T	T
T	F	F	T	F	T
F	T	T	F	T	F
F	F	T	T	T	T

(1) Since columns 3 and 4 are not identical, a conditional statement and its converse are not equivalent.

(2) Since columns 3 and 6 are not identical, a conditional statement and its inverse are not equivalent.

(3) Since columns 3 and 5 are identical, a conditional statement and its contrapositive are equivalent.

In many cases the direct proof of a statement is difficult, but the proof of the contrapositive statement is easy.

EXAMPLE 1. Prove that if n^2 is odd, then n is odd.

Solution. Let p be "n^2 is odd," and let q be "n is odd." We are asked to prove $p \Longrightarrow q$. Since $p \longrightarrow q$ is equivalent to $\sim q \longrightarrow \sim p$, it is sufficient to prove $\sim q \Longrightarrow \sim p$. Now the negation of "$n^2$ is odd" is the statement "n^2 is even." Similarly for n. Hence the contrapositive of the given statement is the statement "If n is even, then n^2 is even." This new statement is much easier to treat than the original one, as we shall now show.

Since n is even, we may set $n = 2k$, where k is some integer. Then $n^2 = (2k)(2k) = 2(2k^2)$, which is also even because it is twice the integer $2k^2$. Hence the statement $\sim q \longrightarrow \sim p$ is indeed an implication. Consequently the equivalent statement $p \longrightarrow q$ is also an implication. We have proved that if n^2 is odd, then n is odd.

The English language offers a large variety of ways of saying the same thing, and these diversities reach a peak with the conditional $p \longrightarrow q$ (as the reader will soon see). Hence it is very essential that we fully understand whether the person making the assertion means the conditional, or its converse, or both (the biconditional). We give below four different statements, which may seem to have diverse meanings but in fact say the same thing.

(a) "If a girl is beautiful, then she is interesting."

In this form, the statement is reasonably clear.

(b) "A girl is beautiful only if she is interesting."

The assertion (b) in fact states "If a girl is not interesting, then she is not beautiful," which is the contrapositive of (a) and hence equivalent to it. Thus "if p, then q" and "p only if q" are equivalent statements.

We say "p is a sufficient condition for q" if q happens whenever p happens. In assertion (a) to know that a girl is interesting, it suffices to know that she is beautiful. We could then express statement (a) as

(c) "A girl being beautiful is a sufficient condition for her to be interesting."

In other words,

(c) "For a girl to be interesting, it is sufficient that she be beautiful."

To say "q is a necessary condition for p" means "p does not take place unless q takes place." That is, "$p \longrightarrow q$." We can express assertion (a) as

(d) "For a girl to be beautiful, it is necessary that she be interesting."

Thus being interesting is a necessary condition for a girl to be beautiful. It is not sufficient, since we have not precluded the possibility that other girls are interesting.

We note that if p is a sufficient condition for q, then q is a necessary condition for p and vice versa.

To summarize, we have the following equivalent forms of a statement:

(a) If p, then q.
(b) p only if q.
(c) p is a sufficient condition for q.
(d) q is a necessary condition for p.

The biconditional $p \longleftrightarrow q$ is simpler because of its symmetry. From the above discussion it is clear that the statements

(a) p if and only if q,
(b) q if and only if p,
(c) p is a necessary and sufficient condition for q, and
(d) q is a necessary and sufficient condition for p

are equivalent to $p \longleftrightarrow q$.

Exercise 8

1. Prove that $(p \longrightarrow q) \longrightarrow r$ and $p \longrightarrow (q \longrightarrow r)$ are not equivalent.
2. Prove that $p \longrightarrow q \equiv \sim p \lor q$.
3. Convert the following into "If ..., then ..." statements:
    (a) The sun is shining only if it is warm.
    (b) It is necessary for you to breathe in order to live.
    (c) All men are mortal.
    (d) He is not lazy or he will fail this course.
4. Consider the following statements:
    (a) If John does not wear a coat, then he is warm.
    (b) John is cold is a sufficient condition for his wearing a coat.
    (c) John is warm or he is wearing a coat.
    (d) His wearing a coat is a necessary condition for John to be cold.
    Which of the above statements are equivalent to "If John is cold, then he is wearing a coat"?

5. Write the converse, contrapositive, and inverse of the following statements:

   (a) If a baby does not cry, then he is happy.
   (b) If a baby cries, then he is sad.
   (c) A baby cries only if he is sad.

6. Which statements in problem 5 are equivalent?

7. Which of the following statements is (are) equivalent to "If n is a natural number, then n is an integer"?

   (a) n is a natural number is a sufficient condition for n to be an integer.
   (b) n is not a natural number or n is an integer.
   (c) n is an integer is a necessary condition for n to be a natural number.

8. Prove that if n^2 is even, then n is even.

9. Among the following statements, which pairs are equivalent?

   (a) $(\sim p \longrightarrow q) \wedge (\sim r \longrightarrow \sim q)$.
   (b) $r \longrightarrow \sim p$.
   (c) $p \longrightarrow \sim r$.
   (d) $\sim[(\sim q \longrightarrow \sim p) \wedge (q \longrightarrow \sim r)]$.

10. Prove that if mn is an odd number, then both m and n are odd.

11. Prove the equivalence of the following statements:

    (a) $p \longrightarrow q$.
    (b) $(p \wedge \sim q) \longrightarrow \sim p$.
    (c) $(p \wedge \sim q) \longrightarrow q$.

9.   on the validity of an argument

An *argument* is an assertion that a given set of statements, called the *premises,* yields another statement, called the *conclusion.* The argument is said to be *valid* if and only if the conjunction of the premises implies the conclusion. In other words if we grant that the statements constituting the premises are all true, then (for a valid argument) the conclusion must be true. If an argument is not valid it is called a *fallacy.*

The validity of an argument can easily be checked by constructing a truth table. All we have to show is that conjunction of premises $\Longrightarrow$ conclusion; i.e., the conclusion is true whenever all the premises are true.

EXAMPLE 1.   Is the following argument valid?

> If I am wealthy, then I am happy.
> I am happy.
> ———————————
> Therefore, I am wealthy.

Solution. This is not a valid argument. Let us analyze why.

Let p be the statement "I am wealthy," and let q be the statement "I am happy." We write the above argument symbolically as

$$p \longrightarrow q$$
$$\underline{q}$$
$$\therefore p$$

(The three dots $\therefore$ are to be read "therefore.")

This argument would be valid if the conditional $[(p \longrightarrow q) \wedge q] \longrightarrow p$ were always true. We construct the truth table for this conditional:

TABLE 16

p	q	$p \longrightarrow q$	$(p \longrightarrow q) \wedge q$	$[(p \longrightarrow q) \wedge q] \longrightarrow p$
T	T	T	T	T
T	F	F	F	T
F	T	T	T	F
F	F	T	F	T

Since in Table 16 there is an F entry in the third line of column 5, while in that case both premises are true, the conditional is not true in all cases. Hence the argument is not valid.

EXAMPLE 2. Check the validity of the following argument:

If Nixon is smart, then the population of India will not explode.
Nixon is smart.
_____
Therefore, the population of India will not explode.

Solution. At first this looks like a ridiculous conclusion. However, the argument above is a valid one. Let p be "Nixon is smart," and let q be "the population of India will not explode." Symbolically the above argument is

$$p \longrightarrow q$$
$$\underline{p}$$
$$\therefore q$$

By constructing the truth table we can easily see that the conditional $[(p \longrightarrow q) \wedge p] \longrightarrow q$ is always true. Hence the argument is valid.

One may be a little surprised that a valid argument gives a ridiculous conclusion. What has the smartness of Nixon to do with the population of

India? The conclusion may be false if one of the premises is false. In this particular case one naturally reexamines the premises, and one must either reject one of the premises or accept the conclusion.

EXAMPLE 3. Show that the following argument is valid:

$$p \longrightarrow q$$
$$\underline{q \longrightarrow r}$$
$$\therefore \; p \longrightarrow r$$

Solution. The reader can verify that the above argument is valid by showing that the conditional $[(p \longrightarrow q) \wedge (q \longrightarrow r)] \longrightarrow (p \longrightarrow r)$ is always true. This is the so-called *law of syllogism*.

EXAMPLE 4. (Lewis Carroll). Prove that the following argument is valid.

(1) All the dated letters in this room are written on blue paper.
(2) None of them are in black ink, except those that are written in the third person.
(3) I have not filed any of those that I can read.
(4) None of those that are written on one sheet are undated.
(5) All of those that are not crossed out are in black ink.
(6) All of those that are written by Brown begin with "Dear Sir."
(7) All of those that are written on blue paper are filed.
(8) None of those that are written on more than one sheet are crossed out.
(9) None of those that begin with "Dear Sir" are written in the third person.

∴ I cannot read any of Brown's letters.

Solution. Let

> p be "the letter is dated,"
> q be "the letter is written on blue paper,"
> r be "the letter is written in black ink,"
> s be "the letter is written in the third person,"
> t be "the letter is filed,"
> u be "I can read the letter,"
> v be "the letter is written on one sheet,"
> w be "the letter is crossed out,"
> x be "the letter is written by Brown,"
> y be "the letter begins with 'Dear Sir.'"

Symbolically the above argument is

(1)	$p \longrightarrow q$	
(2)	$\sim s \longrightarrow \sim r$	
(3)	$u \longrightarrow \sim t$	equivalently $t \longrightarrow \sim u$
(4)	$v \longrightarrow p$	
(5)	$\sim w \longrightarrow r$	equivalently $\sim r \longrightarrow w$
(6)	$x \longrightarrow y$	
(7)	$q \longrightarrow t$	
(8)	$\sim v \longrightarrow \sim w$	equivalently $w \longrightarrow v$
(9)	$y \longrightarrow \sim s$	

$$\therefore x \longrightarrow \sim u$$

The situation becomes clearer if we put the premises in the order **(6)**, **(9)**, **(2)**, **(5)**, **(8)**, **(4)**, **(1)**, **(7)**, and **(3)**. Then the conjunction of the premises has the form $(x \longrightarrow y) \wedge (y \longrightarrow \sim s) \wedge (\sim s \longrightarrow \sim r) \wedge (\sim r \longrightarrow w) \wedge (w \longrightarrow v) \wedge (v \longrightarrow p) \wedge (p \longrightarrow q) \wedge (q \longrightarrow t) \wedge (t \longrightarrow \sim u)$. By repeated use of the law of syllogism we have the valid conclusion $x \longrightarrow \sim u$.

In everyday affairs one frequently encounters fallacious arguments. Sometimes the fallacy is introduced quite by accident in the heat of a discussion, but at other times it is advanced quite purposefully just to confuse the opponent. For example, one might argue as follows:

> If Mr. Wronchy is honest and smart, then he will have an important position with his company.
> Mr. Wronchy is indeed the Treasurer of Major Motors, Inc.
> Therefore, Mr. Wronchy is honest and smart.

An analysis of this argument shows that it has the form

$$\textbf{(1)} \quad \begin{array}{l} p \longrightarrow q \\ \underline{q} \\ \therefore p \end{array}$$

A truth table shows that the argument is fallacious even though if uttered forcefully and quickly it does sound quite convincing.

Other common fallacies can be reduced to one of the following forms.

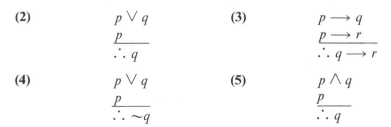

$$\textbf{(2)} \quad \begin{array}{l} p \vee q \\ \underline{p} \\ \therefore q \end{array} \qquad\qquad \textbf{(3)} \quad \begin{array}{l} p \longrightarrow q \\ \underline{p \longrightarrow r} \\ \therefore q \longrightarrow r \end{array}$$

$$\textbf{(4)} \quad \begin{array}{l} p \vee q \\ \underline{p} \\ \therefore \sim q \end{array} \qquad\qquad \textbf{(5)} \quad \begin{array}{l} p \wedge q \\ \underline{p} \\ \therefore q \end{array}$$

(6)
$$p \longrightarrow q$$
$$\sim p$$
$$\therefore q$$

(7)
$$p \longrightarrow q$$
$$r \longrightarrow q$$
$$\therefore p \longrightarrow r$$

(8)
$$p \longrightarrow q$$
$$q \vee r$$
$$\therefore r \longrightarrow \sim p$$

(9)
$$p \longrightarrow q$$
$$\sim p \vee q$$
$$\therefore q \longrightarrow p$$

Exercise 9

An argument is given in problems 1 through 8. In each case determine whether the argument is valid or is a fallacy.

1. p
   q
   $\therefore p \longrightarrow q$

2. p
   $\sim q$
   $\therefore q \longrightarrow p$

3. $p \wedge \sim q$
   $p \longrightarrow q$
   $\therefore p \wedge q$

4. $p \vee \sim q$
   $r \vee \sim p$
   $\therefore q \longrightarrow r$

5. $p \vee \sim r$
   $p \vee q$
   $q \longrightarrow r$
   $\therefore q \vee r$

6. $p \wedge \sim q$
   $\sim q$
   $\therefore p$

7. $q \longrightarrow p$
   $\sim q \longleftrightarrow p$
   $\therefore p$

8. $p \longrightarrow \sim q$
   $p \wedge r$
   $\therefore \sim q \longleftrightarrow \sim r$

In problems 9 through 12 determine the validity of the given argument.

9. If John drinks beer, he is at least 18 years old.
   John does not drink beer.
   $\therefore$ John is not yet 18 years old.

10. If girls are blonde, they are popular with the boys.
    Ugly girls are unpopular with boys.
    Intellectual girls are ugly.
    $\therefore$ The blonde girls are not intellectual.

11. Your troubles start when you get married.
    You have no troubles.
    $\therefore$ You are not married.

12. If I study, then I will not fail this course.
    If I do not play cards too often, then I will study.
    I failed this course.
    $\therefore$ I did play cards too often.

In problems 13 through 19, find the best conclusion using all the hypotheses.

13. (a) All young people are smart.
    (b) No student in this school is old.
    (c) Only stupid people get drunk.
14. (a) Every cab driver is a noisy conversationalist.
    (b) Weak men never wrestle with alligators.
    (c) If a man is strong, he always speaks softly.
15. (a) No army officer with a Purple Cross pushes peanuts with his nose.
    (b) No coward is ever promoted to the rank of a major.
    (c) Every brave man in the 99th regiment has a Purple Cross.
16. (a) One who is not patriotic always fails to vote in the presidential elections.
    (b) Snakes are hopelessly lacking in emotions.
    (c) Anyone without emotions cannot possibly be patriotic.
17. (a) Topologists enjoy only complicated spaces.
    (b) Classical mathematicians like to study Euclidean spaces.
    (c) Kuratowski is a famous topologist.
    (d) Euclidean spaces are essentially simple.
18. (a) No theorem is important unless we are certain that it is true.
    (b) In Isaac Newton's works, the proofs are all based on intuition.
    (c) If the proof of a theorem is not rigorous, we cannot be sure that the theorem is true.
    (d) A first class mathematician proves at least one important theorem.
    (e) A rigorous proof cannot use intuitive arguments.
19. (a) Pushy people are undependable.
    (b) Men are not talkative.
    (c) A shy person is always happy.
    (d) No lady is a concrete worker.
    (e) One can always depend on a quiet person.

★10. design of switching networks

In this section we shall apply the theory of compound statements to develop a theory of simple switching networks. A *switching network* is an arrangement of wires and switches connecting two terminals. A switch can be either "open" or "closed." An open switch prevents the flow of current, while a closed switch permits the flow of current (see Fig. 1).

open switch   closed switch
    (a)   (b)

FIGURE 1

Two switches P and Q may be connected in *series* as in Fig. 2. In Fig. 2, A and B are terminals. In (a) P is open and Q is closed, while in (b) both P and Q are closed.

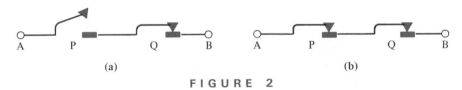

(a)   (b)

FIGURE 2

We let p denote some statement, and we associate with this statement a switch P, with the property that P is closed if p is true and P is open if p is false. Similarly the statement q is true if and only if the corresponding switch Q is closed. For example in Fig. 2 the current will flow if and only if p is true and q is true; i.e., if $p \land q$ is true.

Two switches P and Q may be connected in *parallel* as in Fig. 3. In Fig. 3(a) P is closed while Q is open, while in (b) both P and Q are closed. It is clear that in Fig. 3 the current will flow if and only if either P or Q is closed; i.e., if and only if $p \lor q$ is true.

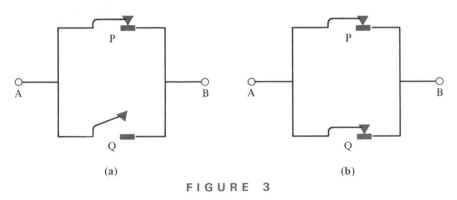

(a)   (b)

FIGURE 3

The two simple networks can be combined to form more complicated networks, and for each such network there is a corresponding compound statement that is true if the current will flow in the network and false if the current will not flow.

Suppose that we have three switches P, Q, and R, and p, q, and r represent the corresponding statements. Then the circuit shown in Fig. 4 corresponds to the statement $(p \land q) \lor r$, since current will flow if P and Q are closed or R is closed. The circuit shown in Fig. 5 corresponds to the statement $(p \lor q) \land r$.

It is possible to couple two or more switches together so that they open and close simultaneously; these switches are said to be equivalent. We shall represent all equivalent switches by the same letter. It is also possible to couple two switches together so that if one switch is open the other is closed

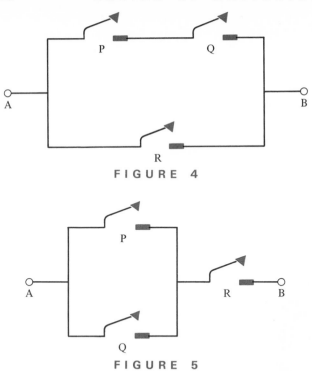

FIGURE 4

FIGURE 5

and vice versa. Such switches will be called complementary. We shall designate the complement of a switch P by P′. Thus if p is the statement that corresponds to the switch P, then $\sim p$ is the statement that corresponds to the switch P′.

We say that two electrical networks are equivalent if they have the same electrical properties with respect to the flow or nonflow of current. To test two networks for equivalence, one merely examines their corresponding statements to see if the statements are equivalent.

EXAMPLE 1. Prove that the circuit shown in Fig. 6 is equivalent to the one shown in Fig. 4.

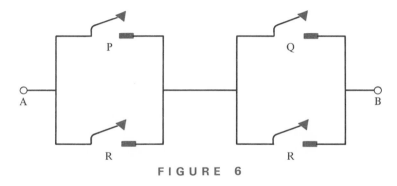

FIGURE 6

Solution.   The network shown in Fig. 6 corresponds to the statement $(p \lor r) \land (q \lor r)$. By the distributive law, equation (5), this statement is equivalent to $(p \land q) \lor r$. Consequently the given network is equivalent to the one shown in Fig. 4.

EXAMPLE 2.   Find a network equivalent to the one shown in Fig. 7.

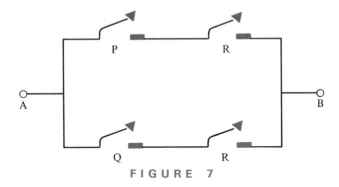

FIGURE 7

Solution.   The network shown in Fig. 7 corresponds to the statement $(p \land r) \lor (q \land r)$. By the distributive law, equation (6), this is equivalent to $(p \lor q) \land r$. But this is the statement for the network shown in Fig. 5.

EXAMPLE 3.   Find the compound statement corresponding to the switching circuit shown in Fig. 8. Then find an equivalent network that is simpler.

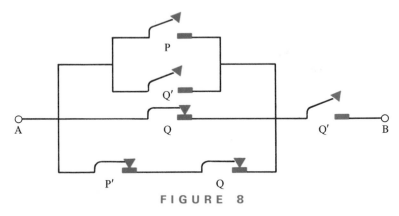

FIGURE 8

Solution.   The corresponding statement is

$$[(p \lor \sim q) \lor q \lor (\sim p \land q)] \land \sim q.$$

By using the laws of algebra of statements, one can show that this compound statement is equivalent to $\sim q$. Hence the simple network shown in Fig. 1, with a switch labeled Q', is equivalent to the complicated circuit shown in Fig. 8.

In problems 1 through 3, draw a pair of switching networks to illustrate each of the pairs of equivalent statements.

1. $p \vee p \equiv p$.
2. $(p \wedge q) \vee p \equiv p$.
3. $p \vee (\sim p \wedge \sim q) \vee (p \wedge q) \equiv p \vee \sim q$.

In problems 4 through 7, draw switching networks corresponding to the compound statements.

4. $[(p \vee \sim q) \vee q \vee (\sim p \vee q)] \vee \sim q$.
5. $[(p \wedge \sim q) \wedge q \wedge (\sim p \wedge q)] \vee \sim q$.
6. $[(p \wedge \sim q) \vee q \vee (\sim p \wedge q)] \wedge \sim q$.
7. $\{[(p \wedge \sim q) \wedge q] \vee (\sim p \wedge q)\} \wedge \sim q$.
8. For each of problems 4 through 7, find a simpler network equivalent to the one drawn.
★9. In a large hall it is desired to turn the lights on or off from any one of three different switches. Design such a circuit.
★10. Assume three people vote for an issue by pressing a button (closing a switch). Design a network such that current will flow if and only if a majority vote in favor of the issue.
★11. Repeat problem 10 with five people voting.

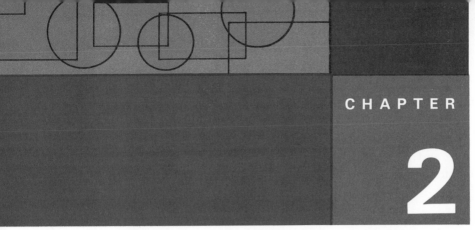

2

the algebra of sets

In Chapter 1 we studied a system of algebra for statements. There is a similar algebra for sets, and the two algebras are closely related, as we shall see in Section 8.

We have just observed the convenience and the usefulness of the logical connectives $\wedge$, $\vee$, and $\sim$. In set theory, these connectives are replaced by $\cap$, $\cup$, and $'$, respectively.

Although the theory of sets is spiritually present in every branch of modern mathematics and is central in the foundations of mathematics, these deeper applications do not concern us here. We use the theory of sets merely to supply a precision to our ideas (definitions, theorems, and proofs) that might otherwise be lacking.

1. the concept of a set

This basic concept is set forth in

> **Definition 1.** A set is any well-defined collection of objects. Any object in the set is called an element of the set or a member of the set.

The principal character of a set is that it is well defined. Given a particular object, it must be clear that either the object is an element of the set or the object is not an element of the set. For example, the collection of all millionaires in the United States might be considered as ill-defined, because it may not be clear how to evaluate a person's property in terms of money.

39

But if we consider instead the collection of all United States citizens who on a certain fixed date have a total of at least one million dollars in one or more savings accounts in United States banks, then this collection is well defined and is a suitable set for a mathematical discussion (although it may not be very interesting).

In general we will use italic letters such as A, B, C, ... to denote various sets and reserve such letters as a, b, c, ... for elements of the set. It is customary to use braces to enclose the elements of a set. Thus the notation

$$A = \{a,b,c,d\}$$

means that A is a set consisting of the four elements a, b, c, and d. We can indicate that a is an element of the set A by writing the symbols

$$a \in A$$

(read "a is an element of A" or "a belongs to A").

The symbol $\notin$ denotes the negation of $\in$. Consequently,

$$e \notin A$$

(read "e is not an element of A" or "e does not belong to A") means that e is not a member of the set A.

A set may contain infinitely many elements. For example, the set of all positive integers (the *natural numbers*)

(1) $$N = \{1,2,3,4,5, \ldots\}$$

contains infinitely many elements. Here, of course, we cannot write them all down, but the three dots that occur in (1) indicate that the enumeration of the elements in N continues in the natural manner.

A set of numbers is often defined by requiring that the elements of the set satisfy certain inequalities. For example, if S is the set of all integers that are greater than -3 and less than 4, then

$$S = \{-2,-1,0,1,2,3\}.$$

There is a convenient shorthand for writing a set that is formed by setting conditions on the elements. Thus the example set just given could be written[1] as

(2) $$S = \{x \mid x \text{ is an integer}, -3 < x < 4\}.$$

---

[1] We assume that the reader is familiar with the inequality symbols: $>$ (read "greater than") and $<$ (read "less than"). A brief treatment of these symbols and their properties is given in Appendix 2.

Here x is a letter which merely represents an element of the set. The vertical bar | can be translated as the phrase "such that" and after the bar we find the conditions on x that must be satisfied for x to be an element of the set S. Equation (2) may be read "S is the set of all elements x such that x is an integer, and x is greater than -3 and less than 4." The reader may amuse himself by thinking of S as a club and the material after the vertical bar as the requirement on x for membership in the club.

A set may not have any elements. For example, the set T defined by

$$T = \{x \,|\, x \text{ is an integer, } x > 10, \text{ and } x < 5\}$$

has no elements, because there is no integer that is greater than 10 and less than 5. A set that has no elements is called the *empty set* and will be denoted by the symbol Ø (to remind us of the number zero). It is also called the *null set* or the *void set*.

EXAMPLE 1. List explicitly the elements of the set

$$A = \{x \,|\, 1 < x < 13, \text{ and } x \text{ is an even integer}\}.$$

Solution. $A = \{2,4,6,8,10,12\}$.

EXAMPLE 2. Baer, Jeyn, and Liekker are to play a chess tournament. Find the set of possible results if there are no ties.

Solution. We use the symbol (L,B,J) to indicate that Liekker won, Baer was second, and Jeyn was third. Then

$$S = \{(L,B,J), (L,J,B), (J,B,L), (J,L,B), (B,L,J), (B,J,L)\}.$$

This example shows that the elements of a set need not be numbers.

EXAMPLE 3. Let N be the set of natural numbers. Give explicitly the elements in each of the following sets:

$$A = \{x \,|\, x = 3n - 2, \, n \in N, \, 1 \leq n \leq 5\},$$
$$B = \{x \,|\, x = 4n + 1, \, n \in N, \, 1 \leq n \leq 5\},$$
$$C = \{y \,|\, y = n^2 + 1, \, n \in N, \, 1 \leq n \leq 7\}.$$

Solution. The notation for A means that we are to replace n by 1, 2, 3, 4, and 5 successively and compute the result. For example, if $n = 2$, then $3n - 2 = 6 - 2 = 4$. In this way we find that

$$A = \{1,4,7,10,13\}.$$

Similarly we find that

$$B = \{5,9,13,17,21\}$$

and

$$C = \{2,5,10,17,26,37,50\}.$$

EXAMPLE 4. Is the set P of all primes well defined?

Solution. We recall that a positive integer $p \geq 2$ is a prime if it has no divisors other than p and 1. We can begin to describe the set P by listing some of the elements:

(3) $$P = \{2,3,5,7,11,13,17, \ldots, p_n, \ldots\},$$

but at present we have no way of giving a meaningful formula that generates all primes, nor does it seem likely that we shall find such a formula soon. It may be very difficult to decide whether an integer such as $x = 2^{1234567} + 17$ is in the set or not. Nevertheless, a positive integer is either a prime or it is not a prime. Hence we must regard the set P as well defined even though we can name only a finite number of elements in this infinite set.

Exercise 1

In problems 1 through 14, a collection is described by a statement. Which of the statements determine a set (a well-defined collection) and which do not?

1. The collection of all famous living Hollywood movie stars.
2. The collection of all good men living in the United States.
3. The collection of all men who are or have been members of the Xyzyvuw Club.
4. The collection of all characters in the Chinese language.
5. The collection of all men in the United States who regard themselves as good.
6. The collection of all good presidents.
7. The collection of all roots of the equation

$$x^{100} + \sqrt{17}\, x^{79} + \pi^2 x^{31} - 57 = 0.$$

8. The collection of all integers that divide n, where

$$n = (12,345,678,910,111,213)^{101} + 1.$$

9. The collection of all pairs (x,y), where x and y are both numbers.

10. The collection of all pairs (x,y), where x is a circle and y is a straight line.

11. The collection of all sets of real numbers.

12. The collection of all beautiful girls in Florida.

13. The collection of all beautiful girls that are registered students in an all-men's college.

14. The collection of all good textbooks on mathematics.

In problems 15 through 22 a set is specified by certain conditions. In each case list explicitly all of the elements in the set. In each of these problems n is an integer.

15. $\{n \mid 4 \leqq n < 9\}$.   **16.** $\{n^2 \mid 0 \leqq n \leqq 7\}$.

17. $\{n \mid -5 < n < 3\}$.   **18.** $\{n \mid n$ divides $1001,\ n \geqq 1\}$.

19. $\{n^4 \mid -4 < n < 4\}$.   **20.** $\{n \mid 1 < n < 30,\ n$ a prime$\}$.

21. $\{n \mid n^2 \leqq 16\}$.   **22.** $\{n \mid n^2 + 13 < 7\}$.

23. The following sets are described in words. Give an alternative description of the same set using a notation similar to that of Equation (2).

    (a) The set of all even integers.

    (b) The set of all odd integers.

    (c) The set of all positive integers which leave a remainder of 2 when divided by 5.

    (d) The set of all numbers that have exactly two distinct prime factors each greater than 1.

24. At a recent Republican convention there were three serious contenders for the presidential nomination, and at the Democratic convention there were two serious contenders for the same honor. A public opinion poll is to determine in advance the voters' preference for each possible pair in the November election. Find the number of different pairs that the voter must be questioned about.

25. Find the number of different pairs as described in problem 24 if there are m possible Republican candidates and n possible Democratic candidates.

★26. If Tahl enters the chess tournament described in Example 2, find the number of possible results.

2.   subsets and supersets

Let

(4) $A = \{2,4,6\}, \qquad B = \{2,3,4,6,7\}, \qquad C = \{1,2,3,4,5,6,7,11,48\}.$

We see immediately that every number in A is also in B and that every number in B is also in C. We describe such an event by saying that A is a *subset* of B and C is a *superset* of B. We also symbolize the situation by writing

(5) $A \subset B$

(read "A is a subset of B" or "A is contained in B") and

(6) $C \supset B$

(read "C is a superset of B" or "C contains B"). Clearly the symbols $\subset$ and $\supset$ were selected because they remind us of the inequality signs $<$ and $>$.

> **Definition 2.** The set A is said to be contained in B and is called a subset of B if
>
> (7) $x \in A$ implies that $x \in B$.
>
> The set C is said to contain B and is called a superset of B if
>
> (8) $x \in B$ implies that $x \in C$.
>
> In the first case we write $A \subset B$, and in the second case we write $C \supset B$.

With this definition it is clear that for every set A we have $A \subset A$ and $A \supset A$. Further, $\varnothing \subset A$ for every A.

> **Definition 3. Equality of Sets.** Two sets A and B are said to be equal if $A \subset B$ and $B \subset A$. We symbolize this by writing $A = B$.

This means that every element of A is an element of B and that every element of B is an element of A. Clearly the equality of two sets means that they are identical. However, the two identical sets might arise from two different sources and at first glance appear to be quite different.

EXAMPLE 1. Let $A = \{0,1,2,3\}$ and let B be the set of roots of the equation $x^4 - 6x^3 + 11x^2 - 6x = 0$. Are these sets equal?

Solution. If $x = 3$, then

$$x^4 - 6x^3 + 11x^2 - 6x = 81 - 162 + 99 - 18 = 0.$$

Hence 3 is a root of the given equation. Similar computations show that

0, 1, and 2 are also roots. Since a fourth-degree equation cannot have more than four roots, we infer that $B = \{0,1,2,3\} = A$.

EXAMPLE 2. Let C be the set of airplanes that made a successful landing on the moon sometime between the years 1850 and 1900. Let D be the set of iceboxes sold last year to people who have green hair and live within a hundred miles of the north pole. Are these two sets equal?

Solution. Since both C and D are empty, we have $C = D$.

These concepts suggest a number of theorems about sets.

Theorem 1. If $A = B$ and $B = C$, then $A = C$.

Theorem 2. If $A \supset B$, then $B \subset A$ and conversely.

Theorem 3. If $A \subset B$ and $B \subset C$, then $A \subset C$.

Each of these theorems is so obvious that no formal proof is required. However, as an example we give a proof of Theorem 3.

Proof. Let x be any element of A. Since $A \subset B$ we see that x is also an element of B (Definition 2). Since $B \subset C$ and x is an element of B, we see that x is also an element of C. Hence $x \in A$ implies that $x \in C$. Hence, by definition, A is contained in C. ∎

3. unions and intersections of sets

Given a number of sets, we may make larger sets by putting them together. When we put two sets A and B together, the resulting set is called the *union* of A and B and is denoted by the symbol $A \cup B$ (read "A union B"). Referring to Fig. 1 we let the elements of A and B be represented by the points inside the two circles, respectively. Then the set $A \cup B$ is represented by the points of the shaded region.

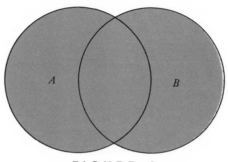

FIGURE 1

Definition 4. Union. The set $C = A \cup B$ consists of all those elements that are either in A or in B or in both A and B. In symbols, $x \in C$ if either $x \in A$ or $x \in B$ or both. The set C is called the union of A and B.

Notice that an element that happens to be in both A and B is still counted just once in the set C.

EXAMPLE 1.   Find $A \cup B$ if

$$A = \{1,2,3,4,5,6\} \quad \text{and} \quad B = \{4,5,6,7,8\}.$$

Solution. $A \cup B = \{1,2,3,4,5,6,7,8\}$. Observe that the numbers 4, 5, and 6 occur in both sets, and yet when we form the union we list each of these numbers only once.

EXAMPLE 2.   Let D be the set of all odd integers, and let E be the set of all even integers. Describe $D \cup E$.

Solution.   Clearly $D \cup E = Z = \{0,\pm 1,\pm 2,\pm 3,\ldots\}$, the set of all integers.

Given two sets A and B, we may be interested in just those elements that belong to both of the sets. Such a set will in general be smaller than either one of the sets A and B. When we take the set of all elements that are common to the two sets A and B, the result is called the *intersection* of A and B and is denoted by the symbol $A \cap B$ (read "A intersection B"). If we let the elements of A and B be represented by the points inside the two circles, as in Fig. 1, then the set $A \cap B$ is represented by the points of the shaded region in Fig. 2.

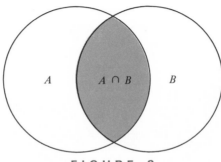

FIGURE 2

Definition 5. Intersection.   The set $C = A \cap B$ consists of all those elements that are in A and in B. In symbols, $x \in C$ if $x \in A$ and $x \in B$. The set C is called the intersection of A and B. If $A \cap B$ is empty the two sets are said to be disjoint.

EXAMPLE 3.   Find $A \cap B$, $D \cap E$, $A \cap D$, $A \cap E$, $B \cap D$, and $B \cap E$ for the sets of Examples 1 and 2.

Solution.   The elements common to A and B are 4, 5, and 6. Hence, by the definition of intersection, $A \cap B = \{4,5,6\}$. Since D consists of all of the odd integers and E consists of all the even integers, we see that $D \cap E$ has no elements because there is no integer that is both odd and even (0 is even). Hence $D \cap E = \varnothing$, where we recall that $\varnothing$ is the symbol for the empty set.

It is easy to check each of the following:

$$A \cap D = \{1,3,5\}, \qquad A \cap E = \{2,4,6\},$$
$$B \cap D = \{5,7\}, \qquad B \cap E = \{4,6,8\}.$$

It is possible to unionize more than two sets or to take the intersection of more than two sets. For example, $(A \cap B) \cap D$ means that one first forms a set by intersecting A and B, and then with this new set one takes the intersection with D. Using for A, B, and D the sets of Examples 1 and 2, we see that $A \cap B = \{4,5,6\}$, and selecting from this set the odd integers we have

$$(A \cap B) \cap D = \{5\},$$

a set that consists of just one element. Such a set is frequently called a *singleton* set.

We have introduced an algebra of sets by defining two operations: $\cup$, union; and $\cap$, intersection. We can now state six simple theorems that are obviously true as soon as the meaning is clear. The student may agree that each of these theorems is obviously true, or he may work out a proof for himself.

Theorem 4.   The union of any two sets contains either one of the sets. In symbols

$$A \cup B \supset A \qquad and \qquad A \cup B \supset B.$$

Thus in general the union of two sets is larger than either of its component parts. However, equality can occur. For example, If Z is the set of all integers, and D and E are sets of odd and even integers, respectively, then

$$Z \cup D = Z \quad \text{and} \quad E \cup Z = Z.$$

Theorem 5. The intersection of any two sets is contained in either one of the sets. In symbols

$$A \cap B \subset A \quad and \quad A \cap B \subset B.$$

Equality can occur. For if Z, D, and E are the sets introduced above, then

$$D \cap Z = D \quad \text{and} \quad Z \cap E = E.$$

Theorem 6. The Commutative Law for Intersections. For any two sets A and B,

$$A \cap B = B \cap A.$$

This means that in forming the intersection of two sets, the order is unimportant. The same is true when forming the union of two sets.

Theorem 7. The Commutative Law for Unions. For any two sets A and B,

$$A \cup B = B \cup A.$$

If we have three sets, the symbol

$$(9) \qquad\qquad S = A \cap B \cap C$$

is at first glance confusing, because it is not clear which pair to consider first. That is, we might form $A \cap B$, and then intersect the resulting set with C. We call the result S_1; that is, we let

$$(10) \qquad\qquad S_1 = (A \cap B) \cap C.$$

If we first form $B \cap C$ and intersect the resulting set with A, we may obtain a set that is different from S_1, so to be safe we call it S_2; that is, we let

$$(11) \qquad\qquad S_2 = A \cap (B \cap C).$$

But the next theorem tells us that for set intersections such care is unnecessary, because $S_1 = S_2$.

> **Theorem 8. The Associative Law for Intersections.** For any three sets A, B, and C,
>
> (12) $$(A \cap B) \cap C = A \cap (B \cap C).$$

Once this theorem is proved, the parentheses can be dropped and we can write expression (9) to represent either (10) or (11), because in fact $S_1 = S_2$, and we make this the definition of S. The perceptive reader will note at once that we could just define S in (9) to be the set of all elements that are in A and in B and in C. The situation is illustrated in Fig. 3, where as usual circles represent the sets, and the intersection of the three sets is shown shaded.

A similar theorem obviously holds for unions.

> **Theorem 9. The Associative Law for Unions.** For any three sets A, B, and C,
>
> (13) $$(A \cup B) \cup C = A \cup (B \cup C).$$

Once this theorem is proved, the parentheses can be dropped, and we can write $A \cup B \cup C$ for the set on either side of (13). This set is shown shaded in Fig. 4. Clearly this set consists of all elements that are in at least one of the three sets.

Of course all of these considerations can be extended to four or more sets, and such material forms its own little domain in mathematics, and is of interest in its own right. But we need not pursue this topic further now.

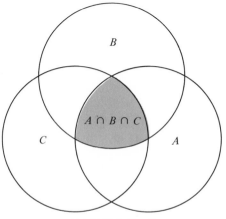

FIGURE 3

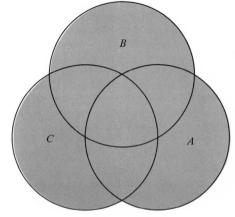

FIGURE 4

When a drawing is used to represent certain relations among sets (as in Fig. 2, 3, and 4), the drawing is called a *Venn diagram*. Further examples of Venn diagrams are given in Figs. 5 through 15.

Exercise 2

1. Let $A = \{1,2,3,4\}$, $B = \{1,2,3,4,5,6\}$, $C = \{2,5,7\}$, $D = \{4,5,6,7\}$, and $E = \{1,3,4\}$. Find
   - (a) $A \cup B$.
   - (b) $A \cap B$.
   - (c) $C \cup D$.
   - (d) $C \cap D$.
   - (e) $C \cap D \cap E$.
   - (f) $C \cup D \cup E$.
   - (g) $C \cap (D \cup E)$.
   - (h) $A \cap (C \cup D)$.
   - (i) $A \cup B \cup C \cup D$.

2. Prove that for any set A,
   - (a) $A \cup A = A$.
   - (b) $A \cup \varnothing = A$.
   - (c) $A \cap A = A$.
   - (d) $A \cap \varnothing = \varnothing$.

3. Using the sets of problem 1, which of the following inclusion relations are true:
   - (a) $A \subset B$.
   - (b) $C \subset B$.
   - (c) $C \subset A \cup D$.
   - (d) $A \cap D \subset E$.
   - (e) $C \subset D \cup E$.
   - (f) $D \cap E \subset B$.

4. (a) Prove that if $A \subset C$ and $B \subset C$, then $A \cup B \subset C$.
   (b) Prove that if $A \supset C$ and $B \supset C$, then $A \cap B \supset C$.

5. Let A be the set of all integers that are divisible by 2, let B be the set of all integers that are divisible by 4, and let C be the set of all integers that are divisible by 6. Is it true that $A \supset B \supset C$?

★6. Find an infinite sequence $C_1, C_2, C_3, \ldots$ of sets of integers such that $C_1 \supset C_2 \supset C_3 \supset \ldots$ and no two of the sets are equal.

★7. In each of the following state whether or not the sets A and B are equal:
   - (a) $A = \{4n \mid n \text{ an integer}\}$, $B = \{2n \mid n \text{ an even integer}\}$.
   - (b) $A = \{2n + 1 \mid n \text{ an integer}\}$, $B = \{2n - 13 \mid n \text{ an integer}\}$.
   - (c) $A = \{4p + 1 \mid p \text{ a positive integer}\}$, $B = \{4n + 1 \mid n \text{ a prime}\}$.
   - (d) $A = \{7n + 2 \mid n \text{ an integer}\}$, $B = \{7n - 26 \mid n \text{ an integer}\}$.
   - (e) $A = \{4n + 1 \mid n \text{ an integer}\}$, $B = \{2n + 1 \mid n \text{ an odd integer}\}$.

★8. For the sets of part (e) of problem 7, find $A \cap B$ and $A \cup B$.

★9. How many different subsets can be formed from the set $A_2 = \{a,b\}$? Write out all of the subsets and observe that both A_2 and the empty set are subsets of A_2. Repeat the problem for $A_3 = \{a,b,c\}$. How many different subsets are there for the set A_n if A_n has n elements?

★★10. Let $A_0 = \{4n \mid n \text{ an integer}\}$, $A_1 = \{4n + 1 \mid n \text{ an integer}\}$, $A_2 = \{4n + 2 \mid n \text{ an integer}\}$, and $A_3 = \{4n + 3 \mid n \text{ an integer}\}$. Prove that these four sets are pairwise disjoint; i.e., that if $i \neq j$, then $A_i \cap A_j = \varnothing$. What can you say about the sets: (a) $A_0 \cup A_2$, (b) $A_1 \cup A_3$, and (c) $A_0 \cup A_1 \cup A_2 \cup A_3$?

★**11.** From problem 9, the set $S = \{m,n,p,q,r,s,t\}$ has 2^7 different subsets. How many of these subsets have an odd number of elements? What is the situation if S has $2n + 1$ elements?

4. the distributive laws for sets

The student is already familiar with the distributive law for real numbers. This law states that for any three numbers a, b, and c,

$$(14) \qquad a(b + c) = ab + ac.$$

For example, $2(3 + 8) = 2 \cdot 3 + 2 \cdot 8$, because $2 \cdot 11 = 6 + 16 = 22$.

In (14) let us replace the numbers by sets, and let us interpret multiplication as set intersection and addition as union. Then equation (14) becomes

$$(15) \qquad A \cap (B \cup C) = (A \cap B) \cup (A \cap C).$$

The remarkable fact is that (15) is true for any three sets, just as (14) is true for any three numbers.

> **Theorem 10. The First Distributive Law for Sets.** For any three sets A, B, and C, equation (15) holds.

Proof. The quickest way to see that Theorem 10 is correct is to use a Venn diagram. In Fig. 5 the three circles representing the three sets divide the plane into eight distinct regions which we have numbered 1, 2, ..., 8. If x is any element whatever, its relation as an element or nonelement of these three sets can be represented by a point in a suitably selected region. For example, If x is in A, B, and C, it is represented by a point in region 1. If x is in B and C but not in A, then it is represented by a point in region 3. If it is in A but not in B and not in C, then region 5 is the proper place for its representative point. If x is not in any of the three sets, then it is represented by a point in region 8. The remaining possibilities are left for the student to check.

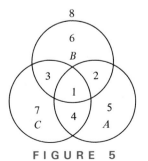

FIGURE 5

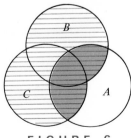

FIGURE 6

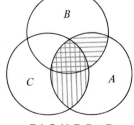

FIGURE 7

Now let R denote the set on the right-hand side of equation (15) and L denote the set on the left-hand side. The set $B \cup C$ is represented by the regions 1, 2, 3, 4, 6, and 7, and these are shown shaded with horizontal lines in Fig. 6. Intersecting this set with A gives the regions 1, 2, and 4. This is the set L and is shown as heavily shaded in Fig. 6.

For the set R we refer to Fig. 7. Here $A \cap B$ is shaded with horizontal lines, and $A \cap C$ is shaded with vertical lines. Putting these regions together yields the region representing the union R. This is clearly the same region that we obtained for L, so $L = R$. ∎

If we wish to avoid reference to a picture the proof would go as follows: Let $x \in A \cap (B \cup C)$. Then x is in A, and it is either in B or in C. Hence it is either in $A \cap B$ or it is in $A \cap C$. In either case it is in the union of these two sets, so that $x \in (A \cap B) \cup (A \cap C)$. This proves that $L \subset R$. We leave it for the student to reverse the argument in order to show that $L \supset R$. It will then follow that $L = R$.

If we interchange the union and intersection signs in equation (15) we obtain a new equation which is also true.

> **Theorem 11. The Second Distributive Law for Sets.** For any three sets A, B, and C,
>
> (16) $\qquad A \cup (B \cap C) = (A \cup B) \cap (A \cup C).$

We leave it for the student to prove this theorem by showing that both sides of (16) yield the shaded portion of Fig. 8.

At the beginning of this section we converted a theorem about numbers, equation (14), into a theorem about sets, equation (15). Suppose now that

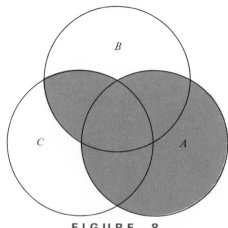

FIGURE 8

we use the same technique on equation (16) to obtain an assertion about numbers. If in (16) we replace union by addition, intersection by multiplication, and the sets A, B, and C by letters a, b, and c representing numbers, we obtain

$$(17) \qquad a + (bc) = (a + b)(a + c).$$

It is obvious that (17) is not always true, and to prove this we need only one example. For $a = 2$, $b = 3$, and $c = 8$, the left-hand side is 26 and the right-hand side is 50. Hence in general (17) is false.

This shows that a theorem may be true in one domain of mathematics but a similar assertion in another domain may be either true or false.

Exercise 3

1. Let $A = \{1,2,3,5,6,7\}$, $B = \{1,2,3,4,8,9\}$, and $C = \{1,2,4,5,10\}$. Use these sets to test Theorems 10 and 11 by finding explicitly the sets on both sides of equations (15) and (16).
2. Complete Table 1, in which the entry "yes" means that the element is in the set at the heading of the column and the corresponding regions (Fig. 5) are listed in the last column. Is it true that this table describes every possible situation with respect to the membership of an element in the sets A, B, and C?

TABLE 1

A	B	C	Region
Yes	Yes	Yes	1
Yes	Yes	No	2
			3
			4
			5
			6
			7
			8

3. Make a table similar to Table 1 for four sets. How many possibilities are there?
4. How many lines would be needed in a table similar to that of Table 1 to show all possible situations with respect to the membership of an element in five different sets? How many for n different sets?
5. Make a Venn diagram showing the various possibilities for four sets, as described in problem 3. Notice that if you try to use circles for all of your sets, such a drawing is impossible.

6. Make a Venn diagram showing each of the following sets:
   (a) $\{x \mid x \in A \text{ and } x \notin B\}$.
   (b) $\{x \mid x \in A,\ x \notin B, \text{ and } x \notin C\}$.
   (c) $\{x \mid x \in A,\ x \in B, \text{ and } x \notin C\}$.
   (d) $\{x \mid \text{either } x \in A \text{ or } x \in B, \text{ and } x \notin C\}$.

7. Subtraction of sets is defined as follows. If A and B are any two sets, then $A - B$ is the set of all x that are in A but not in B. In other words, subtracting B from A is effected by removing from A all of the elements that are also in B [see problem 6(a)]. Prove that $A - B = A$ if and only if $A \cap B = \varnothing$.

8. Use a Venn diagram to prove that
   (a) $(A - B) - C = A - (B \cup C) = (A - C) - (B - C)$.
   (b) $A \cap (B - C) = A \cap B - (A \cap C)$.
   (c) $(A - B) \cup (A - C) = A - (B \cap C)$.
   (d) $(A - B) \cap (A - C) = A - (B \cup C)$.

9. Make a Venn diagram that illustrates each of the following special relations:

(a) $A \subset B$.	(b) $A \cap B = \varnothing$.
(c) $A \cap B \subset C$.	(d) $A \cap B \supset C$.
(e) $A \cap B \cap C = \varnothing$.	(f) $A \cup B \supset C$.

10. For each of the following statements, either prove that the relation is true in every case or give an example to show that it may be false:
   (a) $(A \cup B) \cap C = A \cup (B \cap C)$.
   (b) $A \cup B = A \cup (B - A)$.
   (c) $A \cap (B \cup C) = (A \cap B) \cup C$.
   (d) $(A \cap B) \cup (A \cap C) = A \cap (B \cup C)$.
   (e) $(A \cup B) \cap (A \cup C) = A \cap (B \cup C)$.
   (f) $A \cup B \cup C = A \cup (B - A) \cup [C - (B \cup A)]$.
   (g) $(A \cap B) \cup (A \cap C) = A \cup (B \cap C)$.

★11. Prove that if equation (17) is true, then either $a = 0$ or $a + b + c = 1$, and that conversely when either of these equations is satisfied, then (17) is true.

5.   the complement of a set

Given a set A, we frequently want to consider the collection of all elements that are not in A. This new set is called the *complement* of A and is denoted by A' (read "the complement of A").[1] For example, if A is the set of all numbers less than or equal to 5, it is natural to assume that A' is the set

---

[1] Other notations are also in use. The complement of a set A is sometimes denoted by $\sim A$, by $\bar{A}$, by A^C, or by $C(A)$.

of all numbers greater than 5. But we must observe that this is not quite correct because airplanes, steamships, and oranges are not in A and hence by definition they must be in A'. The way out of this difficulty is reasonably clear. In any sensible discussion there is usually some specified overall set on which the discussion centers. This set is called the *universal set for the discussion* or merely the *universal set* and is denoted by U. It may change as the topic changes, but in general the composition of the universal set is obvious, and in fact one often omits mentioning the set just because it *is* obvious. For example, in a discussion of candidates for the President of the United States, the universal set would certainly be the set of natural-born citizens of the United States who would be at least 35 years of age at the time of the prospective inauguration. In a contest for the prettiest girl in a certain school, the universal set would be the set of all girls who are students at that school.

For the set A cited at the beginning of this section, the universal set would normally be the set of all real numbers. In a course on complex variables, the universal set might well be the set of all complex numbers. In some other context U might be N, the set of natural numbers, or Z, the set of all integers.

It may seem that the changing nature of U, and hence A', might cause trouble. In actual practice it never does, because in any given problem the nature of the universal set U is either obvious, or explicitly stated, and remains constant throughout the discussion of that problem.

Definition 6. Complement. Let U be the universal set, and let A be a subset of U. Then the complement of A is the set of all elements in U that are not in A. This set is symbolized by writing A' or $U - A$.

EXAMPLE 1. Let $U = \{1,2,3,4,5,6,7,8,9\}$, $A = \{2,3,4,5\}$, and $B = \{3,5,7,9\}$. Find A', B', $(A \cup B)'$, and $(A \cap B)'$.

Solution. By definition of the complement,

(18) $\qquad A' = \{1,6,7,8,9\} \qquad$ and $\qquad B' = \{1,2,4,6,8\}$.

Since $A \cup B = \{2,3,4,5,7,9\}$ and $A \cap B = \{3,5\}$, we have

(19) $\quad (A \cup B)' = \{1,6,8\} \qquad$ and $\qquad (A \cap B)' = \{1,2,4,6,7,8,9\}$.

EXAMPLE 2. Let $U = N$ be the set of all positive integers, let P be the set of primes, and let B be the set of perfect squares. Give the first 10 numbers in P' and B'.

Solution. $P = \{2,3,5,7,11,13,17,19,23,\ldots\}$, and $B = \{1,4,9,16,\ldots\}$.
Hence

$$P' = \{1,4,6,8,9,10,12,14,15,16,\ldots\}$$

and

$$B' = \{2,3,5,6,7,8,10,11,12,13,\ldots\}.$$

In making a Venn diagram it is customary to let a box represent the universal set, with the usual circles representing certain subsets of the universal set (see Figs. 9 and 10). For example, in Fig. 9, we have two sets A and B, and the set $(A \cup B)'$ is shown shaded. The set $(A \cap B)'$ is shown

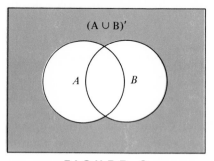

FIGURE 9   FIGURE 10

shaded in Fig. 10. If a suitable Venn diagram is made for A' and B', it will be clear that

(20) $(A \cup B)' = A' \cap B'.$

In a similar manner it is easy to show that for any two sets

(21) $(A \cap B)' = A' \cup B'.$

Theorem 12. DeMorgan's Laws.   For any two sets A and B, equations (20) and (21) are true.

The proof is reserved for the next exercise.

EXAMPLE 3.   Illustrate DeMorgan's laws using the sets of Example 1.

Solution.   From equation set (18)

(22) $A' \cap B' = \{1,6,8\}$ and $A' \cup B' = \{1,2,4,6,7,8,9\}.$

By inspection of equation sets (19) and (22) we see that

$(A \cup B)' = A' \cap B'$ and $(A \cap B)' = A' \cup B'$.

A number of other relations among sets are stated in

Theorem 13. If U is the universal set, $\varnothing$ is the empty set, and A is any subset of U, then

(23) $$U' = \varnothing, \qquad \varnothing' = U,$$
(24) $$A \cup A' = U, \qquad A \cap A' = \varnothing$$
(25) $$A \cup U = U, \qquad A \cap U = A,$$
(26) $$(A')' = A.$$

These are obvious, and the proof is left for the reader.

Exercise 4

1. Let the universal set U be the set of integers $\{0,1,2,3,4,5,6,7,8,9\}$. Find the complement of each of the following sets: $A = \{0,1,2,3\}$, $B = \{$the even integers of $U\}$, and $C = \{$the integers of U that are not divisible by $4\}$.

2. For the sets of problem 1 find
   (a) $(A \cup B)'$.   (b) $A' \cap B'$.
   (c) $(A \cup C)'$.   (d) $A' \cap C'$.
   (e) $(A \cap C)'$.   (f) $A' \cup C'$.

3. Prove Theorem 13.

4. Prove Theorem 12, using a Venn diagram. Next prove Theorem 12 using a logical argument that does not refer to a picture.

5. Prove that if $A \supset B$, then $A' \subset B'$.

6. From DeMorgan's law derive the two alternative forms:
   (a) $(A' \cup B')' = A \cap B$.
   (b) $(A' \cap B')' = A \cup B$.

7. Prove that if A and B are any two sets, then $A - B = A \cap B'$. This shows that subtraction can be defined by complementation.

8. Prove that for any three sets A, B, and C,

$$A \cap (B - C) = B \cap (A - C) = A \cap B \cap C'.$$

9. Prove that if $A \cap B = \varnothing$, then $A \subset B'$ and conversely.

10. Prove that if A and B are any two sets, then

$$(A - B) \cup (B - A) = (A \cup B) - (A \cap B),$$

and make a Venn diagram showing the sets.

Problems 11, 12, 13, and 14 extend some of the earlier results to any finite number of sets. In these problems $A, A_1, A_2, \ldots, A_n, B_1, B_2, \ldots, B_n$ are arbitrary sets.

★11. Prove that

$$A \cap \{B_1 \cup B_2 \cup B_3 \cup \cdots \cup B_n\}$$
$$= (A \cap B_1) \cup (A \cap B_2) \cup (A \cap B_3) \cup \cdots \cup (A \cap B_n).$$

★12. Prove that

$$A \cup \{B_1 \cap B_2 \cap B_3 \cap \cdots \cap B_n\}$$
$$= (A \cup B_1) \cap (A \cup B_2) \cap (A \cup B_3) \cap \cdots \cap (A \cup B_n).$$

★13. A generalization of DeMorgan's laws. Prove that
   (a) $(A_1 \cup A_2 \cup \cdots \cup A_n)' = A_1' \cap A_2' \cap \cdots \cap A_n'$.
   (b) $(A_1 \cap A_2 \cap \cdots \cap A_n)' = A_1' \cup A_2' \cup \cdots \cup A_n'$.

★14. Construct your own sets and illustrate each of the results obtained in problems 11, 12, and 13.

15. Prove that
   (a) $(A \cap B) \cup (A \cap B') = A$.
   (b) $A \cup (B \cap A') = A \cup B$.
   (c) $A \cap (A \cup B) = A$.

16. As in problem 15 express each of the following combinations of sets in a simpler form:
   (a) $(A \cup A') \cap (B \cup B') \cup (A \cap B)$.
   (b) $[(C \cap D)' \cup (C \cap C')] \cap C$.
   (c) $[B \cup (B' \cap U)] \cup [C \cap (C' \cup \varnothing)]$.
   (d) $(A \cup B') \cap (B \cup A')$.

★17. Suppose that $(X \cap A') \cup (X' \cap A) = A$. What can you say about the set X?

★18. For each of the Venn diagrams in Fig. 11, find a simple expression for the set represented by the shaded region.

19. Suppose that in making a Venn diagram we use a circle, rectangle, and triangle to represent the sets C, R, and T, respectively, and that, as shown in Fig. 12, we obtain 12 regions. Do we gain or lose anything essential if we use this diagram in place of the usual three-circle one?

20. In Fig. 12, the set $C \cap R \cap T'$ is represented by the two regions labeled 6 and 8. Find three other such sets, and for each set name the associated pair of regions.

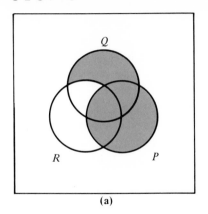

(a)

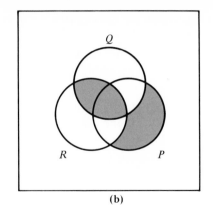

(b)

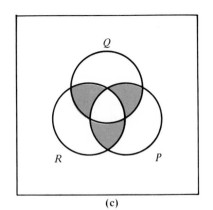

(c)

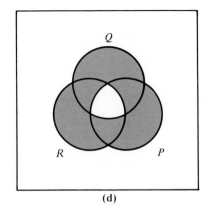

(d)

FIGURE 11

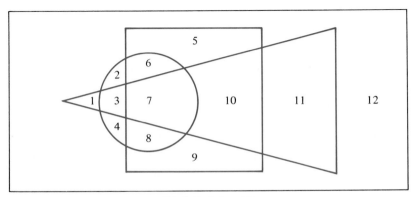

FIGURE 12

6. the number of elements in a set

In some of our earlier examples and exercises, we have determined the number of elements in a set. We now examine this type of problem more closely.

We denote the number of elements in the set A by the symbol $n(A)$, (read "n of A"). If the elements of A are given by a list, one merely counts the elements to find $n(A)$. This is certainly not very interesting, and if A is a large set, then this counting process is dull and tedious. It is more exciting to look for formulas for computing $n(A)$ when the set A is defined in some theoretical way. In some cases it may be easy to find a formula for $n(A)$. In other cases the discovery of a formula for $n(A)$ may require considerable ingenuity. An extreme example is the set P_M consisting of all prime numbers p that are less than or equal to M. To this day no formula for $n(P_M)$ is known.

In this section we shall consider a few simple situations and reserve the more complicated cases for Chapters 3 and 4.

EXAMPLE 1. In a recent election one could vote either Democratic, Independent, or Republican for the president. At the same time, running for superintendent of the local school board (without party affiliation) were Wobbly, Xerxes, York, and Zorro. In how many different ways could a man vote for candidates for these two offices?

Solution. We use the notation (D,W) to indicate a vote for the Democrat for president and a vote for Wobbly for superintendent. Then all of the possibilities can be arranged in a rectangular array, as displayed in Table 2. Since this array has three rows and four columns, it is clear that the number of different ways of voting is $3 \times 4 = 12$.

TABLE 2

	W	X	Y	Z
D	(D,W)	(D,X)	(D,Y)	(D,Z)
I	(I,W)	(I,X)	(I,Y)	(I,Z)
R	(R,W)	(R,X)	(R,Y)	(R,Z)

In solving this problem it was natural to consider pairs (x,y), where x is from a certain set $A = \{D,I,R\}$ and y is from the set $B = \{W,X,Y,Z\}$. This formation of a set of pairs (x,y) occurs so frequently (although often in disguise) that it is worthwhile to give the set a special name and symbol. We do this in

Definition 7. Cartesian Product. Let A and B be two sets. The set of all pairs (x,y) such that $x \in A$ and $y \in B$ is called the Cartesian product of the sets A and B and is denoted by $A \times B$. It is also called the direct product of A and B.

For example, if $A = \{D,I,R\}$ and $B = \{W,X,Y,Z\}$, then $A \times B$ is just the set of 12 pairs displayed in Table 2.

Suppose that A and B are the same sets or have elements in common. To be specific, suppose that $A \cap B$ contains the numbers 4 and 5 as elements. Then both (4,5) and (5,4) are in $A \times B$. In such a case we regard the pair (4,5) as different from (5,4). In other words the order of the elements in the pair is important. To be precise we give

Definition 8. Equality in $A \times B$. Let (s,t) and (u,v) be elements of $A \times B$. If $s = u$ and $t = v$, then we say that $(s,t) = (u,v)$. Conversely if $(s,t) = (u,v)$, then $s = u$ and $t = v$. Any pair (x,y) in $A \times B$ is called an ordered pair.

EXAMPLE 2.   If $C = \{0,1,2,3,4,5,6,7\}$ and $D = \{3,4,5,6,7,8,9\}$, how many elements are in $C \times D$?

Solution.   Without bothering to list all such pairs, it is clear that the elements of $C \times D$ can be arranged in a rectangular array (as in Table 2) with eight rows and seven columns. Hence $n(C \times D) = 8 \times 7 = 56$.

The principle illustrated in Examples 1 and 2 is perfectly general. This is stated in

Theorem 14.   If A and B are finite sets, then

(27) $$n(A \times B) = n(A)n(B).$$

For example, if $n(A) = 170$ and $n(B) = 1300$, then [by equation (27)] the set $A \times B$ contains $170 \times 1300 = 221{,}000$ elements. Certainly the formula of Theorem 14 is superior to counting the elements of $A \times B$.

It is easy to find $n(A \times B)$ because there are no "duplicates" in the set $A \times B$. This rather mystic remark is clarified in the next three examples where "duplicates" do appear.

EXAMPLE 3.   In a survey of students of foreign languages it was found that 77 were studying French, 44 were studying Russian, and 13 were studying both French and Russian. How many students were studying at least one foreign language?

Solution. We observe that a student who is studying both French and Russian is counted in all three groups. Thus the sum $77 + 44$ counts such students twice, and to correct this error we must subtract 13. Thus we find

(28) $$77 + 44 - 13 = 108$$

for the number of students studying at least one foreign language.

This situation may be clarified by the Venn diagram in Fig. 13.

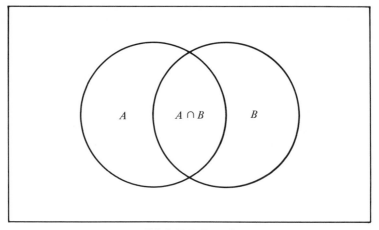

FIGURE 13

If A represents the set of French students and B represents the set of Russian students, then clearly

(29) $$n(A \cup B) = n(A) + n(B) - n(A \cap B).$$

Using $n(A) = 77$, $n(B) = 44$, and $n(A \cap B) = 13$, equation (29) yields equation (28).

The reasoning used in this example applies to any pair of sets. Consequently we have

Theorem 15. If A and B are any pair of finite sets, then equation (29) holds.

If $A \cap B = \varnothing$, then the last term in (29) is zero, and in this special case we obtain the obvious formula

$$n(A \cup B) = n(A) + n(B).$$

Suppose that we want to compute $n[(A \cup B)']$, the number of elements that are neither in A nor in B. If we know $n(U)$, where U is the universal set, then clearly

(30) $$n[(A \cup B)'] = n(U) - n(A \cup B).$$

Using equation (29) in equation (30), we find

(31) $$n[(A \cup B)'] = n(U) - n(A) - n(B) + n(A \cap B).$$

Theorem 16. If A and B are any pair of finite sets, then equation (31) holds.

EXAMPLE 4.   Let U be the set of integers n such that $1 \leq n \leq 200$. Let C be the integers from U that are not divisible by either 2 or 3. Find $n(C)$.

Solution.   It would certainly be boring to count the elements in $C = \{1,5,7,11,\ldots,199\}$. To avoid this, we let $A = \{2,4,6,\ldots,200\}$ be the set of even integers in U, and we let $B = \{3,6,9,\ldots,198\}$ be the set of integers in U that are divisible by 3. Clearly $n(A) = 100$, and since $198 = 3 \times 66$, we find that $n(B) = 66$. Finally $A \cap B = \{6,12,18,\ldots,198\}$, and since $198 = 6 \times 33$, we see that $n(A \cap B) = 33$. Using these results in equation (31) we find that

$$n(C) = n[(A \cup B)'] = 200 - 100 - 66 + 33 = 67.$$

Some labor was involved in finding $n(C) = 67$. Nevertheless, it was certainly simpler than counting the elements in C, and far more interesting.

The formulas of Theorems 15 and 16 are extended to three sets in

Theorem 17. If A, B, and C are any three finite sets, then

(32) $$n(A \cup B \cup C) = n(A) + n(B) + n(C)$$
$$- n(A \cap B) - n(B \cap C) - n(C \cap A)$$
$$+ n(A \cap B \cap C)$$

and

(33) $$n[(A \cup B \cup C)'] = n(U) - n(A) - n(B) - n(C)$$
$$+ n(A \cap B) + n(B \cap C) + n(C \cap A)$$
$$- n(A \cap B \cap C).$$

The generalization of Theorem 17 to an arbitrary finite number of sets is rather complicated and the proof is even more so. We shall give this generalization and proof in Chapter 4, after suitable preparations have been made.

Although the formulas given in Theorem 17 are useful, it is often more convenient to use a Venn diagram, placing in each region the number of elements that belong to the set represented by that region. This is illustrated in

EXAMPLE 5. A survey of 100 students gave the following data:

> 41 students taking Spanish.
> 29 students taking French.
> 26 students taking Russian.
> 15 students taking Spanish and French.
> 8 students taking French and Russian.
> 19 students taking Spanish and Russian.
> 5 students taking all three languages.

Find the number of students among the 100 that are not taking any of the three languages. How many take just one of the languages?

Solution. It is simpler to use the data in just the reverse of the order in which it is given. We first use the fact that 5 students are taking all three languages and place this number in the proper region as indicated in Fig. 14. Since 19 students are taking Spanish and Russian and 5 are taking all three languages, we see that $19 - 5$ or 14 students are taking Spanish and

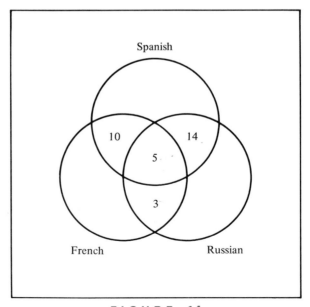

FIGURE 14

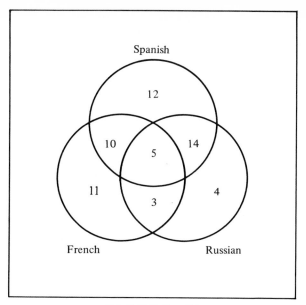

FIGURE   15

Russian only. This is also represented by placing the number 14 in the appropriate region. Similarly we find that $8 - 5 = 3$ students are taking French and Russian only and that $15 - 5 = 10$ students are taking Spanish and French only. These entries complete Fig. 14.

To find the number of students who have Russian as their only language, we subtract from 26, the sum $14 + 5 + 3 = 22$, as dictated by Fig. 14, to find 4, and this number is entered in the proper place in Fig. 15. Similarly we locate $29 - (10 + 5 + 3) = 11$ and $41 - (10 + 5 + 14) = 12$ in their respective regions. Using Fig. 15 it is now easy to find the total number of students studying at least one language. Indeed we find $12 + 10 + 11 + 3 + 4 + 14 + 5 = 59$, and consequently $100 - 59 = 41$ students are not taking any foreign language.

If we use equation (33) to find the same number, the computation would run

$$n[(A \cup B \cup C)'] = 100 - 41 - 29 - 26 + 15 + 8 + 19 - 5 = 41.$$

Finally, the Venn diagram tells us immediately that $12 + 11 + 4 = 27$ students are studying exactly one language.

Exercise 5

1. Let P_M be the set of prime numbers that are less than or equal to M. List the elements of P_{10}. Find $n(P_{10})$, $n(P_{20})$, $n(P_{30})$, $n(P_{40})$, and $n(P_{50})$.

2. Suppose that A is the empty set. What can you say about the set $A \times B$? What is $n(A \times B)$?

3. A man can select any one of three different airlines in traveling from St. Louis, Mo., to Atlanta, Ga., and any one of four different airlines in continuing from Atlanta to Miami, Fla. In how many different ways can he fly from St. Louis to Miami? In how many different ways can he make the round trip by air?

4. A large TV set with an outdoor antenna receives 13 channels, while a smaller set in the den can be tuned only to the 5 local channels. In how many different ways can these two sets be receiving programs?

5. Two scholars agree to meet promptly at 7:00 P.M. for dinner at a Chinese restaurant. They are both punctual, and both remember the date agreed upon, but they forget to specify which restaurant. If there are seven Chinese restaurants in the area, find the number of ways in which they can miss each other.

6. If A, B, and C are three sets, the Cartesian product $A \times B \times C$ is defined to be the set of all ordered triples (x,y,z) such that $x \in A, y \in B$, and $z \in C$. Prove that $n(A \times B \times C) = n(A)n(B)n(C)$. If A has 4 elements, B has 7 elements, and C has 11 elements, what is $n(A \times B \times C)$?

7. Give a suitable definition for the Cartesian product of any finite number of sets. Give a formula for computing $n(A \times B \times C \times D)$. Give a suitable definition for equality in $A \times B \times C \times D$.

8. The chief designer for a large automobile company is considering four different radiator grilles, two different styles of headlights, and five different rear fender designs. With respect to these items alone how many different style cars can be made? *Hint:* Use the formula developed in problem 6.

9. How many integers are there between 100 and 1000 for which each digit is odd? How many of these are between 100 and 500?

10. A college freshman is to have a program of four courses. He is to take either one of two English courses, a beginning course in any one of four different foreign languages, any one of three different mathematics courses, and any one of four different science courses. How many different programs of study are possible? *Hint:* Use the formula developed in problem 7.

11. How many five-letter words can be formed from the letters of the English alphabet if the second and fourth letters must be one of the vowels, A, E, I, O, and U, while the other letters must be consonants and the letter Q is not to be used? By definition, a *word* is any sequence of letters and need not have a meaning.

12. A manufacturer makes shirts in five different patterns, each in seven different neck sizes, and each neck size with three different sleeve lengths. How many different shirts must a store stock in order to carry a complete line of these shirts (one of each type)?

13. In example 5, find
   (a) The number of students studying exactly two languages.
   (b) The number of students who take French if and only if they take Russian.

14. Among 70 freshman students a survey showed that

> 23 were taking physics.
> 25 were taking biology.
> 22 were taking chemistry.
> 6 were taking physics and biology.
> 7 were taking biology and chemistry.
> 8 were taking chemistry and physics.
> 2 were taking all three sciences.

How many of the freshmen were taking none of the three sciences? How many were taking just one of the three sciences?

15. A survey of Neurosisville showed that

> 33 percent read magazine X.
> 29 percent read magazine Y.
> 22 percent read magazine Z.
> 13 percent read magazines X and Y.
> 6 percent read magazines Y and Z.
> 14 percent read magazines Z and X.
> 6 percent read all three magazines.

What percent read none of the three magazines? What percent read magazines X and Y and not Z? What percent read magazine X if and only if they read magazine Z?

16. The girls in Alpha Alpha were classified as indicated in Table 3.

TABLE 3

	Beautiful and clever	Plain and clever	Beautiful and dumb	Plain and dumb
Tall	22	13	11	4
Short	17	18	7	5

Let T denote the set of tall girls, B the set of beautiful girls, etc. Find the number of Alpha Alpha girls in each of the following sets:
   (a) $T \cup S$.   **(b)** $B \cup P$.   **(c)** $T \cap P$.
   (d) $T \cup (P \cap C)$.   **(e)** $T' \cup (P \cap T)$.
   (f) $(P \cap C) \cup (B \cap D) \cup S'$.
   (g) $(S \cap B \cap C) \cup (S' \cap B' \cap D)$.

17. A random sample showed that 33 students preferred the quarter system over the semester system. If the set consisted of 50 undergraduates and 35 graduate students, how many favored the semester system? If 23 undergraduates preferred the semester system, how many graduate students favored the quarter system?

18. A recent party was attended by 9 Socialists, 27 Republicans, and 27 Democrats. There were twice as many Republican men as there were Republican women, but the ratio was just the reverse for the Socialists. If there were 37 men present, how many of the ladies were Democrats?

★19. Let U be the set of integers m such that $1 \leq m \leq 300$. Find $n(C)$, where C is the set of integers from U that are
   (a) Not divisible by 2 or 3.
   (b) Not divisible by 2 or 5.
   (c) Not divisible by 3 or 5.

★20. With U as in problem 19, find the number of integers in U that are not divisible by 2, or 3, or 5.

7. tree diagrams

In finding $n(A)$ it is frequently convenient to think of the set A as created by a sequence of operations. Whenever this is the case, the construction of the set can be visualized by drawing a *tree diagram*. This is illustrated in

EXAMPLE 1. Service and Lob agree to play a tennis match in which the player who first wins three games is the winner of the match. Find the number of possible outcomes for the games.

Solution. We can list the elements of this set A by using symbols such as (S,L,S,L,S) to indicate a particular sequence in which Service won the match. However, the sequence $(S,S,S,L,L) \notin A$, because, according to the rules, the match ended after the first three games. Rather than count the sequences in A, we make the tree diagram shown in Fig. 16.

In this diagram the point labeled "start" is the *root* of the tree. The two "branches," which start at the root, represent the two possible results of the first game; the vertex labeled L indicates that Lob won and the vertex labeled S indicates that Service won. Any sequence of branches that moves upward from the root to a "tip" of the tree is called a *path*. Notice that some paths have only three branches and some have four branches. For example, the path marked in color corresponds to the sequence (L,S,L,L) and represents a match that Lob won in four games. It is now an easy matter to count the tips (the ends of the paths) and to decide that there are 20 different sequences of games.

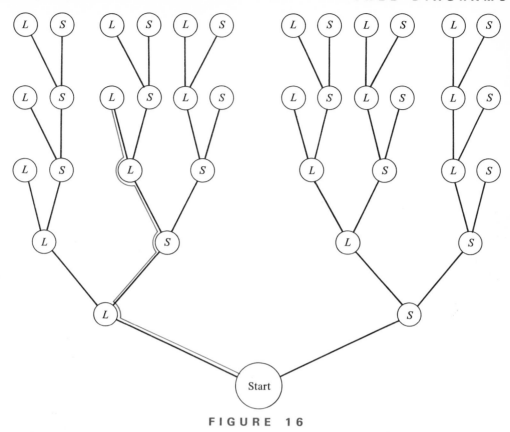

FIGURE 16

In Fig. 16 the tree appears in an upright position, but in general a tree diagram can be drawn to suit the reader's taste: upright, left to right, or downward. When the tree must represent a large number of possibilities, then the drawing becomes unwieldy and overcrowded. Nevertheless, one can often obtain the information that is needed by drawing a small part of the tree or by drawing half of the tree and appealing to symmetry.

EXAMPLE 2. An interviewer wants to file his completed forms in accordance with the sex, political affiliation, and age of the subject. How many different folders will be needed if he classifies each person as Republican, Democrat, or Independent, and the age groups are *Y* (under 30), *MA* (30 to 60), and O (over 60).

Solution. We use a tree to help solve this problem. In Example 1, there was a natural sequence, the order in which the games are played. Here there is no natural sequence, but we can create one by asking the questions in the particular order already specified in the problem. Then it is easy to draw a tree diagram, and this is given in Fig. 17.

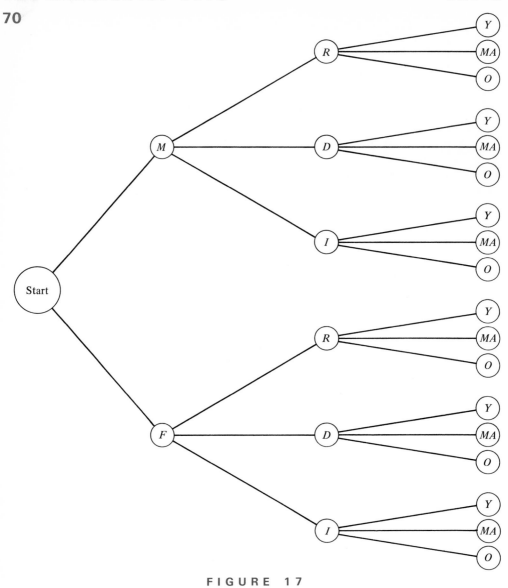

FIGURE 17

By counting the tips of the tree, it is clear that 18 different categories are needed to classify people with respect to these three items.

We can also use Cartesian products to obtain the same numerical result. If we let $A = \{M,F\}$, $B = \{R,D,I\}$, and $C = \{Y,MA,O\}$, then the number of categories is also the number of elements in the Cartesian product $A \times B \times C$. By the theorem of problem 6, Exercise 5, we find $n = 2 \times 3 \times 3 = 18$.

1. For the match described in Example 1, find the number of ways that the match can end (a) in three games and (b) in four games. In how many ways can Lob win the first game and lose the match?

2. Construct a tree for Example 2 if the people are classified first with respect to their political affiliation, second with respect to their age, and third with respect to their sex.

3. The interviewer in Example 2 finds that all of his 18 folders are bulging and that a further refinement is necessary. He decides to further classify the people as smokers or nonsmokers and in accordance with their annual income using four different ranges (below $8000, etc.). How many folders will be needed to file the completed forms? Should you make a tree diagram for this problem?

4. Referring to the problems in Exercise 5, make a tree diagram for each of the situations described in problems 3, 5, and 8.

5. The World Series (baseball) is played between N and A, the two champions of the National and American Leagues, respectively. The first team to win four games is declared the world champion. Draw that portion of the tree for which N wins the first game. How many paths are there for this part of the tree? In how many ways can N win the first two games and lose the series?

6. There are three urns; the first contains three red balls and one green ball, the second contains two red balls and three green balls, and the third contains five green balls. One urn is selected and two balls are removed from that urn successively. Draw a tree representing the selection of the urn and the color of the balls that are removed. How many paths does the tree have? For how many paths is the second ball green?

★7. Suppose that in problem 6 the balls are numbered as well as colored, so that two balls of the same color are distinguishable. Again, one urn is selected and two balls are removed successively. Draw a tree that represents all of the possible results. How many paths does this tree have? For how many paths is the first ball drawn a green ball? On how many paths is the second ball red?

8. A coin is tossed three times. Draw a tree representing the various possible outcomes, assuming that at each toss the result is either heads (H) or tails (T). How many paths does the tree have? In how many cases does H occur twice and T once? If the coin is tossed four times, how many paths does the corresponding tree have? How many, if the coin is tossed n times?

9. In how many different ways can we make 45 cents using only quarters, dimes, and nickels? In how many ways can we make 75 cents?

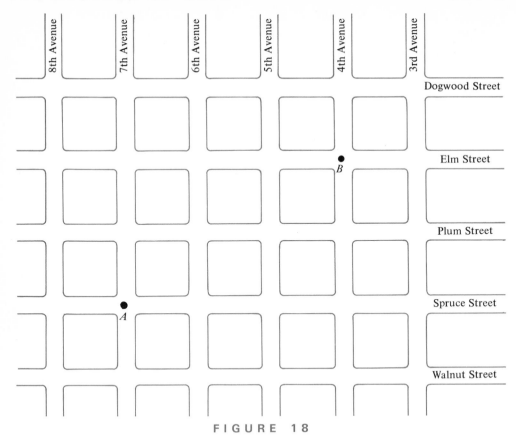

8th Avenue 7th Avenue 6th Avenue 5th Avenue 4th Avenue 3rd Avenue

Dogwood Street

Elm Street

B

Plum Street

Spruce Street

A

Walnut Street

FIGURE 18

★**10.** In how many different ways can a taxi driver go from point A to point B on the streets shown in Fig. 18 and still take a shortest route?

11. The head of a telephone committee phones a message to each of m helpers. In turn each of these helpers phones the message to n helpers, and each of these in turn passes the message on to p more persons. If no person is informed twice, how many people receive the message?

12. Find m, n, and p in problem 11 if they are all equal and the head wishes to reach at least 75 people.

★**8.   the relation between set theory and logic**

We begin with a universal set U and make some statement p concerning the elements of U. This statement creates a set $P \subset U$ in a natural manner. Indeed, P is just the collection of those elements of U for which the statement p is true. The set P is called the *truth set* of p. For example, if U is a certain collection of girls and p is the statement "the girl has red hair," than P is the collection of redheaded girls in U. If q is "the girl is pretty," then Q is the collection of pretty girls in U.

The concept of a truth set allows us to pass from a statement p to its corresponding set P. In the reverse direction given a set P we can always find some corresponding statement p such that P is the truth set for p.

The connectives $\sim$, $\vee$, $\wedge$, $\longrightarrow$, and $\longleftrightarrow$ and the relations $\Longrightarrow$ and $\Longleftrightarrow$ also have a natural image in set theory. If p and q are the statements given above, then the compound statement $p \wedge q$ has for its truth set the set $P \cap Q$, that is, the set of girls that are both redheaded and pretty. In Table 4 we display a "dictionary" which gives the translation from the language of logic to the language of set theory.

TABLE 4

	Logic	Sets
1	p	P
2	q	Q
3	$p \wedge q$	$P \cap Q$
4	$p \vee q$	$P \cup Q$
5	$p \longrightarrow q$	$P' \cup Q$
6	$p \longleftrightarrow q$	$(P' \cup Q) \cap (Q' \cup P)$
7	$p \Longrightarrow q$	$P \subset Q$
8	$q \Longrightarrow p$	$P \supset Q$
9	$p \Longleftrightarrow q$	$P = Q$
10	$\sim p$	P'

EXAMPLE 1.   How does the relation

(34) $$\sim p \vee \sim q \Longleftrightarrow \sim(p \wedge q)$$

translate into the language of set theory?

Solution.   As always we let P be the truth set of p and let Q be the truth set of q. Using Table 4, we find that (34) translates into

(35) $$P' \cup Q' = (P \cap Q)'.$$

This is one of DeMorgan's laws (see Theorem 12) and is true for any two sets P and Q.

EXAMPLE 2.   Let $P = \{2,4,8,16,32\}$. Find a statement for which P is the truth set.

Solution.   Let p be "$x = 2^n$, where $n \in N$ and $1 \leq n \leq 5$". Then P is the set of x for which p is true; i.e., P is the truth set of p. Notice that other statements could be used. Indeed, "x is an integer between 1 and 57 with no odd divisors greater than 1" has the same truth set.

EXAMPLE 3. In Chapter 1, Exercise 9, problem 13, we were asked to draw a correct conclusion from the statements (a) all young people are smart, (b) no student in this school is old, and (c) only stupid people get drunk. Solve this problem by considering suitable truth sets.

Solution. Let U be the set of all people (or if we wish a smaller set, those living in the area served by this school). Let Y be the set of young people, let S be the set of smart people, let T be the set of students in this school, and let D be the set of people who get drunk (at least once). Then the three statements give the following relations among the truth sets:

$$(36) \qquad\qquad Y \subset S, \qquad T \subset Y, \qquad D \subset S'.$$

From $D \subset S'$ we have $S \subset D'$. Hence (36) gives

$$(37) \qquad\qquad T \subset Y, \qquad Y \subset S, \qquad S \subset D'.$$

This gives $T \subset D'$, and hence we conclude that students in this school never get drunk.

Exercise 7

In problems 1 through 8, find statements p such that P is the truth set of p.

 1. $\{11,13,17,19,23,29\}$.
 2. $\{2,5,8,11,14,17,20,23\}$.
 ★3. $\{6,10,14,15,21,35\}$.
 ★4. $\{$Roosevelt, Fairbanks, Sherman, Marshall, Coolidge, Dawes, Curtiss$\}$.
 5. $\{K,Q,B,N,R,P\}$.
 6. $\{$Wilbur, Thomas, Orville$\}$.
 7. $\{$point, line, two intersecting lines, circle, parabola, ellipse, hyperbola$\}$.
 ★8. $\{B,E,H,P,X,Y\}$.
 9. Let P, Q, and R be the truth sets of p, q, and r, respectively. Find truth sets for each of the following statements.
   (a) $p \vee q$.   (b) $\sim p \wedge q$.   (c) $p \wedge q \wedge r$.
   (d) $(p \wedge q) \vee r$.   (e) $(p \vee q) \wedge \sim r$.   (f) $(p \vee q) \wedge (r \vee q)$.

In problems 10 through 13, interpret the given logical relations as relations among sets.

10. $p \Longrightarrow q \vee r$.   11. $q \Longrightarrow r \wedge p$.
12. $p \vee q \Longleftrightarrow r$.   13. $p \vee q \Longleftrightarrow (p \wedge q) \vee r$.

In problems 14 through 18 we give a number of tautologies. Translate these tautologies into theorems about sets.

14. $[(p \longrightarrow q) \wedge (q \longrightarrow r)] \Longrightarrow (p \longrightarrow r).$

15. $p \wedge q \Longleftrightarrow \sim(\sim p \vee \sim q).$

16. $(p \longrightarrow q) \Longleftrightarrow (\sim p \vee q).$

17. $(p \longrightarrow q) \Longleftrightarrow (\sim q \longrightarrow \sim p).$

18. $[p \vee (q \wedge r)] \Longleftrightarrow [(p \vee q) \wedge (p \vee r)].$

19. Follow the method used in Example 3 of this section to solve problems 14 through 19 of Exercise 9, Chapter 1.

★9. voting coalitions

For brevity we shall call any decision-making body, or policy-making body, a *committee*. Such a body might be the city council, the state legislature, or the Supreme Court. In the usual situation each member of a committee has one vote and a majority of the votes is enough to pass a measure or to defeat it. However, there are cases where passage requires 2/3 majority (or some other fraction) and some members of the committee may have more votes than other members.

If the members of a subset A of the committee have enough votes among them to pass a measure, then A is said to form a *winning coalition*. It may happen that the set A cannot carry a measure but has enough votes to block the passage of a measure. In this case A is called a *blocking coalition*. Finally, if the set A is powerless to prevent the passage of a measure, it is called a *losing coalition*.

It follows from these definitions that if A is a winning coalition, then A' is a losing coalition. Further, if A is a blocking coalition, then A' is also a blocking coalition.

A winning coalition may have more members than it needs to win. In such a case some members may be dropped and the smaller set will still form a winning coalition. If this is not the case, then the set is called a *minimal winning coalition*. In other words A is called a minimal winning coalition if it is a winning coalition, and if any one member is removed from A, then the remaining set is no longer a winning coalition. In a similar manner we define a minimal blocking coalition.

EXAMPLE 1. A committee has an odd number of members and each member has exactly one vote. Describe the various possible coalitions if a simple majority is sufficient to pass a measure.

Solution. Suppose that the committee has $2m + 1$ members. Then $m + 1$ members just barely form a majority. If $n(A) \geqq m + 1$, then A is a winning coalition. If $n(A) = m + 1$, then A is a minimal winning coalition. All other coalitions are losing coalitions. There are no blocking coalitions.

EXAMPLE 2. A committee to pass judgment on the amount of violence in a TV show consists of the president, vice-president, treasurer, and manager of the TV station. If these members have five, four, three, and two votes, respectively, and a simple majority is sufficient for approval of a particular program, find the minimal winning and minimal blocking coalitions.

Solution. We use p, v, t, and m to denote the members of the committee. There are 14 votes in all, so 8 votes or more are required for passage. It is easy to see that $\{p,v\}$, $\{p,t\}$, and $\{v,t,m\}$ are minimal winning coalitions. Further $\{v,t\}$ and $\{p,m\}$ each have 7 votes, so each forms a minimal blocking coalition.

Exercise 8

In problems 1 through 7, find the minimal winning coalitions and the minimal blocking coalitions under the given conditions.

1. A committee consists of $2m$ members, each member has one vote, and a simple majority is sufficient to pass (or defeat) a measure.
2. A committee has $3m$ members, each member has one vote, and at least 2/3 of the total votes is required for passage of a measure.
3. Suppose that in Example 2 each member of the committee is given two more votes, so that p, v, t, and m have seven, six, five, and four votes, respectively.
4. Suppose that in Example 2 the number of votes of each member of the committee is tripled, so that p, v, t, and m have 15, 12, 9, and 6 votes, respectively.
5. Suppose that in Example 2 the members p, m, v, and t have 16, 8, 4, and 2 votes, respectively.
6. The committee consists of a, b, c, d, and e, and the members have 6, 5, 4, 3, and 2 votes respectively. Passage of a bill requires 13 or more votes.
7. The Security Council of the United Nations has 15 members. These include the 5 permanent members[1] and 10 others that are each elected for a two-year term. For a measure to pass, 9 votes are required, but these 9 votes must include all of the 5 permanent members.
8. If a committee member x forms a winning coalition by himself, then x is called a *dictator*. If y is not a member of any minimal blocking coalition then y is said to be *powerless*. Prove that if x is a dictator, then all other members are powerless.
9. Let A and B be winning coalitions. Prove that $A \cap B \neq \emptyset$.

[1] These are the United States, United Kingdom, France, Russia, and China.

★10. Suppose that each member of a committee has a certain number of votes and that a measure passes if it receives more than pS votes, where p is a fixed fraction between 0 and 1 and S is the total number of votes in the committee. Now suppose that the number of votes available to each member is doubled, thereby doubling S. Prove that under the new arrangement winning coalitions are the same as under the old arrangement. What about blocking coalitions?

★11. Prove that the theorem stated in problem 10 is still true if instead of doubling the vote of each member the vote is multiplied by k, where k is some positive number, the same for each member of the committee.

★★12. By a well-chosen example, show that the theorem stated in problem 10 becomes false if in place of doubling the vote of each member we add m votes to his vote, where m is the same for each member.

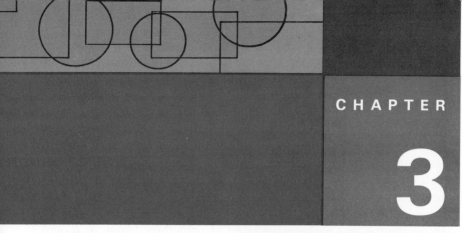

3

elementary combinatorial analysis

This chapter is devoted to counting. Although counting the elements of a set may sound trivial and dull, it can be rather complicated and perhaps exciting.

Certain sets which at first glance seem quite different may have an underlying similar structure. When this is the case, one formula will suffice to count any one of the similarly structured sets.

Two structures occur quite frequently in counting, and these are described briefly by the titles (1) permutations and (2) combinations. The difference in these two structures is explained in Section 1.

1. two simple examples

The following two examples appear to be identical and we might at first glance expect the same answer for both. However, a little thought reveals that they are fundamentally different.

EXAMPLE 1.   How many three-letter arrangements can be made from the four letters A, B, C, and D if repetitions of a letter in a given arrangement are not allowed?

EXAMPLE 2.   Four points A, B, C, and D are given in a plane, with no three on a straight line. How many different triangles can be drawn using these points as vertices?

Solution. For brevity we call each three-letter arrangement a *word*. In solving the first problem the words *ABC*, *BAC*, and *CAB* are all admissible and must be regarded as distinct in our counting. However, in the second problem the triangles *ABC*, *BAC*, and *CAB* are all the same and should not be regarded as distinct in our counting.

We now have a clear picture of the objects to be counted in each case. After a general theory has been formulated, these two problems will be trivial. But until we have such a theory, the simplest procedure is to make a list of the objects and count them directly. Further, it is instructive to do this because it suggests the theory and helps us to understand it.

In solving the first problem we select first the three letters *A*, *B*, and *C* and then make all possible arrangements of these three letters. We then do the same with letters *A*, *B*, and *D*. Continuing in this systematic way, we eventually arrive at the list of Table 1, which contains 24 words. Therefore the answer to the first problem is 24.

If we turn to the second problem we notice at once that all six of the words in the first column represent one and the same triangle. The same assertion is true for each of the other columns. Thus, using vertices from among the four points *A*, *B*, *C*, and *D*, one can draw four different triangles, one for each column in Table 1.

TABLE 1

Words without *D*	Words without *C*	Words without *B*	Words without *A*
ABC	*ABD*	*ACD*	*BCD*
ACB	*ADB*	*ADC*	*BDC*
BAC	*BAD*	*CAD*	*CBD*
BCA	*BDA*	*CDA*	*CDB*
CAB	*DAB*	*DAC*	*DBC*
CBA	*DBA*	*DCA*	*DCB*

In problems similar to the first one, the objects counted are called *permutations*, while those in the second problem are called *combinations*. To be exact we state

Definition 1. Permutation. Suppose that in an arrangement of *k* things selected from a set of *n* different things, the order of arrangement is important, so that different orderings of the same set of *k* things are regarded as different. In this case each such arrangement is called a permutation.

We speak of each such arrangement as a permutation of *n* things taken *k* at a time, or more briefly as a *k-permutation*. We shall use the symbol

$P(n,k)$ to denote the number of different permutations of n things taken k at a time. In the next section we shall obtain a formula for $P(n,k)$.

Definition 2. Combination. Suppose that in an arrangement of k things selected from a set of n different things, the order of arrangement is not important so that different orderings of the same set of k things are regarded as being the same. In this case each arrangement is called a combination.

We speak of each such arrangement as a combination of n things taken k at a time, or more briefly as a k-*combination*. We shall use the symbol $C(n,k)$ to denote the number of different combinations of n things taken k at a time. In Section 3 we shall obtain a formula for $C(n,k)$.

We use the above examples to illustrate our new terminology and notation.

In Example 1, the number of different three-letter words that can be formed from the letters A, B, C, and D is just the number of permutations of four things taken three at a time. From Table 1 we see that $P(4,3) = 24$.

In Example 2, the number of different triangles that can be formed with vertices selected from four points A, B, C, and D in a general position is just the number of combinations of four things taken three at a time. From Table 1 we see that $C(4,3) = 4$.

It is instructive to rephrase the definitions of combination and permutation. Let U be a set of n elements, and let A and B be two combinations of k elements from U. Since the order of the elements in a combination is unimportant, the combinations A and B are the same if and only if A and B are the same as sets. This gives the alternative

---

Definition 3. Combination. A combination of n things taken k at a time is a subset of k elements selected from a set U of n elements. The number of different combinations of n things taken k at a time is the number of different subsets of k elements contained in the set U.

To rephrase the definition of permutation, we must first make the idea of order in an arrangement a little more precise. We have already encountered ordered pairs in Definition 8 of Chapter 2. The natural extension of this concept is given in

---

Definition 4. Ordered k-tuple. The Cartesian product $U \times U \times \cdots \times U$, where the set U occurs k times, is the collection of all elements of the form

(1) $$P = (a_1, a_2, a_3, \ldots, a_k)$$

where each a_i is in U. Each such P is called an ordered k-tuple. Let Q be an ordered k-tuple,

$$(2) \qquad\qquad Q = (b_1, b_2, b_3, \ldots, b_k),$$

from the same Cartesian product $U \times U \times \cdots \times U$. We say that $P = Q$ if and only if

$$(3) \qquad\qquad a_1 = b_1, \quad a_2 = b_2, \quad a_3 = b_3, \ldots, a_k = b_k.$$

Now we may replace Definition 1 by

Definition 5. Permutation. A permutation is an ordered k-tuple $P = (a_1, a_2, \ldots, a_k)$ in which no two of the a_i's in P are the same. P is a permutation of n things taken k at a time if the a_i's are from U, a set of n elements.

For example, suppose that $U = \{0,1,2,3,4,5,6\}$. Then each of the ordered 4-tuples

$$(4) \qquad\qquad P = (6,4,0,3) \qquad \text{and} \qquad Q = (6,0,4,3)$$

is a permutation of seven things taken four at a time. In this example $P \neq Q$ because $a_2 \neq b_2$ and $a_3 \neq b_3$. However, regarded as a combination, P and Q are the same because as sets

$$P = \{6,4,0,3\} = \{6,0,4,3\} = Q.$$

Finally, we should observe that $R = (6,4,3,3)$ is an ordered k-tuple (with $k = 4$) but it is not a permutation because the element 3 occurs twice in R.

2. permutations and the factorial notation

We now derive a formula for the number of permutations $P(a_1, a_2, \ldots, a_k)$ of n things taken k at a time. We observe that a_1, the first element in P, can be selected in n different ways. Once it has been selected and fixed, then there are only $n - 1$ ways of selecting a_2, the second element of P, because the second element in P must be different from the first. These two selections for the first two elements of P can be represented by a tree diagram (see Fig. 1). The precise nature of the elements in U is of no importance

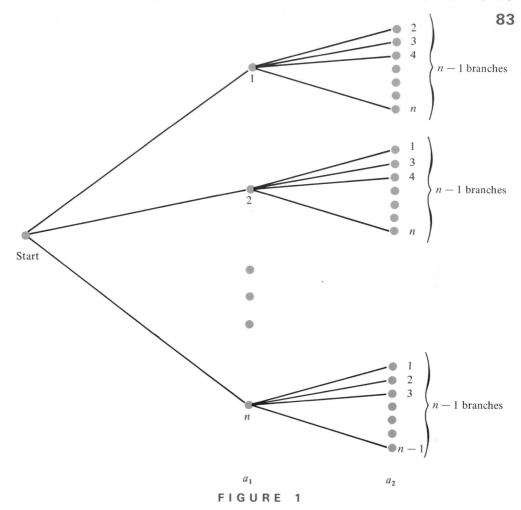

FIGURE 1

when we are finding the number of permutations so we may select U to be the set $\{1,2,3,\ldots,n\}$.

The number of ways of selecting the first two elements of P is just the number of paths in the tree. From the start to the first vertex there are n different branches representing the n different choices for a_1. At each of these vertices there are $n-1$ different branches which represent the $n-1$ different ways of selecting a_2. Consequently, the number of paths through the tree in Fig. 1 is $n(n-1)$. If we are to select a third element a_3 in P, then the tree in Fig. 1 is extended by adding to each vertex on the right $n-2$ branches representing the $n-2$ different choices for a_3. The extension of the two top paths in Fig. 1 is indicated in Fig. 2. In this case each of the $n(n-1)$ paths in Fig. 1 gives rise to $n-2$ paths, and hence there are $n(n-1)(n-2)$ paths in the tree that represents the number of different ways of selecting a_1, a_2, and a_3 to form the permutation (a_1,a_2,a_3). This

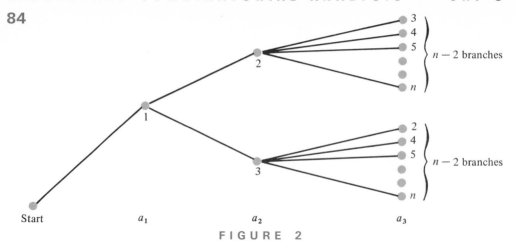

FIGURE 2

process can be continued until all k of the elements in $P = (a_1, a_2, \ldots, a_k)$ have been selected. We thus arrive at

Theorem 1. The formulas

(5) $\qquad P(n,1) = n, \qquad\qquad P(n,2) = n(n - 1),$

(6) $\qquad P(n,k) = n(n - 1) \cdots (n - k + 1), \qquad\qquad k > 2,$

give the number of permutations of n things taken k at a time.

EXAMPLE 1. Use Theorem 1 to solve the first example of section 1.

Solution. If repetitions are not allowed, then the number of three-letter words that can be made from the four letters A, B, C, and D is just the number of permutations of four things taken three at a time. If $n = 4$ and $k = 3$, then $n - k + 1 = 4 - 3 + 1 = 2$ and equation (6) gives

(7) $\qquad\qquad\qquad P(4,3) = 4 \cdot 3 \cdot 2 = 24.$

EXAMPLE 2. First, second, third, and fourth prizes are to be awarded in a beauty contest in which 16 very pretty girls are entered. In how many different ways can the prizes be awarded?

Solution. Certainly the order of selection is important (at least to the girls involved) and repetitions are not allowed (no girl can receive more than one prize). Hence the answer is just the number of permutations of 16 things taken 4 at a time. Since $n - k + 1 = 16 - 4 + 1 = 13$, we have

(8) $\qquad\qquad\qquad P(16,4) = 16 \cdot 15 \cdot 14 \cdot 13 = 43{,}680.$

No wonder the judges are either gray-haired or bald!

The formulas of Theorem 1 (and other formulas) can be expressed concisely if we have at hand a special symbol to represent the product of all the positive integers from 1 to n inclusive. The symbol universally used is $n!$ (read "n factorial"). Thus by definition we have

$$1! = 1 \qquad\qquad 5! = 1 \cdot 2 \cdot 3 \cdot 4 \cdot 5 = 120$$
$$2! = 1 \cdot 2 = 2 \qquad\qquad 6! = 1 \cdot 2 \cdot 3 \cdot 4 \cdot 5 \cdot 6 = 720$$
$$3! = 1 \cdot 2 \cdot 3 = 6 \qquad\qquad 7! = 1 \cdot 2 \cdot 3 \cdot 4 \cdot 5 \cdot 6 \cdot 7 = 5040$$
$$4! = 1 \cdot 2 \cdot 3 \cdot 4 = 24 \qquad\qquad 8! = 1 \cdot 2 \cdot 3 \cdot 4 \cdot 5 \cdot 6 \cdot 7 \cdot 8 = 40{,}320$$

and in general for any positive integer n,

$$(9) \qquad\qquad n! = 1 \cdot 2 \cdot 3 \cdots (n - 2)(n - 1)n,$$

or what is the same thing

$$(10) \qquad\qquad n! = n(n - 1)(n - 2) \cdots 3 \cdot 2 \cdot 1.$$

We observe that in general

$$(11) \qquad\qquad (n + m)! \neq n! + m!.$$

For example, if $n = 3$ and $m = 5$, then the left-hand side of (11) is $8! = 40{,}320$, while the right-hand side of (11) is $3! + 5! = 6 + 120 = 126$. Certainly these are not equal.

Similarly we observe that $(2n)!$ and $2 \cdot n!$ have different meanings. In the first expression the factorial applies to $2n$, while in the second expression we are to double $n!$. For example if $n = 3$, then $(2n)! = 6! = 720$, while on the other hand $2 \cdot n! = 2 \cdot 3! = 2 \cdot 6 = 12$. Quite a difference!

With our new notation we can put the formulas of Theorem 1 in the form

$$(12) \qquad\qquad P(n,k) = \frac{n!}{(n - k)!},$$

for each integer k, such that $1 \leqq k \leqq n - 1$.

For example,

$$P(8,5) = \frac{8!}{(8 - 5)!} = \frac{8 \cdot 7 \cdot 6 \cdot 5 \cdot 4 \cdot 3 \cdot 2 \cdot 1}{3 \cdot 2 \cdot 1} = 8 \cdot 7 \cdot 6 \cdot 5 \cdot 4.$$

The reader may test his understanding of the factorial notation by proving the following assertions:

If $n > 1$, then $n \cdot (n - 1)! = n!$
If $n > 2$, then $(n^2 - n) \cdot (n - 2)! = n!$
If $n \geqq k$, then $(n - k + 1) \cdot n! + k \cdot n! = (n + 1)!$
If $n > 2$, then $n! - (n - 1)! = (n - 1)^2 \cdot (n - 2)!$

1. Find all positive integers n for which $(2n)! = 2 \cdot n!$.
2. Are there any positive integers n such that $(3n)! = 3 \cdot n!$?
3. Find all pairs of positive integers m and n such that $(m + n)! = m! + n!$.
4. Solve the equation $(n + 2)! = 90 \cdot n!$ for n.
5. Find a positive integer n, such that $n^2! \neq (n!)^2$.
6. Prove that if $1 \leq k \leq n - 1$, then $P(n,k) = n!/(n - k)!$.
7. Give an interpretation for the number of permutations of n things taken k at a time when $k = 0$. Does Theorem 1 make sense when $k = 0$? What about equation (12)?
8. How many three-letter words can be made using the first 10 letters of the alphabet if no letter is repeated in any one word. How many, using all the letters of the alphabet?
(9.) Three flags are to be displayed on a vertical flagpole in order to transmit a message from one boat to another by a prearranged code. If the three flags can be selected from 10 different flags, how many different messages can be transmitted?
10. How many fraternity names can be made using the letters of the Greek alphabet (there are 24 letters) if the name consists of three letters and no repetitions are allowed? How many if repetitions are allowed?
11. In how many ways can six people line up for a group photograph?
12. The population of Fort Lauderdale, Fla., is 83,000. Prove that if each resident has three initials, then there are at least two with the same initials.
13. A Scrabble player with seven different letters on his rack decides to test all possible five-letter permutations before making his next play. If he tests one permutation each second, how long will it take before he is ready to play?
14. A liberal organization decides to rank the nine Supreme Court Justices in accordance with their liberalism. They select 20 crucial cases and give each justice 2 points if he votes "right," 1 point if he abstains, and 0 points if he votes "wrong." Assuming that there are no ties, how many different rankings are possible?
15. In the test described in problem 14, how many different scores are possible for a particular justice?
★★16. A second liberal organization rates the nine justices using the same test material as in problem 14 but gives x points if he votes "right," y points for an abstention, and z points if he votes "wrong," where $x > y > z$. Is it possible that the two rankings might be different when the votes are weighted differently?
17. A congressman sends out questionnaires asking his constituents to rank in order of importance 10 different possible improvements in

the present tax laws. He then instructs his secretary to put the replies in the same folder if and only if they give the same ranking to the proposed reforms. What is the largest number of folders that she might need? Is it possible that no folder will contain more than one reply?

In problems 18 through 20 solve for n.

18. $P(n,4) = 1680$.

19. $5P(n,3) = 2P(n-1,4)$.

20. $15P(n,1) + P(n,2) = P(n,3)$.

21. Solve the problem posed in Example 1 by making a suitable tree diagram.

22. Find the number of permutations of k things taken k at a time.

23. If we apply equation (12) in solving problem 22, we obtain 0! in the denominator. What definition for 0! does this suggest?

3.   combinations

It is now easy to prove

> **Theorem 2.** If $1 \leq k \leq n-1$, then the number of combinations of n things taken k at a time is given by the formula

(13) $$C(n,k) = \frac{P(n,k)}{k!} = \frac{n!}{k!(n-k)!}.$$

Proof. Each permutation of n things taken k at a time can be obtained in two steps: (1) We select k things from the universal set, and (2) we then rearrange these k things to form the desired permutation.

Now we can do step 1 in $C(n,k)$ different ways because $C(n,k)$ is just the symbol for the number of combinations of n things taken k at a time.

Once we have selected a particular set of k things, these can be permuted in $P(k,k) = k!$ different ways. Hence step 2 can be performed in $k!$ different ways.

We can picture this two-step process by a tree. Step 1 is represented by $C(n,k)$ different branches starting from the root. Step 2 is represented by a set of $k!$ branches drawn from the ends of each branch constructed in step 1. Consequently, the number of paths through the tree is just the product $C(n,k) \cdot k!$. Since the number of paths is also the number of permutations of n things taken k at a time we have

(14) $$C(n,k) \cdot k! = P(n,k).$$

Dividing both sides of (14) by $k!$ gives equation (13). ∎

In equation (13) the special case $k = 0$ or $k = n$ will be treated in problem 13 of the next exercise.

EXAMPLE 1.   Use Theorem 2 to solve the second example of Section 1.

Solution.   We are to compute the number of triangles with vertices among four given points, and this is just the number of combinations of four things taken three at a time.

From Theorem 2, equation (13), we find that

$$(15) \qquad C(4,3) = \frac{4!}{3!1!} = \frac{4 \cdot 3 \cdot 2 \cdot 1}{3 \cdot 2 \cdot 1 \cdot 1} = 4.$$

EXAMPLE 2.   How many different poker hands are possible? A poker hand consists of five cards dealt from a deck of 52 cards.

Solution.   The arrangement of the given cards in the hand is unimportant, so this is a problem on combinations. We want the number of combinations of 52 things taken five at a time. By equation (13),

$$(16) \qquad C(52,5) = \frac{52!}{5!47!} = \frac{52 \cdot 51 \cdot 50 \cdot 49 \cdot 48}{5 \cdot 4 \cdot 3 \cdot 2 \cdot 1} = 2{,}598{,}960.$$

The numbers $C(n,k)$ have many interesting properties. One useful property is set forth in

Theorem 3.   If $1 \leqq k \leqq n - 1$, then

$$(17) \qquad C(n + 1, k + 1) = C(n,k) + C(n, k + 1).$$

Proof.   Let the universal set of $n + 1$ things be the set $U = \{a_1, a_2, \ldots, a_n, x\}$. The reader should note that we have singled out x for special consideration by giving that element of U a distinctive name. Let S be the set of all combinations of $k + 1$ things taken from U. Now some of these combinations will contain x and others will not. If we let S_1 be the subset of those combinations of S that contain x and let S_2 be the subset of those that do not, then obviously

$$(18) \qquad S = S_1 \cup S_2, \quad \text{and} \quad S_1 \cap S_2 = \emptyset.$$

Hence

$$(19) \qquad n(S) = n(S_1) + n(S_2).$$

Regarding the set S_1, once x is selected to be in a combination of $k + 1$

things from U, one can select the remaining k elements from $\{a_1, a_2, \ldots, a_n\}$ in $C(n,k)$ different ways. Therefore

$$(20) \qquad\qquad n(S_1) = C(n,k).$$

Since x is not present in the combinations in S_2, these combinations are obtained by selecting $k + 1$ elements from $\{a_1, a_2, \ldots, a_n\}$. Consequently

$$(21) \qquad\qquad n(S_2) = C(n, k + 1).$$

Now $n(S) = C(n + 1, k + 1)$ by definition. If we use equations (21) and (20) in (19) we obtain (17). ∎

Exercise 2

1. Compute $C(5,2)$ and $C(5,3)$.
2. Compute $C(10,3)$ and $C(10,7)$.
3. The preceding problems illustrate the fact that $C(n,k) = C(n, n - k)$. Prove that this assertion is always true in two different ways: **(a)** by using equation (13), and **(b)** by considering the fact that every selection of k elements from a set of n elements leaves $n - k$ elements.
4. How many triangles are formed by six lines in the plane if no two are parallel and no three are concurrent? Note that some triangles may be inside others. Generalize this problem to n lines.
5. In how many ways can eight astronauts be divided into two parties of four each, if one party is to explore the moon while the other party remains at the base? Find the number of ways if there is at least one astronaut in each party.
6. In how many different ways can the Supreme Court give a five to four decision upholding a lower court? In how many ways can it give a majority decision reversing a lower court?
7. Given 10 points in space with no 4 lying in the same plane, how many different planes are there containing 3 of the given points? Generalize this problem to n points.
8. A salesman represents a shoe company that is currently pushing shoes in 15 different styles. His sample case will hold only 7 different shoes. In how many different ways can he select the shoes to put in his case? Of course he carries the maximum number, and no two of the shoes selected are of the same style.
★9. Suppose that in problem 17 of Exercise 1 the senator asks his constituents to select the 5 most important tax reforms from the suggested list of 10, instead of asking for a ranking. Find the maximum number of folders his secretary might need if replies are filed together if and only if they make the same selection?

10. Given n points in a plane, all possible lines are drawn connecting pairs of these points. How many lines are there if no three of the given points are collinear?

11. A mathematics department with 20 assistant professors is granted the right to promote 3 of them to the rank of associate professor. In how many different ways can the men be selected for promotion?

12. A newly formed Young Republican Club has 20 members. In how many different ways can the club elect a president, vice-president, and publicity director if each member is eligible for each position but no one can hold two posts?

★13. From Exercise 1, problem 23, we recall that $0! = 1$, by definition. Attach a meaning to the symbol $C(n,k)$ when $k = 0$ or $k = n$. Does Theorem 2 seem reasonable when $k = 0$ or $k = n$? Is it true that $C(n,k) = C(n, n - k)$ when $k = 0$ or $k = n$?

★14. Is Theorem 3 still true when $k = 0$?

★15. Give a second proof of Theorem 3 using equation (13) instead of the combinatorial proof given in the text. Which proof do you prefer?

4. more difficult problems

We have developed a formula for the number of permutations of n things taken k at a time and another formula for the number of combinations of n things taken k at a time. These formulas will handle the standard problems of counting. We can continue to develop more formulas to meet more complex situations, and in Chapter 4 we shall do this. However, in many problems we need only a slight modification of the two formulas already available. To avoid an excess of formulas it is often better to treat a problem individually on its own merits.

EXAMPLE 1.   In how many ways can five people line up for a group picture if two of the five are not on speaking terms and refuse to stand next to each other?

Solution.   Let A, B, C, D, and E denote the various members of the group and suppose that A and B are the cantankerous ones. If we ignore the wishes of A and B, then the number of ways of lining up is just the number of permutations of five things taken five at a time and is $5! = 120$. We must subtract the number of permutations in which A and B stand side by side. We denote the pair A,B (in either order) by a single letter X and observe that the number of permutations of X, C, D, E is just $4! = 24$. In these permutations A and B stand next to each other. But we can interchange A and B in each such permutation. For example, $CDXE$ represents either one of the two permutations $CDABE$ and $CDBAE$ in which A and B are side by side. Therefore, the number of permutations of the set $U = \{A,B,C,D,E\}$ with A and B next to each other is $2 \times 4!$. Hence the number of ways of lining up for a photograph with A and B separated is

$$(22) \qquad\qquad 5! - 2 \times 4! = 120 - 48 = 72.$$

EXAMPLE 2. There are 10 boys and 6 girls at the tennis courts. In how many ways can a doubles game be arranged if each side consists of a boy and a girl?

Solution. The two boys can be selected in $C(10,2) = 10 \times 9/2$ different ways. Similarly the two girls can be selected in $C(6,2) = 6 \times 5/2$ different ways. After the four have been selected to play, one of the boys may select either of the two girls for his partner. Hence, the number of ways of arranging a doubles game is

$$(23) \qquad\qquad 2 \times \frac{10 \times 9}{2} \times \frac{6 \times 5}{2} = 1350.$$

EXAMPLE 3. Find the number of permutations that can be formed with the letters of the word *Nevada*.

Solution. First we regard the two *a*'s as different and and in order to observe the difference we might paint one red and the other blue. Then the number of permutations of these six letters all different is $6!$. Now with any one permutation in which the red *a* appears first there is a similar permutation with the red *a* and the blue *a* interchanged so that the blue *a* appears first. When the paint is removed these two permutations are indistinguishable. Hence only half of the $6!$ permutations are different and the answer is $6!/2 = 360$.

Suppose that we are to find the number of permutations of the letters of the word *almanac*. If the *a*'s are all different, then there are $7!$ such permutations. But the three *a*'s can be permuted in $3!$ ways in each such arrangement, so that if the *a*'s are indistinguishable, then there are only $7!/3! = 840$ permutations.

EXAMPLE 4. How many different sums of money can be made using a penny, a nickel, a dime, a quarter, and a half dollar?

Solution. We have five coins, so taking just one coin there are $C(5,1) = 5$ different sums. Using two coins there are $C(5,2) = 10$ different sums. In general if we use k of the coins, the number of different sums is the number of combinations of five things taken k at a time. Then the total number of possible sums of money is

$$(24) \quad \begin{array}{c} C(5,1) + C(5,2) + C(5,3) + C(5,4) + C(5,5) \\ = \ 5 \ + \ 10 \ + \ 10 \ + \ 5 \ + \ 1 \ = 31. \end{array}$$

We leave for the reader the task of proving that two different selections of coins have different sums.

EXAMPLE 5. Find the number of different poker hands that contain exactly three aces, while the remaining two cards do not form a pair.

Solution. There are $C(4,3)$ different ways of selecting the three aces from the four that are in a standard deck. The fourth card must not be an ace but may be any one of the 48 remaining cards. The fifth card should not form a pair with the fourth card and hence must be selected from the $52 - 4 - 4 = 44$ remaining cards. However in selecting the last two cards the pairs (a,b) and (b,a) give the same hand. Hence the number of poker hands with three aces, and nothing else of interest, is

$$(25) \quad C(4,3) \times \frac{48 \times 44}{2} = 4224.$$

Exercise 3

1. In how many ways can six persons A, B, C, D, E, and F line up for a picture if A and B refuse to stand next to each other, while C and D insist on standing next to each other?
2. Do problem 1 under the assumption that A, B, and C insist on being together with B always between A and C.
3. In how many ways can four boys and four girls stand in line if the boys and the girls are to alternate in line?
4. There are seven boys and four girls at a dance. In how many ways can they form couples to dance if all of the girls have partners?
5. Do problem 4 if one couple is going steady and refuses to split.
6. Find the number of different permutations of the letters in each of the following words: (a) *monotone*, (b) *banana*, and (c) *ukulele*.
7. Find the number of different weights (greater than zero) that can be formed using weights of 1, 2, 4, 8, 16, and 32 lb.
8. Find the number of different poker hands with exactly two aces that can be obtained from a regulation 52-card deck
    (a) If the other cards are unrestricted.
    (b) If the other three cards do not contain a pair.
★9. In the design of a four-engine plane consideration must be given to the possibility that some of the engines may fail to operate. How

many different design conditions must be considered if at least one engine is running and if symmetrical situations are regarded as the same?

★10. In how many different ways can four people be seated around a circular table? Two ways are counted as different only if someone has a different neighbor in the two arrangements. Generalize to find a formula for n people.

★11. Two tables of bridge are to be arranged among eight players. In how many different ways can this be done if two arrangements are regarded as the same when each player has the same partner and the same pair of opponents?

★12. In how many different ways can the opening day baseball games be arranged in a league consisting of eight teams, if we disregard the place where the game is played? Find the number of ways, if at least one team from each pair is playing on its home ground and the location of the game is regarded as important.

★★13. Let each of p points lying on a line be joined by line segments to each of q points lying on a second line parallel to the first. Find the number of points of intersection of these line segments that lie in the strip bounded by the given parallel lines, if no three of the segments are concurrent.

★14. Given n points on a circle, all chords with these points as end points are drawn. If no three chords are concurrent, find the number of intersection points of these chords in the circle.

★15. Given n points in a plane, we join them by straight lines in all possible ways. Assuming that no two lines are parallel and no three are concurrent, find the number of points of intersection of these lines not counting the originally given points.

★16. Given n points in a plane with k of them on a straight line but the others in general position, find the number of straight lines that can be formed by joining pairs of these points.

★17. A rectangle is cut by p lines parallel to one side and q lines parallel to the second side. Find the number of different rectangles in the resulting figure. Note that some of the rectangles may be contained in other rectangles.

18. A true-false examination has 10 statements which the student may mark T or F. In how many different ways can the student guess at the answers?

19. Suppose that in a certain multiple-choice examination a student must select any one of three possible answers as the correct one. If the examination has 10 questions, in how many different ways can the student guess at the answers?

20. Assume that each correct answer in the examination described in

problem 18 is worth 10 points. Find the number of ways that the student may score

(a) 70.

(b) 70 or better.

21. Repeat problem 20 for the examination described in problem 19.
22. Find the number of ways of making a score less than 70
    (a) For the exam in problem 20.
    (b) For the exam in problem 21.

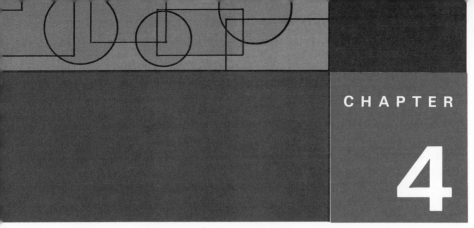

the binomial theorem and related topics

The binomial theorem gives a formula for the expansion of $(x + y)^n$. In the most general form of the theorem, n may be an arbitrary real or complex number, but here we are concerned only with the simple case in which n is a positive integer. Aside from its intrinsic interest, the binomial theorem plays an important role in combinatorial analysis, probability, and statistics.

★1.   the Σ and Π notation

We recall from elementary algebra (or Appendix 3) the formulas:

(1) $\qquad 1 + 3 + 5 + \cdots + (2n - 1) = n^2,$

(2) $\qquad 1 + 2 + 3 + \cdots + n = \dfrac{n(n + 1)}{2},$

(3) $\qquad 1^2 + 2^2 + 3^2 + \cdots + n^2 = \dfrac{n(n + 1)(2n + 1)}{6},$

and

(4) $\qquad 1^3 + 2^3 + 3^3 + \cdots + n^3 = \dfrac{n^2(n + 1)^2}{4},$

each one true for an arbitrary $n \in N$.

95

In these formulas the three dots suffice to indicate the nature of the terms to be summed on the left side. A purist might object to the three dots on the grounds that they do not specify the terms with sufficient clarity. It is now time to introduce a new notation which replaces the three dots and offers greater precision. Moreover, as we shall see, the new notation is more compact, is more efficient, and has greater flexibility.

The symbol Σ is a capital sigma in the Greek alphabet and corresponds to our English S. Thus it naturally reminds us of the word sum. The symbol

$$\sum i^2$$

means that we are to sum the numbers i^2 for various integer values of i. The range for the integers is indicated by placing them below and above the Σ. For example,

$$\sum_{i=1}^{4} i^2$$

(read "the sum of i^2 as i runs from 1 to 4") means

$$1^2 + 2^2 + 3^2 + 4^2 = 1 + 4 + 9 + 16 = 30.$$

Thus we substitute in i^2 successively all of the integers between and including the lower and upper limits of summation, in this case $i = 1, 2, 3$, and 4, and then add the results.

The sum need not start at 1 or end at 4. Further, any letter can be used instead of i. Finally, any function may be used in place of i^2 and we indicate the sum $f(1) + f(2) + \cdots + f(n)$ by writing

$$\sum_{i=1}^{n} f(i).$$

The following examples should indicate the various possibilities. The new shorthand notation is on the left-hand side, and its meaning is on the right-hand side in each of these equations.

$$\sum_{i=1}^{4} i^3 = 1 + 8 + 27 + 64 = 100.$$

$$\sum_{k=3}^{8} k^2 = 9 + 16 + 25 + 36 + 49 + 64 = 199.$$

$$\sum_{j=3}^{5} j(j + 2) = 3 \cdot 5 + 4 \cdot 6 + 5 \cdot 7 = 74.$$

$$\sum_{j=1}^{8} 1 = 1 + 1 + 1 + 1 + 1 + 1 + 1 + 1 = 8.$$

$$\sum_{k=1}^{n} f(k) = f(1) + f(2) + f(3) + \cdots + f(n).$$

$$\sum_{t=1}^{n} f(t) = f(1) + f(2) + f(3) + \cdots + f(n).$$

$$\sum_{\theta=1}^{n} g(\theta) = g(1) + g(2) + g(3) + \cdots + g(n).$$

Sometimes the terms to be added involve subscripts or combinations of functions with subscripts. These possibilities are illustrated below.

$$\sum_{n=3}^{7} a_n = a_3 + a_4 + a_5 + a_6 + a_7.$$

$$\sum_{n=1}^{5} nb_n = b_1 + 2b_2 + 3b_3 + 4b_4 + 5b_5.$$

$$\sum_{k=1}^{n} \frac{a_k}{k} = a_1 + \frac{a_2}{2} + \frac{a_3}{3} + \cdots + \frac{a_n}{n}.$$

EXAMPLE 1. Find $\sum_{k=2}^{6} k^2$.

Solution. By the meaning of the symbols, we have

$$\sum_{k=2}^{6} k^2 = 4 + 9 + 16 + 25 + 36 = 90.$$

EXAMPLE 2. Use the Σ notation to rewrite formulas (1), (2), (3), and (4).

Solution. With the Σ notation, (1), (2), (3), and (4) become

(1⋆) $$\sum_{k=1}^{n} (2k - 1) = n^2.$$

(2⋆) $$\sum_{k=1}^{n} k = \frac{n(n + 1)}{2}.$$

(3⋆) $$\sum_{k=1}^{n} k^2 = \frac{n(n + 1)(2n + 1)}{6}.$$

(4⋆) $$\sum_{k=1}^{n} k^3 = \frac{n^2(n + 1)^2}{4}.$$

EXAMPLE 3. Prove that

(5)
$$\sum_{j=0}^{n} x^{n-j+1}y^j = \sum_{k=-1}^{n-1} x^{n-k}y^{k+1}$$

Solution. If we examine the exponent of y in (5) it appears that j has been replaced by $k + 1$, in the transition from the left-hand to the right-hand side. We set $j = k + 1$. As j runs through the integers 0, 1, 2, ..., n, the index $k = j - 1$ runs through the integers $-1, 0, 1, \ldots, n - 1$. The exponent $n - j + 1$ of x on the left-hand side becomes $n - (k + 1) + 1 = n - k$, the exponent of x on the right-hand side of (5). Consequently, both sides of equation (5) contain exactly the same terms.

As one might expect, there is a similar notation for products. The symbol Π is a capital pi in the Greek alphabet and corresponds to our English P. Thus it naturally reminds us of the word "product." The symbol

(6)
$$\prod_{i=1}^{n} f(i)$$

(read "the product of f of i as i runs from 1 to n") means the product

(7)
$$f(1)f(2)f(3) \cdots f(n).$$

The Π notation is not quite so useful as the Σ notation, but it is still worthwhile to have it available when the need arises.

EXAMPLE 4. Find $\prod_{i=1}^{4} i^2$.

Solution. By the meaning of the symbols we have

$$\prod_{i=1}^{4} i^2 = 1^2 \times 2^2 \times 3^2 \times 4^2 = 576.$$

EXAMPLE 5. Find a simpler form for

$$Q = \prod_{k=1}^{10} x^k.$$

Solution. Clearly we have

$$\prod_{k=1}^{10} x^k = x x^2 x^3 x^4 \cdots x^{10} = x^{1+2+3+\cdots+10} = x^s,$$

(8) $$s = 1 + 2 + 3 + \cdots + 10 = \sum_{k=1}^{10} k.$$

But by equation (2) or (2*), we find that $s = 10(10 + 1)/2 = 55$. Hence $Q = x^{55}$.

Exercise 1

1. Show by direct computations that each of the following assertions is true.

(a) $\displaystyle\sum_{k=5}^{8} k = 26.$ (b) $\displaystyle\sum_{k=1}^{5} k^3 = 225.$

(c) $\displaystyle\sum_{n=1}^{10} 2n = 110.$ (d) $\displaystyle\sum_{n=0}^{5} \frac{1}{2} n(n - 1) = 20.$

(e) $\displaystyle\sum_{n=1}^{5} \frac{1}{n} = \frac{137}{60}.$ (f) $\displaystyle\sum_{n=1}^{7} \frac{1}{n(n + 1)} = \frac{7}{8}.$

In problems 2 through 12 a number of assertions are given with the summation notation. In each case write out both sides of the equation in full and decide whether the given assertion is always true or sometimes may be false.

2. $c \displaystyle\sum_{k=1}^{n} k^4 = \sum_{k=1}^{n} ck^4.$

3. $c \displaystyle\sum_{k=1}^{n} a_k = \sum_{k=1}^{n} ca_k.$

4. $\displaystyle\sum_{k=1}^{N} b_k = \sum_{j=1}^{N} b_j.$

5. $\displaystyle\sum_{k=1}^{n} f(k) = \sum_{k=2}^{n+1} f(k).$

6. $\displaystyle\sum_{k=1}^{n} f(k) = \sum_{k=2}^{n+1} f(k - 1).$

7. $\left(\displaystyle\sum_{k=1}^{n} a_k\right)\left(\sum_{k=1}^{n} b_k\right) = \sum_{k=1}^{n} a_k b_k.$

8. $\displaystyle\sum_{k=1}^{n} b_k + \sum_{k=n+1}^{N} b_k = \sum_{k=1}^{N} b_k,$ if $1 < n < N.$

9. $\left(\displaystyle\sum_{k=1}^{n} a_k\right)^2 = \displaystyle\sum_{k=1}^{n} a_k^2 + \displaystyle\sum_{k=1}^{n} 2a_k + \displaystyle\sum_{k=1}^{n} 1.$

10. $\displaystyle\sum_{k=1}^{n} a_k + \displaystyle\sum_{k=1}^{n} b_k = \displaystyle\sum_{k=1}^{n} (a_k + b_k).$

11. If $a_k \leq b_k$ for each positive integer k, then $\displaystyle\sum_{k=1}^{n} a_k \leq \displaystyle\sum_{k=1}^{n} b_k.$

12. $\displaystyle\sum_{k=0}^{n} a_k = \displaystyle\sum_{k=0}^{n} a_{n-k}.$

★13. Solve the equation $\displaystyle\sum_{k=1}^{5} k^2 = \displaystyle\sum_{k=1}^{5} t^2.$

14. Give explicitly the meaning of $\displaystyle\sum_{k=0}^{5} x^k y^{5-k}.$

15. As in problem 14, give explicitly $\displaystyle\sum_{i=2}^{6} A^i B^{7-i}.$

16. Define $n!$ using the Π symbol.

17. Evaluate each of the following:

(a) $\displaystyle\prod_{k=1}^{3} 2^k.$
(b) $\displaystyle\prod_{k=1}^{5} 1^k.$

(c) $\displaystyle\prod_{k=3}^{7} (k - 2).$
(d) $\displaystyle\prod_{k=5}^{20} (k - 7).$

(e) $\displaystyle\prod_{k=1}^{9} k^2.$
(f) $\displaystyle\prod_{j=1}^{5} x^{-j}.$

In problems 18 through 23, find simpler forms for the given quantity.

18. $\displaystyle\prod_{i=1}^{5} \frac{i}{i+1}.$
19. $\displaystyle\prod_{j=1}^{6} \frac{j+3}{j+1}.$

★20. $\displaystyle\prod_{k=1}^{n} \frac{(k+1)(k+3)}{k+2}.$
★21. $\displaystyle\prod_{k=1}^{n-1} \frac{k}{n-k}.$

★★22. $\dfrac{\displaystyle\prod_{k=1}^{n} k^3}{\displaystyle\prod_{j=3}^{n+1} (j-1)^2}$
★★23. $\displaystyle\prod_{n=1}^{5} \left(\displaystyle\sum_{k=1}^{n} k\right).$

★24. Prove that

$$(x+y) \displaystyle\sum_{k=0}^{n} a_k x^{n-k} y^k = a_0 x^{n+1} + a_n y^{n+1} + \displaystyle\sum_{k=0}^{n-1} (a_k + a_{k+1}) x^{n-k} y^{k+1}.$$

★25. For each of the following, state whether the assertion is always true or sometimes may be false:

(a) If $1 \leq k < n$, then $\prod_{i=1}^{k} a_i \prod_{i=k+1}^{n} a_i = \prod_{i=1}^{n} a_i.$

(b) If $1 \leq k$, then $\prod_{i=1}^{k} a_i + \prod_{i=2}^{k+1} a_i = (a_1 + a_{k+1}) \prod_{i=2}^{k} a_i.$

(c) If $1 \leq k < n$, and all a_i's are different from zero, then

$$\frac{\prod_{i=1}^{n} a_i}{\prod_{i=1}^{k} a_i} = \prod_{i=k+1}^{n} a_i.$$

★26. If $k = n$ in part (c) of problem 25, then the left-hand side equals 1. But the right-hand side is

$$\prod_{i=n+1}^{n} a_i,$$

an empty product. How would you define this empty product?

2. the binomial theorem

The expression $(x + y)^n$ occurs so frequently in mathematics that it pays to have a systematic method of writing the expansion of this expression. By direct (and for higher values of n laborious) computation, we find that

$$(x + y)^0 = \qquad\qquad 1$$
$$(x + y)^1 = \qquad\qquad x + y$$
$$(x + y)^2 = \qquad\qquad x^2 + 2xy + y^2$$
$$(x + y)^3 = \qquad\qquad x^3 + 3x^2y + 3xy^2 + y^3$$
$$(x + y)^4 = \qquad\qquad x^4 + 4x^3y + 6x^2y^2 + 4xy^3 + y^4$$
$$(x + y)^5 = \qquad\qquad x^5 + 5x^4y + 10x^3y^2 + 10x^2y^3 + 5xy^4 + y^5$$
$$(x + y)^6 = \qquad x^6 + 6x^5y + 15x^4y^2 + 20x^3y^3 + 15x^2y^4 + 6xy^5 + y^6.$$

Can we predict the expansion of $(x + y)^7$ from a study of these expansions?

The pattern of the powers of x and y is clearly discernible. Indeed in going from left to right the powers of x decrease from n to 0 and the powers of y increase from 0 to n. If we ignore the coefficients (the numbers that multiply x and y), then the structure of $(x + y)^n$ has the form

$$x^n + (\text{---})x^{n-1}y + (\text{---})x^{n-2}y^2 + \cdots + (\text{---})x^{n-k}y^k + \cdots + y^n,$$

where (———) represents the missing coefficient. Consequently, to write the expansion of $(x + y)^n$ we need to know only the coefficients.

We shall soon see that the coefficient of $x^{n-k}y^k$ is simply $C(n,k)$, the number of combinations of n things taken k at a time. However, custom dictates that a different symbol $\binom{n}{k}$ (read "n above k") be used for this quantity. Since both of the symbols $C(n,k)$ and $\binom{n}{k}$ are in common use, we shall employ both interchangeably.

Definition 1. For each positive integer n and each integer k, such that $0 \leq k \leq n$, the symbol $\binom{n}{k}$ is defined by

(9)
$$\binom{n}{k} = C(n,k) = \frac{n!}{k!(n-k)!}.$$

We observe that if $k = 0$ or $k = n$, the quantity $0!$ appears on the right-hand side of (9) and we recall[1] that $0! = 1$. Consequently, for these special values of k we have

(10)
$$\binom{n}{0} = \frac{n!}{0!n!} = 1 \quad \text{and} \quad \binom{n}{n} = \frac{n!}{n!0!} = 1.$$

We also recall from Chapter 3, Theorem 3, (page 88) that

(11)
$$C(n,k) + C(n, k+1) = C(n+1, k+1).$$

Restated in our new notation, equation (11) gives

Theorem 1. For each positive integer n and each integer k, such that $0 \leq k \leq n-1$,

(12)
$$\binom{n}{k} + \binom{n}{k+1} = \binom{n+1}{k+1}.$$

The important result for this section is

Theorem 2. For each positive integer n, and each integer k, such that $0 \leq k \leq n$, the coefficient of $x^{n-k}y^k$ in the expansion of $(x + y)^n$ is $\binom{n}{k}$.

---

[1] See Chapter 3, Exercise 1, problem 23.

In other words

(13)
$$(x + y)^n = \sum_{k=0}^{n} \binom{n}{k} x^{n-k} y^k$$

or

(14)
$$(x + y)^n = \binom{n}{0} x^n + \binom{n}{1} x^{n-1} y + \binom{n}{2} x^{n-2} y^2 + \cdots +$$

$$\binom{n}{k} x^{n-k} y^k + \cdots + \binom{n}{n-1} xy^{n-1} + \binom{n}{n} y^n.$$

Proof. We use mathematical induction. It is sufficient to begin with $n = 1$, but this case is so simple that we prefer to examine the more interesting case $n = 2$. Certainly, we already know that

(15)
$$(x + y)^2 = x^2 + 2xy + y^2.$$

But from (9) we also have

(16) $\binom{2}{0} = \dfrac{2!}{0!2!} = 1, \qquad \binom{2}{1} = \dfrac{2!}{1!1!} = 2, \qquad \binom{2}{2} = \dfrac{2!}{2!0!} = 1,$

and these are indeed the coefficients in (15). Hence Theorem 2 is true for $n = 2$.

We now assume that the theorem is true for index n. Thus we assume that equation (13) holds. Using equation (13) and the definition of $(x + y)^{n+1}$ we have

$$(x + y)^{n+1} = (x + y)(x + y)^n = (x + y) \sum_{j=0}^{n} \binom{n}{j} x^{n-j} y^j$$

$$= x \sum_{j=0}^{n} \binom{n}{j} x^{n-j} y^j + y \sum_{j=0}^{n} \binom{n}{j} x^{n-j} y^j$$

(17)
$$(x + y)^{n+1} = \sum_{j=0}^{n} \binom{n}{j} x^{n-j+1} y^j + \sum_{j=0}^{n} \binom{n}{j} x^{n-j} y^{j+1}.$$

To find the coefficient of the term involving y^{k+1} on the right-hand side of (17) we set $j = k + 1$ in the first sum and $j = k$ in the second sum. Combining these, we find that the coefficient of $x^{n-k} y^{k+1}$ on the right-hand side of (17) is

(18)
$$\binom{n}{k+1} + \binom{n}{k}.$$

By Theorem 1, this new coefficient is

(19)
$$\binom{n}{k+1} + \binom{n}{k} = \binom{n+1}{k+1}.$$

But the right-hand side of (19) has exactly the form required for the coefficient of $x^{n-k}y^{k+1}$ in the expansion of $(x+y)^{n+1}$. This argument holds if $k = 0, 1, 2, \ldots, n-1$. The extreme terms, x^{n+1} and y^{n+1} in (17) are not covered by the preceding discussion, but these terms obviously have 1 as coefficients, as required by Theorem 2. ∎

The binomial theorem in its most general form gives the expansion of $(x+y)^n$ if n is an arbitrary real or complex number. We have proved the binomial theorem in the special case that n is a positive integer but this is enough for our needs.

EXAMPLE 1. Expand $(A - 3B)^4$.

Solution. By Theorem 2 the coefficients are

$$\frac{4!}{0!4!} = 1, \qquad \frac{4!}{1!3!} = 4, \qquad \frac{4!}{2!2!} = 6, \qquad \frac{4!}{3!1!} = 4, \qquad \frac{4!}{4!0!} = 1.$$

We can write $(A - 3B)^4 = [A + (-3B)]^4$. Hence

$$(A - 3B)^4 = A^4 + 4A^3(-3B) + 6A^2(-3B)^2 + 4A(-3B)^3 + (-3B)^4$$
$$= A^4 - 12A^3B + 54A^2B^2 - 108AB^3 + 81B^4.$$

EXAMPLE 2. Find the coefficient of x^7y^3 in the expansion of $(x+y)^{10}$.

Solution. Here $n = 10$ and $k = 3$. By Theorem 2 the coefficient is

$$\binom{10}{3} = \frac{10!}{3!(10-3)!} = \frac{10 \cdot 9 \cdot 8 \cdot 7 \cdot 6 \cdot 5 \cdot 4 \cdot 3 \cdot 2 \cdot 1}{3 \cdot 2 \cdot 1 \cdot 7 \cdot 6 \cdot 5 \cdot 4 \cdot 3 \cdot 2 \cdot 1} = 120.$$

★3. the Pascal triangle

When the coefficients in the expansion of $(x+y)^n$ are arranged as shown in Fig. 1, the triangle of numbers that we obtain is known as Pascal's triangle.[1]

[1] Blaise Pascal (1623–1662), at the age of 16, made important discoveries in geometry. Together with Pierre Fermat, he founded the theory of probability.

$(x + y)^0$

$(x + y)^1$

$(x + y)^2$

$(x + y)^3$

$(x + y)^4$

$(x + y)^5$

$(x + y)^6$

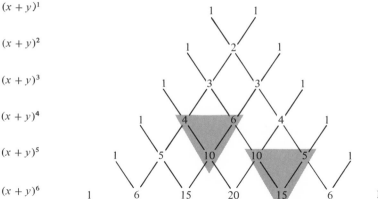

FIGURE 1

An alternative rule of formation for this triangle is almost obvious. First, the border of the triangle is formed of 1's. Second, each number inside the triangle is the sum of the two numbers which appear in the line above just to the left and to the right of the number. This is indicated in Fig. 1 by the little colored triangles. The first one shows that 10 is obtained as $4 + 6$, and the second little triangle shows that 15 is obtained as $10 + 5$. We have already proved in Theorem 1 that this rule holds for all of the entries inside Pascal's triangle, no matter how far downward the triangle is extended. This is quite useful because we can now obtain the expansion of $(x + y)^7$ very quickly. Indeed, we use the rule just mentioned to find the next row in Fig. 1. According to the rule these new entries are

$$1, \quad 1 + 6 = 7, \quad 6 + 15 = 21, \quad 15 + 20 = 35,$$
$$20 + 15 = 35, \quad 15 + 6 = 21, \quad 6 + 1 = 7, \quad 1.$$

Consequently (by Theorems 1 and 2),

$$(x + y)^7 = x^7 + 7x^6y + 21x^5y^2 + 35x^4y^3 + 35x^3y^4 + 21x^2y^5 + 7xy^6 + y^7.$$

When n is large, it is impractical to use Pascal's triangle, and it is better to use equation (9) to obtain the binomial coefficients. However, if n is small (say $n \leq 10$), then Pascal's triangle gives the results more rapidly. Of course if one must use the binomial coefficients very often, then it is best to have at hand a reliable table. A brief table of binomial coefficients is given in Appendix 4.

Using either Pascal's triangle or the factorial representation in Definition 1, it is clear that

(20)
$$\binom{n}{k} = \binom{n}{n-k}.$$

This symmetry allows us to cut the table in half.

Exercise 2

1. Extend Pascal's triangle and find the coefficients in the expansion of $(x + y)^8$ and $(x + y)^9$.
2. Find the coefficient of x^3y^2 in the expansion of $(3x - 2y)^5$.
3. Find the coefficient of x^2y^5 in the expansion of $[(x/2) - (y/3)]^7$.
4. Find the coefficient of u^7v^2 in the expansion of $(2v - u)^9$.

In problems 5 through 16, compute explicitly the given quantity.

5. $\binom{10}{5}$.

6. $\binom{11}{8}$.

7. $\binom{11}{2}$.

8. $\binom{55}{54}$.

9. $\binom{99}{0}$.

10. $\dfrac{5!}{0!}$

11. $\binom{5}{4}\binom{6}{5}$.

12. $\binom{6}{5}\binom{7}{6}$

13. $\dfrac{8}{\binom{6}{4}}$.

14. $\dfrac{\binom{8}{6}}{4}$.

15. $\binom{6}{\binom{4}{2}}$.

16. $\binom{\binom{6}{4}}{2}$.

★17. Prove that if $0 \leqq k \leqq n - 2$, then

$$\binom{n+2}{k+2} = \binom{n}{k+2} + 2\binom{n}{k+1} + \binom{n}{k}.$$

Hint: Either use equation (12) twice or use the factorial expression in Definition 1.

★18. Prove that for each positive integer n,

$$\sum_{k=0}^{n}\binom{n}{k} = \binom{n}{0} + \binom{n}{1} + \binom{n}{2} + \cdots + \binom{n}{n} = 2^n.$$

Hint: Let $x = y = 1$ in equation (14).

★19. Prove that for each positive integer n,

$$\sum_{k=0}^{n} (-1)^k \binom{n}{k} = 0 \quad \text{and} \quad \sum_{k=0}^{n} 2^k \binom{n}{k} = 3^n.$$

In each case write explicitly a few terms of the sum indicated on the left. Check the assertion for small values of n using Pascal's triangle. For example, if $n = 3$, then $1 - 3 + 3 - 1 = 0$ and $1 + 6 + 12 + 8 = 27$.

★20. Give an alternative proof of Theorem 2 by checking the details in the following argument. By definition

$$(x + y)^n = (x + y)(x + y) \cdots (x + y),$$

where the factor $(x + y)$ on the right-hand side is repeated n times. When the multiplication indicated on the right is performed, the term $x^k y^{n-k}$ occurs once for each selection of k x's. Since there are n factors, the number of such terms is the number of combinations of n things taken k at a time. Hence the coefficient of $x^k y^{n-k}$ is $C(n,k)$.

In problems 21 through 25, find the coefficient of the specified term in the expansion of the given binomial.

21. $\left(x^3 - \dfrac{2}{x^2}\right)^4$, coefficient of x^2 and x^4.

22. $(x^2 + 3x^3)^5$, coefficient of x^9, x^{11}, and x^{13}.
23. $(1 - u^3)^{11}$, coefficient of u^{10}, u^{20}, and u^{30}.
24. $[3A + (B - 2A)]^7$, coefficient of $A^3 B^4$ and $A^4 B^3$.
★25. $[2w + (w - 1)^2]^9$, coefficient of w^6, w^9, and w^{12}.

4. partitions

Let us look at combinations in a new light, as illustrated in

EXAMPLE 1.   Eleven senators are to be divided into two subcommittees. The first set, consisting of 5 members, is to discuss foreign affairs, and the remaining 6 are to discuss domestic matters. In how many different ways can these subcommittees be formed?

Solution.   Once the first 5 are selected, the remaining 6 are automatically determined. Consequently we need to find only the number of ways of selecting the first 5. But this is exactly the number of combinations of 11 things taken 5 at a time. Hence the number of ways is

(21) $C(11,5) = \dbinom{11}{5} = \dfrac{11!}{5!6!} = 462.$

To extend this problem we now consider three subcommittees. Of course we are not concerned with the subject that they choose to discuss, so we reword the problem slightly in

EXAMPLE 2. Eleven persons are to be distributed in three rooms, A, B, and C, so that 5 are in room A, 2 are in room B, and 4 are in room C. In how many ways can this be done?

Solution. The number of ways of selecting 5 persons for room A is $C(11,5)$ as before. There are now 6 persons left to distribute among rooms B and C. Clearly there are $C(6,2)$ ways of placing 2 of those remaining in room B, and the rest are left for room C. Hence the number of ways is the product

$$(22) \qquad C(11,5)C(6,2) = \frac{11!}{5!6!} \cdot \frac{6!}{2!4!} = \frac{11!}{5!2!4!} = 6930.$$

The distribution described above is an example of a *partition*. In this case we are forming a partition of 11 things into three sets, with 5 in the first set, 2 in the second set, and 4 in the third set. In this example the order of the sets is regarded as important (5 in room A and 2 in room B is not the same as 2 in room A and 5 in room B). When the order of the subsets is important, we call the distribution an *ordered partition*. In the general case we may have n things (all different or distinguishable) that are to be divided into r subsets. If we wish we may regard the subsets as placed in rooms or boxes: B_1, B_2, ..., B_r, and we are to put n_1 of the things into box B_1, n_2 of the things into box B_2, etc. Of course if we do not run short of things or have some left over, then we must have

$$(23) \qquad n_1 + n_2 + \cdots + n_r = n.$$

Definition 2. Ordered Partition. Any distribution of n different things into r sets with n_1 in the first set, n_2 in the second set, ..., and n_r in the rth set is called an ordered partition. The number of such ordered partitions into r sets with n_k elements in the kth set is denoted by the symbol

$$(24) \qquad \binom{n}{n_1, n_2, \ldots, n_r}.$$

A formula for computing the number of ordered partitions is the theme of

> **Theorem 3.** For any set of positive integers that satisfies equation (23),

$$(25) \qquad \binom{n}{n_1, n_2, \ldots, n_r} = \frac{n!}{n_1! n_2! \cdots n_r!}.$$

Proof. We use induction on r, the number of subsets. We begin with $r = 2$ (but the reader should observe that the theorem is true when $r = 1$).

If $r = 2$, we set $n_1 = k$, and then by equation (23) we must have $n_2 = n - k$.

In this special case the theorem asserts that

$$(26) \qquad \binom{n}{k, n - k} = \frac{n!}{k!(n - k)!}.$$

But (26) is indeed true, because the number of ordered partitions of n things into two sets with k things in the first set is just the number of combinations of n things taken k at a time, and this is the right-hand side of (26).

We next assume that (25) is true for r sets, and we shall prove that then it is true for $r + 1$ sets. In this induction, equation (23) must be modified to read

$$(27) \qquad n_1 + n_2 + \cdots + n_r + n_{r+1} = n$$

because we now have $r + 1$ sets, and we use $n\star$ to denote the number of elements that are to be distributed into the first r sets. Thus

$$(28) \qquad n_1 + n_2 + \cdots + n_r = n\star$$

and we observe that $n - n\star = n_{r+1}$.

We make the distribution in two steps. We select $n\star$ things from the set of n things, leaving n_{r+1} things for the $(r + 1)$th set. This can be done in $C(n, n\star)$ different ways, where

$$(29) \qquad C(n, n\star) = \frac{n!}{n\star!(n - n\star)!} = \frac{n!}{n\star! n_{r+1}!}.$$

The $n\star$ things selected are now distributed into the r sets putting n_1 into the first set, n_2 into the second set, $\ldots$, and n_r into the rth set. According to the induction hypothesis, the number of ways in which this can be done is

$$(30) \qquad \binom{n\star}{n_1, n_2, \ldots, n_r} = \frac{n\star!}{n_1! n_2! \cdots n_r!}.$$

The number of ways in which the two steps can be taken is the product (see Theorem 14 in Chapter 2, page 61). Using equations (29) and (30) we obtain

$$(31) \qquad \frac{n!}{n\star! n_{r+1}!} \cdot \frac{n\star!}{n_1! n_2! \cdots n_r!} = \frac{n!}{n_1! n_2! \cdots n_r! n_{r+1}!}.$$

This is equation (25) for $r + 1$ sets. ∎

EXAMPLE 3. In how many different ways can nine children be divided into sets of three to be sent to three different summer camps?

Solution. By Definition 2 and Theorem 3,

$$(32) \qquad \binom{9}{3,3,3} = \frac{9!}{3!3!3!} = 1680.$$

The numbers denoted by the symbol in (24) are called *multinomial* coefficients, and we shall see why in the next section.

Unordered partitions are a little more complicated than ordered partitions. As before, we partition a set of n distinguishable things into r subsets A_1, $A_2, \ldots, A_r$, but if A_1 and A_2 have the same number of elements, then the two partitions $(A_1, A_2, \ldots, A_r)$ and $(A_2, A_1, \ldots, A_r)$ are regarded as the same for an unordered partition, while they are regarded as different for an ordered partition. A general formula for the number of unordered partitions is too cumbersome, but the following examples should suffice.

EXAMPLE 4. In how many different ways can nine sales executives be divided into three discussion groups of three each to discuss methods of promoting sales?

Solution. This example is similar in structure to Example 3, but with one important difference. In Example 3 the camps were different, so that Abel, Bozo, and Cory at camp X is not the same as Abel, Bozo, and Cory at camp Y. Thus in Example 3 we are to count the number of *ordered* partitions.

In this example Adams, Brown, and Cutler presumably have the same discussion, no matter which conference room they use, and hence we are to count the *unordered* partitions. Now each unordered partition into sets A, B, C gives rise to $3! = 6$ ordered partitions by permuting A, B, amd C. Thus if U is the number of unordered partitions, then

$$6U = \binom{9}{3,3,3} = 1680,$$

or $U = 280$.

EXAMPLE 5. We modify Examples 3 and 4 as follows:
(a) Two children go to camp X, four children go to Camp Y, and three children go to camp Z.
(b) The nine executives break up into three groups, one with four members, one with three members, and one group with just two members.

Solution. Both the number of ordered partitions in (a) and the number of unordered partitions in (b) are given by

$$\binom{9}{2,3,4} = \frac{9!}{2!3!4!} = 1260.$$

Exercise 3

In problems 1 through 9, compute the given multinomial coefficient.

1. $\binom{7}{2,2,3}$.

2. $\binom{8}{1,2,5}$.

3. $\binom{6}{1,1,2,1,1}$.

4. $\binom{8}{2,4,2}$.

5. $\binom{9}{1,1,7}$.

6. $\binom{11}{7,2,1,1}$.

7. $\binom{6}{4,2,0}$.

8. $\binom{9}{7,0,1,1}$.

9. $\binom{6}{0,2,4,0,0}$.

10. Prove that if $h + i + j = n$, then

$$\binom{n}{h}\binom{n-h}{i} = \binom{n}{h,i,j}.$$

11. Prove that if $h + i + j + k = n$, then

$$\binom{n}{h}\binom{n-h}{i}\binom{n-h-i}{j} = \binom{n}{h,i,j,k}.$$

12. Prove that if $n \geq r \geq k \geq s \geq 0$, then

$$\binom{n}{r}\binom{r}{k}\binom{k}{s} = \binom{n}{n-r,r-k,k-s,s}.$$

13. Convert the following product to a suitable multinomial coefficient:

$$\binom{7}{6}\binom{6}{5}\binom{5}{4}\binom{4}{3}\binom{3}{2}\binom{2}{1}.$$

14. In how many ways can eight different toys be divided among three children if the oldest gets two toys and the others get three toys each?

15. The Carcinoma College football team is scheduled to play nine games. In how many different ways can the season end with four wins, three losses, and two ties?

16. In how many ways can the Carcinoma team win six games, lose two, and tie one?

17. Twenty fraternity brothers pile into three distinguishable cars to drive to a picnic. In how many different ways can this be done if the Dunebuggy holds 6, the Mustang holds 7, and the Dart holds 7, and nobody knows or cares who is driving? What is the situation if each owner insists on driving his own car?

18. Grades A, B, C, and F are to be assigned to a class of 12 students. Find the number of ways this can be done if
    (a) There are no restrictions.
    (b) The teacher is instructed to give two A's, two B's, six C's, and two F's.
    (c) The teacher wants a bell-shaped curve with three A's, six B's, three C's and no F's.
    (d) The number of A's and C's is the same, and the number of B's is just one more than the number of F's.
    (e) There are three A's, five B's, and more C's than F's.

19. Ten mutual strangers at a resort register for a doubles tennis tournament. In how many different ways can they be assigned partners (arranged in five pairs)?

20. The 10 players in problem 19 insist on playing singles. In how many ways can the first round of the tournament be arranged?

21. In how many ways can the 10 players in problem 20 break up into groups of 3, 3, and 4 to discuss the results of the first round?

22. A gambler decides always to play the favorite at a dog race. If there are 11 races, in how many different ways can he play the favorite to win two times, to place three times, and to show six times?

In problems 23 through 26, give answers using multinomial coefficients. Do not compute.

23. In how many ways can 40 people be divided into four discussion groups with 10 in each group?

24. Four men are playing poker. How many different games (deals) are possible? Two games are considered different if one of the players has a different hand in the two games.

25. Find the number of different bridge deals.

26. A large business has trained 31 young executives. In how many different ways can these ambitious young men be distributed among five branch offices, if at least 6 are sent to each office? How many ways are there for 32 trainees?

★27. Many problems can be reduced to the following type of problem on permutations with repetitions. Given n_1 letters of type A_1, n_2 letters of type A_2, ..., and n_r letters of type A_r, prove that if $n_1 + n_2 + \cdots + n_r = n$, then the number of different n-letter words that can be made with these letters is

$$\binom{n}{n_1, n_2, \ldots, n_r}.$$

Hint: Number the position of the letters in the word from 1 to n. Then consider the ordered partition of these n numbers into sets $A_1^\star, A_2^\star, \ldots,$ $A_r^\star$ (Compare this type of argument with that given in Chapter 3, Section 4, Example 3, page 91).

28. Use the formula from problem 27 to find the number of different permutations of the letters in each of the following words: **(a)** *mimic,* **(b)** *parallel,* **(c)** *Tennessee,* and **(d)** *Mississippi.*

29. The first row of pieces in a chess game consists of two rooks, two knights, two bishops, one queen, and one king (at the beginning of the game). In how many wrong ways can this first row be set up to start the game?

5.   the multinomial theorem

We begin with

EXAMPLE 1.   Find the expansion of $(x + y + z)^4$.

Solution.   The energetic student who performs the indicated multiplication will find that

(33)
$$\begin{aligned}
(x + y + z)^4 = {} & x^4 + y^4 + z^4 \\
& + 4(x^3y + x^3z + xy^3 + xz^3 + y^3z + yz^3) \\
& + 6(x^2y^2 + x^2z^2 + y^2z^2) \\
& + 12(x^2yz + xy^2z + xyz^2).
\end{aligned}$$

Surely there must be an easier way. First we observe that in each term, the sum of the exponents is 4. Further every nonnegative collection of exponents that has the sum 4 does occur in (33); i.e.,

$$4 + 0 + 0 = 4, \quad 3 + 1 + 0 = 4, \quad 2 + 2 + 0 = 4, \quad 2 + 1 + 1 = 4.$$

Consequently, if we use (——) as a placeholder for the coefficients, then (33) has the form

(34) $$(x + y + z)^4 = \sum_{n_1+n_2+n_3=4} (\text{---}) x^{n_1} y^{n_2} z^{n_3}.$$

We examine more closely the formation of the product

(35) $$P = (x + y + z)(x + y + z)(x + y + z)(x + y + z)$$
$$= \quad A \quad \times \quad B \quad \times \quad C \quad \times \quad D.$$

The expansion of P can be obtained by selecting one term from each of the factors A, B, C, and D, forming the product of the four selected, and then taking a sum of all such products. Now the term $x^2 yz$ will occur once for every selection of x from two of the factors A, B, C, D; y from one of the two remaining factors; and z from the remaining factor. The number of such selections is the coefficient of $x^2 yz$ in (34). Hence the coefficient must be

$$\binom{4}{2, 1, 1} = \frac{4!}{2!1!1!} = 12,$$

and this is consistent with equation (33). The same type of argument will apply to any term of the expansion. Hence (33) can be written in the compact form

(36) $$(x + y + z)^4 = \sum_{n_1+n_2+n_3=4} \binom{n}{n_1, n_2, n_3} x^{n_1} y^{n_2} z^{n_3},$$

where the sum includes every sequence n_1, n_2, n_3 of three nonnegative integers with sum 4.

A careful analysis of the argument reveals that it is valid for any exponent $n \geq 1$ and for any number of variables. In stating the general result we replace the three variables x, y, and z by the r variables x_1, x_2, ..., x_r.

Theorem 4. The Multinomial Theorem. For any integer $n \geq 2$, and $r \geq 3$,

(37) $$(x_1 + x_2 + \cdots + x_r)^n$$

$$= \sum_{n_1+n_2+\cdots+n_r=n} \binom{n}{n_1, n_2, \ldots, n_r} x_1^{n_1} x_2^{n_2} \cdots x_r^{n_r},$$

where the sum is over every sequence $n_1, n_2, \ldots, n_r$, of nonnegative integers such that $n_1 + n_2 + \cdots + n_r = n$.

Proof. From the n factors on the left-hand side of (37) we select n_1 of these to form a set A_1. We select n_2 of the remaining factors to form a set A_2, ..., and the remaining n_r factors to form A_r. We obtain the term

(38) $$x_1^{n_1} x_2^{n_2} \cdots x_r^{n_r}$$

by selecting x_1 from each of the factors in A_1, x_2 from each of the factors in A_2, etc. Since the ordered partition of the factors into sets A_1, A_2, ..., A_r can be done in

$$\binom{n}{n_1, n_2, \ldots, n_r}$$

different ways, this is the coefficient of the term (38) on the right-hand side of (37). ■

EXAMPLE 2. In the expansion of $(x + y + z + w)^{10}$, find the coefficient of
(a) $x^4 y^3 z^2 w$.
(b) $x^2 y^3 z^4 w^5$.

Solution.

(a) $\dbinom{10}{4, 3, 2, 1} = \dfrac{10!}{4!3!2!1!} = 12{,}600.$

(b) Since $2 + 3 + 4 + 5 = 14 \neq 10$, the coefficient is 0.

Exercise 4

1. By direct multiplication find the expansion of

(a) $(x + y + z)^2$,   (b) $(x + y + z)^3$,   (c) $(x + y + z + w)^2$,

and in each case check that your result is identical with that given in Theorem 4.
2. Using the result in problem 1(b), check equation (33).
3. List the solutions of $n_1 + n_2 + n_3 + n_4 = 6$, where to shorten the labor and avoid duplications we assume that $n_1 \geq n_2 \geq n_3 \geq n_4 \geq 0$.

In problems 4 through 11, find the term in the expansion of the given quantity that contains the given variable with the indicated exponent.

4. Term containing $xy^3 z^2$ in $(2x - y + 4z)^6$.
5. Term containing $x^4 y^3 z$ in $(x - 2y + 3z)^8$.

6. Term containing xy^5zw in $(x - y + z - w)^8$.

★7. Term containing x^6 in $(x^2 + y + zx)^4$.

★8. Term containing x^9 in $(x^3 + xy + z)^3$.

★★9. Term containing y^3 in $(xy + 2yz - 5zx)^6$.

★★10. Term containing z^5 in $(xz + 8yz - 2z^2x^3)^5$.

★★11. Term containing x^7 in $(x + x^2y + x^3z + w)^4$.

12. Give the proof of Theorem 4, in your own words.

★13. Using the factorial representation, equation (25), prove that

$$\binom{n+1}{i,j,k} = \binom{n}{i-1,j,k} + \binom{n}{i,j-1,k} + \binom{n}{i,j,k-1},$$

where $i + j + k = n + 1$.

★14. Prove the relation in problem 13 by using a combinatorial argument similar to that used to prove Theorem 3, Chapter 3 (page 88).

★15. Prove the relation in problem 13 by using Theorem 4, equation (37), and the fact that

$$(x + y + z)^{n+1} = (x + y + z)(x + y + z)^n.$$

Of the three proofs in problems 13, 14, and 15, which do you prefer?

★★16. From example 1, equation (33), we see that there are 15 terms in the expansion of $(x + y + z)^4$. Let $E(n, r)$ denote the number of terms in the expansion of $(x_1 + x_2 + \cdots + x_r)^n$. Fill in the details of the following argument to prove that

(39)
$$E(n,r) = \binom{n+r-1}{n} = \frac{(n+r-1)!}{n!(r-1)!}.$$

(a) Equation (39) is true when $n = 1$ and $r \geq 1$.

(b) Equation (39) is true when $n \geq 1$ and $r = 1$.

(c) $\binom{n-1+r-1}{n-1} + \binom{n+r-2}{n} = \binom{n+r-1}{n}$.

(d) In the expansion of $(x_1 + x_2 + \cdots + x_r)^n$, show that $E(n, r - 1)$ is the number of terms that do not contain x_r and that $E(n - 1, r)$ is the number of terms that do contain x_r. Hence

$$E(n - 1, r) + E(n, r - 1) = E(n, r).$$

(e) Use mathematical induction on the sum $n + r$. Combining parts (c) and (d), permit the induction from $n + r - 1$ to $n + r$.

17. Find the number of terms in the expansion of each of the following:

(a) $(x + y + z)^7$.

(b) $(A + B - C - D)^8$.

(c) $(u + 2v - 3w + 4x - 5y + 7z)^5$.

★6.   a general counting theorem

We shall need

> **Definition 3. The Characteristic Function.** Let U be a universal set and let S be a subset of U. The characteristic function X_S for the set S is defined by
>
> (40) $$X_S(a) = \begin{cases} 1, & \text{if } a \in S, \\ 0, & \text{if } a \notin S. \end{cases}$$

For example, if S is the set of natural numbers and U is the set of real numbers, then

$$X_S(5) = 1, \qquad X_S(\sqrt{3}) = 0, \qquad X_S(\sqrt{49} + 17) = 1, \qquad X_S\left(\frac{42}{24}\right) = 0.$$

Suppose that $S = S_1 \cap S_2$. How is the characteristic function of S related to the characteristic functions of S_1 and S_2? If $a \in S_1$ and $a \in S_2$, then $a \in S$ and we find that

(41) $$X_S(a) = 1 = 1 \times 1 = X_{S_1}(a)X_{S_2}(a).$$

In any other case the two extreme terms in (41) give zero. Hence the characteristic function satisfies a product rule stated in

> **Theorem 5.** If $S = S_1 \cap S_2$, then
>
> (42) $$X_S = X_{S_1}X_{S_2}.$$

The same type of argument gives immediately

Theorem 6. If $S = S_1 \cap S_2 \cap \cdots \cap S_k$, then

(43) $X_S = X_{S_1}X_{S_2} \cdots X_{S_k}.$

We now consider the product

(44) $P = \prod_{i=1}^{r} (1 - X_{S_i}) = (1 - X_{S_1})(1 - X_{S_2}) \cdots (1 - X_{S_r}).$

Suppose that $a \in U$. If a belongs to just one of the sets $S_1, S_2, \ldots, S_r$, then one of the factors on the right-hand side of (44) is zero and hence $P = 0$. If a does not belong to any of these sets, then $P = 1$. Hence P is the characteristic function for C, where

(45) $C = (S_1 \cup S_2 \cup \cdots \cup S_r)'.$

On the other hand we can expand the right-hand side of (44). This gives

(46) $X_C = 1 - \sum_{i=1}^{r} X_{S_i} + \sum_{i<j} X_{S_i}S_j$

$$- \sum_{i<j<k} X_{S_i}X_{S_j}X_{S_k} + \cdots + (-1)^r \prod_{i=1}^{r} X_{S_i}.$$

Here the notation $\sum_{i<j}$ on the right-hand side means that we are to sum all pairs $X_{S_i}X_{S_j}$ for which $i < j$ with $1 \leq i$ and $j \leq r$. A similar and obvious interpretation is to be made for $\sum_{i<j<k}$ and the dots indicate that the number of indices increases until all are included in the last term.

For example, if $r = 4$, then equation (46) is

$X_C = 1 - (X_{S_1} + X_{S_2} + X_{S_3} + X_{S_4})$
$\quad + X_{S_1}X_{S_2} + X_{S_1}X_{S_3} + X_{S_1}X_{S_4} + X_{S_2}X_{S_3} + X_{S_2}X_{S_4} + X_{S_3}X_{S_4}$
$\quad - (X_{S_1}X_{S_2}X_{S_3} + X_{S_1}X_{S_2}X_{S_4} + X_{S_1}X_{S_3}X_{S_4} + X_{S_2}X_{S_3}X_{S_4})$
$\quad + X_{S_1}X_{S_2}X_{S_3}X_{S_4}.$

Now let $a \in U$ and evaluate each of the characteristic functions at a. If $a \in C$, the left-hand side of (46) is 1 and the right-hand side is a sum that must also give 1. Similarly if $a \notin C$, both sides of (46) give zero. We perform this evaluation for each $a \in U$, obtaining $n(U)$ equations. When we add these equations, each of the products of characteristic functions is replaced by the number of elements in the intersection of the corresponding sets (see Theorem 6). In this way equation (46) yields

(47)
$$n(C) = n(U) - \sum_{i=1}^{r} n(S_i) + \sum_{i<j} n(S_i \cap S_j)$$

$$- \sum_{i<j<k} n(S_i \cap S_j \cap S_k)$$

$$+ \sum_{i<j<k<l} n(S_i \cap S_j \cap S_k \cap S_l) - \cdots + (-1)^r n\left(\bigcap_{i=1}^{r} S_i\right).$$

We have proved

Theorem 7.[1]   Let U be any finite universal set, let $S_1, S_2, \ldots, S_r$ be subsets of U and let $C = (S_1 \cup S_2 \cup \cdots \cup S_r)'$. If $n(S)$ denotes the number of elements in S, then equation (47) holds.

EXAMPLE 1.   Let $r = 2$, $S_1 = A$, and $S_2 = B$. Prove that (47) gives equation (31) of Chapter 2. Let $r = 3$, $S_1 = A$, $S_2 = B$, and $S_3 = C$. Prove that (47) gives equation (33) of Chapter 2.

Solution.   This simple task is left for the reader.

EXAMPLE 2.   An army regiment desperately needs a typist, a cook, an electrician, and a doctor. Four newly arrived draftees happen to fit the needs exactly. In how many different ways can the captain give all four of the new men the wrong assignment.

Solution.   It is just as easy to formulate and solve a more general problem. We consider n distinguishable objects $x_1, x_2, \ldots, x_n$ to be arranged in places numbered 1, 2, ..., n. We say that x_i is in the *right* (correct) place if it occupies the ith place. (For example, the 26 letters of the alphabet "naturally" belong in their usual places.) In how many ways can these n objects be placed so that none of the objects occupies the right place.

We apply Theorem 7 as follows. Each assignment of places for $x_1, x_2, \ldots, x_n$ is a permutation of these objects. Let U be the set of all such permutations. Then $n(U) = n!$ Let S_i be the number of permutations in which x_i is in the right place, the ith place. Since x_i is located, there are $n - 1$ other objects left to distribute. Hence $n(S_i) = (n - 1)!$ and therefore $\Sigma n(S_i) = n \times (n - 1)!$. Similarly we find that $n(S_i \cap S_j) = (n - 2)!$ and

$$\sum_{i<j} n(S_i \cap S_j) = \frac{n(n - 1)}{2} \times (n - 2)!.$$

[1] This theorem was known to Euler and may have been known even earlier. The proof here is due to John Freeman of Florida Atlantic University.

In general if k of the x_i's are first placed correctly, then there are $(n - k)!$ ways of distributing the remaining x_i's. The number of ways of selecting the k x_i's to be correctly assigned is $C(n,k)$, so the corresponding term in (47) is

$$\Sigma n(S_{i_1} \cap S_{i_2} \cap \cdots \cap S_{i_k}) = C(n,k)(n - k)! = \frac{n!}{k!(n - k)!}(n - k)! = \frac{n!}{k!}.$$

Using this in (47) we find that

$$n(C) = n! - n(n - 1)! + \frac{n(n - 1)}{2}(n - 2)! - \cdots$$

$$+ (-1)^k\frac{n!}{k!} + \cdots + (-1)^n\frac{n!}{n!}.$$

On factoring $n!$, we have

(48) $$n(C) = n!\left[1 - 1 + \frac{1}{2!} - \frac{1}{3!} + \frac{1}{4!} - \frac{1}{5!} + \cdots + (-1)^n\frac{1}{n!}\right]$$

$$= n! \sum_{k=0}^{n} \frac{(-1)^k}{k!},$$

where C is the set of assignments in which every x_i is in the wrong place.

For the typist, cook, electrician, and doctor, $n = 4$. Hence equation (48) asserts that there are

$$n(C) = 4!\left[1 - 1 + \frac{1}{2} - \frac{1}{6} + \frac{1}{24}\right] = 12 - 4 + 1 = 9$$

different ways of misplacing all of them.

Exercise 5

1. Consolidated Enterprises, Inc. owns controlling stock in 71 smaller companies, some of which produce oil, gas, electricity, drugs, and a few other items. The Systems-Analysis Machine (SAM) punched out the following symbols:

(O − 26), (G − 37), (E − 34),   (D − 28)
(O, G − 13),   (O, E − 14),   (O, D − 12),
(G, E − 19),   (G, D − 20),   (E, D − 13),

$(O, G, E - 9)$, $(O, G, D - 8)$, $(O, E, D - 7)$,
$(G, E, D - 11)$, $(O, G, E, D - 6)$.

Here the symbol $(O - 26)$ means that 26 of the companies produce oil, and possibly other commodities. How many of the 71 companies did not produce any of the four commodities?

2. Monster Motors produced 1000 cars during the first month of its fiscal year. The important accessories were A, air conditioning; B, power brakes; S, power steering; and T, automatic transmission. SAM (the machine used in problem 1) punched out the symbols

$(A - 500)$, $(B - 480)$, $(S - 570)$,

$(T - 540)$, $(A, B - 270)$, $(A, S - 300)$,
$(A, T - 270)$, $(B, S - 270)$, $(B, T - 270)$,

$(S, T - 330)$, $(A, B, S - 140)$, $(A, B, T - 130)$,
$(A, S, T - 150)$, $(B, S, T - 160)$.

If 110 cars had no accessories, how many cars had all four of the above items?

3. Find the number of terms on the right-hand side of equation (47) when $r = 2$, $r = 3$, $r = 4$, and in general. *Hint:* See problem 18 of Exercise 2.

4. Prove Theorem 6.

5. Prove that if $C = S'$, then $X_C = 1 - X_S$.

6. If $A = S_1 \cup S_2 \cup \cdots \cup S_r$, prove that

(49) $$X_A \leqq X_{S_1} + X_{S_2} + \cdots + X_{S_r}.$$

What is the largest value that the right-hand side of equation (49) may have?

7. With the notation of problem 6, is it true that

$$X_A \geqq X_{S_1} X_{S_2} \cdots X_{S_r}?$$

8. A typist has five letters and five correctly addressed envelopes. In how many different ways can she put every letter in the wrong envelope?

9. A child trying to name the days of the week in order gets every one in the wrong place. In how many different ways can this happen?

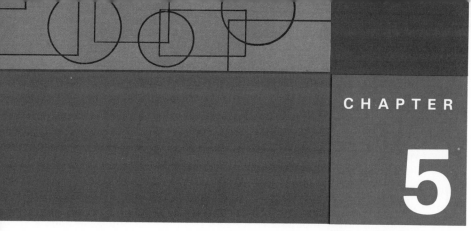

5

probability

Nearly everyone is familiar with the ideas of probability in an intuitive way. Such statements as "It will probably rain today," "A man who is 20 years old will very likely live another 30 years," and "It is silly to expect to have all of the aces and kings in a bridge hand" indicate just a few of the areas in which probability enters in our ordinary affairs. One objective of the theory of probability is to make these statements more precise by introducing a numerical measure for the probability of the event described.

In Section 1 we treat the theory of probability in a highly intuitive manner. The reader should be warned that a continuation along this path will lead to pitfalls that are hidden in the innocent looking phrase "equally likely." When these pitfalls are avoided, we obtain a theory that represents the real world very well and that is extremely useful when properly applied.

1. some simple examples

We begin with a problem that will also illustrate the terms we use.

EXAMPLE 1.   A card is drawn from a standard deck that has been well shuffled. Find the probability that the card is either a queen, a king, or an ace.

Of course we are in no position to work this problem because we have not yet defined the term *probability*. But this example may suggest a suitable definition.

Drawing a card from a deck is an example of an experiment. Other examples of experiments are "roll a die (one of a pair of dice)," "measure the temperature of a particular substance," and "select a chess piece from a box containing a standard set." An experiment may be rather complicated in that it may be compounded from several simple experiments. For example, we may "roll two dice or draw three cards from a deck," or we may "draw two cards and roll three dice." The only requirement is that in an experiment the set of possible outcomes (results of the experiment) must be well defined. In each of the examples mentioned above there are only a finite number of possible outcomes.

The set *U* of all possible outcomes for an experiment is called *the sample space for the experiment,* or just the *sample space.* The use of the word *space* in this connection may seem a little unusual at first glance. Nevertheless the phrase "sample space" is the standard term for the set *U* of all possible outcomes.

Returning to the problem posed in Example 1, we realize that in drawing a card any one of the 52 cards in the standard deck may appear. Hence there are 52 different possible outcomes in this experiment. If the card is a queen, king, or ace we call this a *favorable case,* or a *favorable outcome* in this problem. We also call it a *success.* When we ask for the probability of an event, we really mean the probability that one of the favorable outcomes will occur. Thus by definition *an event is a certain well-defined subset E of U.* An outcome that is in *E* is called a *favorable outcome* or a *success.* An outcome that is not in *E* is called an *unfavorable outcome* or a *failure.*

Notice that this use of the words *favorable* and *success* may not always coincide with the usual meaning of these words. For example, if we speak about the probability of Mr. X dying during the next year, then the event *E* is the death of Mr. X in the specified interval of time. If Mr. X does die, this is regarded as a "favorable case" or as a "success" mathematically, although usually Mr. X, Mrs. X, and *x* would take the opposite point of view.

In Example 1 we are asked to find the probability of the event that the card drawn is a queen, king, or ace. Now the standard deck contains 4 queens, 4 kings, and 4 aces, so in 12 cases the outcome is favorable. Since there are 52 different outcomes, of which 12 are favorable, the fraction 12/52 may be assigned as a measure of the probability of drawing a queen, king, or ace. We use the symbol $P[E]$ to denote the probability of the event *E*.

Definition 1. Probability of an Event. Suppose that in an experiment there are *n* different possible outcomes and that these outcomes are all equally likely. Suppose further that the event *E* occurs in *k* of these outcomes. Then $P[E]$, the probability of the event *E*, is given by

(1) $$P[E] = \frac{k}{n} = \frac{\text{number of successes}}{\text{number of possible outcomes}} = \frac{n(E)}{n(U)}.$$

It is difficult to make the phrase "equally likely" precise. It is here that intuition enters. We must use our feelings about the matter. In Example 1, each card does seem "equally likely" if the deck is well shuffled. In fact we take this as the meaning of the phrase "well shuffled."

Solution to Example 1. Since $k = 12$ and $n = 52$,

$$P[E] = \frac{12}{52} = \frac{3}{13}.$$

Theorem 1. Suppose that an experiment has only a finite number of different outcomes all equally likely. If E is an event, then

(2) $$0 \leqq P[E] \leqq 1.$$

The event is certain to occur if and only if $P[E] = 1$. The event cannot occur if and only if $P[E] = 0$.

Proof. Since E is a subset of U, we have $0 \leqq k = n(E) \leqq n(U)$. Hence

(3) $$0 \leqq \frac{n(E)}{n(U)} \leqq 1.$$

Further, $n(E) = n(U)$ if and only if $E = U$ (the event is certain to occur). Finally, $n(E) = 0$ if and only if $E = \varnothing$ (the event cannot occur). ∎

EXAMPLE 2. Two standard dice are rolled.
  (a) Find the probability of throwing a 17.
  (b) Find the probability of throwing a number between 1 and 25.
  (c) Find the probability of throwing the sum S, where $5 \leqq S \leqq 9$.

Solution.
  (a) The faces of a standard die are numbered from 1 to 6. Hence the maximum value of S is $6 + 6 = 12$. Hence $k = 0$ in equation (1) and the probability of throwing a 17 is zero. We can symbolize this by writing[1] $P[S = 17] = 0$, where we place inside the brackets a description of the event E.

---

[1] It does no harm to replace E in $P[E]$ by the condition $S = 17$ that defines the set E. In fact we can even condense the notation further and write $P[17]$. We shall employ such condensations whenever the meaning is clear.

(b) With two dice the largest value for S is $6 + 6 = 12$, and the smallest is $1 + 1 = 2$. Hence $1 < S < 25$ always occurs. Hence $k = n$ and $P[1 < S < 25] = 1$.

(c) We give three different solutions to this problem, and we obtain three different answers.

(c_1) There are 11 numbers which can actually occur as the sum of the dots when two dice are thrown, and hence $U = \{2,3,4,\ldots,12\}$. Of these $E = \{5,6,7,8,9\}$ represents the favorable cases. Hence

(4) $$P[5 \leqq S \leqq 9] = \frac{n(E)}{n(U)} = \frac{5}{11} = 0.4545 \ldots .$$

(c_2) We list the possible ways of forming the sums:

$$
\begin{array}{lll}
2 = 1 + 1 & \text{1 way.} \\
3 = 1 + 2 & \text{1 way.} \\
4 = 1 + 3 = 2 + 2 & \text{2 ways.} \\
5 = 1 + 4 = 2 + 3 & \text{2 ways.} \\
6 = 1 + 5 = 2 + 4 = 3 + 3 & \text{3 ways.} \\
7 = 1 + 6 = 2 + 5 = 3 + 4 & \text{3 ways.} \quad \text{favorable cases} \\
8 = 2 + 6 = 3 + 5 = 4 + 4 & \text{3 ways.} \\
9 = 3 + 6 = 4 + 5 & \text{2 ways.} \\
10 = 4 + 6 = 5 + 5 & \text{2 ways.} \\
11 = 5 + 6 & \text{1 way.} \\
12 = 6 + 6 & \text{1 way.}
\end{array}
$$

By direct count we see that there are 21 different outcomes for the experiment. Since 13 are favorable, equation (1) gives

(5) $$P[5 \leqq S \leqq 9] = \frac{n(E)}{n(U)} = \frac{13}{21} = 0.6190 \ldots .$$

(c_3) Suppose that we paint one die red and the other die green so that we can tell them apart. Let (a,b) denote the case in which the red die shows an a and the green die shows a b. Then all possible results for one throw of these colored dice are shown in the 6 by 6 array of Table 1 and those for which $5 \leqq S \leqq 9$ are enclosed by the colored lines.

TABLE 1

(1,1)	(1,2)	(1,3)	(1,4)	(1,5)	(1,6)
(2,1)	(2,2)	(2,3)	(2,4)	(2,5)	(2,6)
(3,1)	(3,2)	(3,3)	(3,4)	(3,5)	(3,6)
(4,1)	(4,2)	(4,3)	(4,4)	(4,5)	(4,6)
(5,1)	(5,2)	(5,3)	(5,4)	(5,5)	(5,6)
(6,1)	(6,2)	(6,3)	(6,4)	(6,5)	(6,6)

A direct count shows that $n(E) = 24$ and $n(U) = 36$. Hence

(6) $$P[5 \leqq S \leqq 9] = \frac{n(E)}{n(U)} = \frac{24}{36} = \frac{2}{3} = 0.666 \ldots .$$

We have worked this problem in three different ways and obtained three different answers, namely 5/11, 13/21, and 2/3. Which answer is correct? This problem brings to light the difficulty in the phrase "equally likely" which occurs in Definition 1. We obtained three different answers because in the solutions we made three different selections for the outcomes that are regarded as equally likely.

It is easy to reject the first solution because, as any gambler knows, it is much easier to roll a 7 than it is to roll a 2. Hence the outcomes counted in the first solution are not all equally likely.

The second solution is a little more attractive, and in fact many students feel that it is correct. In this solution the dice are regarded as indistinguishable. By painting the dice as in the third solution we are able to distinguish them. Certainly, if the dice are not loaded, then it seems as though each one of the six faces is equally likely to be showing after a roll. Since the third solution gives an accurate count of the 36 equally likely results, the third solution is correct when the dice are distinguishable. Returning to the second solution, we observe that the dice will behave the same whether they are painted or not. Hence the counting in the second solution is wrong. For example, the sum 6 can be obtained in five different ways:

$$1 + 5, \quad 2 + 4, \quad 3 + 3, \quad 4 + 2, \quad 5 + 1,$$

and not in three different ways.

We notice that the difference in the two answers obtained is $2/3 - 13/21 = 1/21$, a rather small number. Suppose that we decided to check the correctness of our argument by actually performing the experiment and observing the result. To be specific, suppose that we throw a pair of dice 7200 times and actually observe the number of successes; i.e., the number of times that $5 \leqq S \leqq 9$. If equation (5) gives the correct solution, one might hope (foolishly) to observe 4457 successes. If equation (6) gives the correct solution, then one might hope for 4800 successes. But the truth is that neither of these results is very likely. Suppose in fact that on 7200 tries of the experiment we find that $5 \leqq S \leqq 9$ in 4751 cases. Will this tell us which solution is correct? This is indeed a difficult question which we must postpone for the moment. We shall return to it in Chapter 6, Section 6.

EXAMPLE 3.   An integer between 1 and 50 (inclusive) is selected at random. Find the probability that the integer is divisible by 4.

Solution.   Here we must look more carefully at the phrase "selected at random." If a person is asked to select an integer at random from the set $U = \{1,2,\ldots,50\}$, it is not at all obvious that the numbers are all equally likely. Indeed he is more likely to select one from the subset $\{1,2,3,5,50\}$ than from the subset $\{17,23,37,43,47\}$. To obtain a meaningful problem we must be specific about the method of selecting an integer.

Suppose that 50 balls, numbered from 1 to 50, are placed in an urn. After thoroughly mixing, one ball is selected from the urn. Find the probability that n, the number on the ball, is divisible by 4.

It now appears that all of the numbers $1,2,3,\ldots,50$ are equally likely. Since there are 12 numbers in the set $E = \{4,8,12,\ldots,48\}$, we see that

$$P[E] = \frac{n(E)}{n(U)} = \frac{12}{50} = \frac{6}{25}.$$

This device of replacing a given problem by an equivalent problem on drawing balls from an urn is a very useful one that is frequently employed in probability. Just as in the example, it changes an ambiguous problem into a clear one. Many probability problems are stated in terms of drawing balls from urns. As stated, such problems may sound shallow and uninspiring. But the reader should keep in mind that the results can be transferred to a variety of other situations, and in the new settings the problems may be deep and fascinating.

When the probability of an event is determined by some form of reasoning, the number obtained is called an *a priori probability*. This is the type of probability presented in Examples 1, 2, and 3. In each case $P[E]$ was determined before performing the experiment (drawing a card, throwing a pair of dice, selecting a ball from an urn).

On the other hand, suppose that the dice in Example 2 were loaded (or perhaps magnetized on one side and thrown on a steel table). Clearly the probability, $P[5 \leqq S \leqq 9] = 2/3$, computed in solution ($c_3$) would have no meaning for an experiment with these modified dice. In such a case, the usual procedure is to rely on observation. The experiment is performed n times (where n is large), and if k is the number of successes actually observed, then the number k/n is often accepted as a reasonable approximation for the probability. When the probability of an event is estimated in this way, by direct observation, the number obtained is called an *a posteriori proba-* *bility*, or a *statistical probability*, or an *empirical probability*.

Exercise 1

1. When a coin is thrown it will land either heads (H) or tails (T). Unless otherwise stated, we assume that the coin is good and that either outcome is equally probable. Hence $P[H] = P[T] = 1/2$. Two coins are thrown. Find the probability of obtaining
   (a) Two heads.   (b) One head and one tail.
   (c) At least one head.   (d) No heads.
2. Three coins are thrown. Find the probability of obtaining
   (a) Three heads.   (b) One head and two tails.
3. A card is drawn from a standard deck that has been well shuffled. Find the probability that the card is
   (a) A spade.   (b) A seven.
   (c) The seven of spades.   (d) Not an ace.
4. Two dice are thrown. Using the three methods, as in Example 2, find the three possible values for the probability of throwing a 7.
5. Nine black balls, numbered from 1 to 9, are placed in an urn. Nine red balls similarly numbered are placed in a second urn. Two balls are drawn one from each urn. Find $P[S = 10]$, where S is the sum of the numbers on the two balls drawn.
6. Suppose that in problem 5, the urn containing the black balls is unchanged but the other urn contains 15 red balls, numbered from 1 to 15. Two balls are drawn, one from each urn. Find (a) $P[S = 10]$, (b) $P[S = 15]$, and (c) $P[S = 18]$.
7. Under the conditions of problem 6 find (a) $P[S$ is an odd integer] and (b) $P[S$ is even].
★8. Suppose that in the two urns in problem 5, each ball is labeled with an integer. In the first urn d_1 of the integers are odd and e_1 of the integers are even, and in the second urn d_2 of the integers are odd and e_2 of the integers are even. Two balls are drawn, one from each urn. Find $P[S$ is an odd integer].
★9. Under the conditions of problem 8 find a necessary and sufficient condition that $P[S$ is odd$] < 1/2$.
10. Two dice are thrown. Find the probability that the sum is 9 *and* at least one of the two faces is even.

11. Two dice are thrown. Find the probability that the sum is 9, *or* at least one of the two faces is even.

★12. An integer x from 1 to 20 inclusive is selected at random. Find $P[x$ is a prime$]$. Find the probability, if x is selected from the integers between 1 and 100 inclusive. *Note:* 1 is not a prime.

★13. Three standard dice are thrown. Let S be the sum of the dots showing on the three dice. Find **(a)** $P[S = 3]$, **(b)** $P[S = 5]$, **(c)** $P[S = 7]$, and **(d)** $P[S = 25]$.

★14. Continuation of problem 13. Let

$$q(x) = (x + x^2 + x^3 + x^4 + x^5 + x^6)^3 = \sum_{k=3}^{18} c_k x^k.$$

Verify that c_k is the number of different ways that the sum (in problem 13) can be equal to k. Hence in problem 13 we have $P[S = k] = c_k/216$.

★★15. If $q(x)$ is the polynomial of problem 14, it can be proved that

$$q(x) = \frac{x^3(1 - x^6)^3}{(1 - x)^3} = (1 - 3x^6 + 3x^{12} - x^{18}) \sum_{n=2}^{\infty} \frac{n(n - 1)}{2} x^{n+1}.$$

Use this result to find $P[S = 10]$ for the throw of three standard dice.

2. sample spaces

We now examine a little more closely the sample space and the definition of probability given in Section 1. If the individual outcomes are not "equally likely," then Definition 1 must be modified.

If U is a sample space, the individual outcomes are often regarded as points in the space U. If U has only a finite number of points, it is called a *finite sample space*; otherwise it is called an *infinite sample space*. The real difficulties with the foundations of probability theory begin to appear only when we consider infinite sample spaces.

Any event E is a subset of U. By assigning a number as a measure of the probability of E for each subset E of U we are really defining a function on the set of all subsets of U. This can be done in many ways, but for the function to serve as a measure of the probability for the subsets it must have certain special properties. These are listed in

Definition 2. Probability Function. Let U be a finite sample space. A function P, defined for all subsets of U, is a probability function if it has the following three properties:

I. For every subset E of U

(7) $$0 \leqq P[E] \leqq 1.$$

II. $$P[U] = 1.$$

III. If A and B are disjoint sets in U, then

(8) $$P[A \cup B] = P[A] + P[B].$$

If the set U has n elements, then one way to define a probability function is to set

(9) $$P[E] = \frac{n(E)}{n}$$

for every set $E \subset U$. It is easy to see that with this choice for P, the function does satisfy the conditions for a probability function specified in Definition 2. But the reader must realize that there are many other ways of defining a probability function and that in many real situations the one specified by equation (9) does not "fit the facts."

Definition 3. Equiprobable Space. If the probabilities for a finite sample space are defined by equation (9), then the space is said to be an equiprobable space.

To avoid lengthy descriptions for the probability function on a space, we agree that the terms "natural probability" or "usual probability" will mean that the sample space is an equiprobable space; i.e., that $P[E] = n(E)/n$. In fact if we make no other specification for P, then we understand that P is the natural probability function.

This idea of a natural probability function is conveyed by a variety of common terms. For example, in a single toss of a coin, the sample space is {H,T}. If we say that the coin is a true coin, we mean that $P[H] = P[T] = 1/2$. The terms "good coin," "unbiased coin," and "fair coin" all mean the same thing. Any one of these phrases means that the sample space {H,T} is an equiprobable space. In the throw of dice, we may use the same terms, or we may speak of "standard dice" or "balanced dice" or "dice that are not loaded." If the experiment involves selecting a card, then the term "well-shuffled deck" implies the natural probability. In drawing balls from an urn (or any object from any container) each one of the phrases "well mixed," "drawn at random," and "selected at random" implies the natural probability.

The student may have the feeling that in every experiment the natural probability is the one to use. To dispel this feeling, he may consider the probability of a victory in an athletic contest between two teams when one team is much stronger than the other.

Statements about events can be expressed quite efficiently with the notation from set theory. For example, if A and B are events (subsets of the sample space), the statement "A and B both occur" can be condensed to $A \cap B$. Other possibilities with their corresponding symbolism are listed below:

Either A or B occurs:	$A \cup B$.
Neither A nor B occurs:	$A' \cap B'$.
If A occurs, then B occurs:	$A \subset B$.
A and B are equally likely:	$P[A] = P[B]$.
A is more likely than B:	$P[A] > P[B]$.
A is less likely than B:	$P[A] < P[B]$.
If A occurs, then B cannot occur:	$A \cap B = \varnothing$.

(*A* and *B* are mutually exclusive events.)

A set of mutually exclusive and exhaustive events corresponds to a decomposition of U into subsets as described in

Definition 4. Partition. The collection of sets $\{A_1, A_2, \ldots, A_r\}$ is called a partition of U if the collection is exhaustive and pairwise disjoint. The term *exhaustive* means that

$$A_1 \cup A_2 \cup \cdots \cup A_r = U.$$

The phrase *pairwise disjoint* means that

$$A_i \cap A_j = \varnothing,$$

for all $i, j = 1, 2, 3, \ldots, r$ with $i \neq j$.

Exercise 2

An experiment is described in problems 1 through 5. In each case describe an associated sample space and state how many points are in it.

1. A coin is tossed n times in a row, and the result, heads or tails, is noted.

2. A coin is tossed once, then a die is rolled, and finally a card is drawn from a standard deck.

3. An urn contains n balls each bearing one of the numbers from 1 to n inclusive with no two balls having the same number. Three balls are drawn in sequence from the urn.

4. The same experiment as described in problem 3, except that after each ball is drawn and the number is noted, that ball is replaced in the urn before the next drawing.

5. A bridge hand of 13 cards is selected from a standard deck of 52 cards.

6. Explain why n does not cancel on the right-hand side of equation (9). Prove that the function defined by equation (9) satisfies the conditions of Definition 2.

7. Using the language of an experiment, describe each of the following events:

(a) A'. (b) $A \cup B = U$.

(c) $(A \cap B) \cup C$. (d) $A \cap (B \cup C)$.

In problems 8 through 19, assume that P is a probability function defined on a finite sample space U.

8. Prove that $P[\varnothing] = 0$. *Hint:* $\varnothing \cup \varnothing = \varnothing$. Then use property III of Definition 2.

9. If $P[E] = 0$, then $E = \varnothing$. Is this statement always true?

★10. Prove that if $\{A_1, A_2, \ldots, A_r\}$ is any collection of pairwise disjoint sets, then

$$P[A_1 \cup A_2 \cup \cdots \cup A_r] = \sum_{i=1}^{r} P[A_i].$$

Hint: Use mathematical induction.

In problems 11 through 17, prove the given assertion.

11. $P[A'] = 1 - P[A]$.

12. $P[A - B] = P[A] - P[A \cap B]$.

13. $P[A \cup B] = P[A] + P[B] - P[A \cap B]$.

14. $P[A \cup B] \leq P[A] + P[B]$.

15. If $A \subset B$, then $P[A] \leq P[B]$.

16. If $A \cup B \cup C = U$, then $P[A] + P[B] + P[C] \geq 1$.

★17. $P[A \cup B \cup C] = P[A] + P[B] + P[C] - P[A \cap B] - P[B \cap C] - P[C \cap A] + P[A \cap B \cap C]$.

18. Prove that if we know $P[x_i]$ for each point $x_i \in U$, then $P[E]$ is uniquely determined for every $E \subset U$.

19. Is the converse of the result in problem 15 always true?

20. Suppose that a die is loaded in such a way that $P[1] = P[3] = P[5] = 1/12, P[2] = P[6] = 1/8$, and $P[4] = 1/2$. Find (a) $P[x$ is odd], (b) $P[x$ is even], (c) $P[x$ is prime], and (d) $P[x \neq 5]$.

21. Suppose that we wish to simulate the die of problem 20 by an urn loaded with numbered balls using the fewest possible balls. Assume that a ball is drawn at random from the urn, so that for this urn $P[x_k] = 1/n$ for each point in the sample space. If the number on the ball is to correspond to the number on the face of the die, how many balls of each type should be placed in the urn?

★22. A die is loaded so that the probability of throwing an even number is twice the probability of throwing an odd number. All of the odd numbers are equally likely, and all of the even numbers are equally likely. If this die is thrown once, find the probability of getting (a) 2, (b) 2 or 4, and (c) 5 or 6.

★23. A die is loaded so that $P[1] = 1/12$ and the probabilities of throwing 1, 2, 3, 4, 5, and 6 form an arithmetic progression. Find (a) $P[2]$ and (b) $P[$an even number$]$.

★24. K, L, M, and N are candidates for mayor in an election. A newspaper poll gives M and N equal chances. Further, K is only 1/2 as likely to win as M, but L is three times more likely to win than K. Find the probability of winning for each contestant. What is the probability that either L or M will win?

★25. During the campaign, in problem 24, a scandal is uncovered concerning the popular L and he withdraws from the race. Assuming that the ratios of the probabilities for the other candidates remain the same, find the increase in the probability that M will win.

26. A, B, and C are running for district attorney. The probability that either A or B wins is 1/2. The probability that either A or C wins is 5/6. Can you find the probability that A wins?

27. Prove that if $A_1, A_2, \ldots, A_r$ form a partition of U, then

$$P[A_1] + P[A_2] + \cdots + P[A_r] = 1.$$

3. games

The equiprobable spaces almost never occur in a "live" situation. For example, in a horse race, a basketball game, a football game, or a chess tournament, it is seldom the case that the probability function is the natural one. However, in a highly mechanized situation such as the roll of dice, the draw of several cards, or the turn of a wheel, the assumption of an equiprobable sample space works extremely well. In fact, as everyone knows, millions of dollars have been invested in the construction of elegantly furnished seminar rooms for continuous testing of this assumption, and in every case the investors have discovered (to their great satisfaction) that the equiprobable assumption fits the experimental results.

We devote this section to some examples from poker and bridge. Dice is a little more complex and will be considered in Section 10.

EXAMPLE 1. Find the probability of drawing the various hands that are of interest in a poker game.

Solution. A hand consists of 5 cards drawn from a deck of 52 cards. Hence the number of possible hands is

$$(10) \qquad n(U) = \binom{52}{5} = \frac{52 \times 51 \times 50 \times 49 \times 48}{5 \times 4 \times 3 \times 2 \times 1} = 2{,}598{,}960.$$

In Table 2 we give the number of different ways of drawing the hands of interest, together with an approximate value for the probability.

TABLE 2

Rank	E, Type of hand	$n(E)$	$P[E] = n(E)/n(U)$
1	Royal flush: Ace, king, queen, jack, ten in the same suit	4	0.0000015
2	Straight flush: Five cards in a sequence, in the same suit, but not a royal flush	36	0.000014
3	Four of a kind (for example, four jacks or four sevens)	624	0.00024
4	Full house: Three of a kind together with a pair	3744	0.0014
5	Flush: Five cards in a single suit, but not a straight	5108	0.0020
6	Straight: Five cards in a sequence not all in the same suit	10,200	0.0039
7	Three of a kind	54,912	0.0211
8	Two pairs	123,552	0.0475
9	One pair	1,098,240	0.4226
10	Nothing of interest	1,302,540	0.5012

Computations for the first three ranking hands are trivial. We give the computation for the straight, and leave the other cases for the energetic student.

Consider, for example, the straight composed of 7, 8, 9, 10, and jack. The 7 can be drawn in 4 different ways, the 8 can be drawn in 4 different ways, etc. Hence the number of ways of drawing this straight is 4^5. Now a straight may run from ace to 5, from 2 to 6, ..., and from 10 to ace, thus giving 10 different types of straights. The number 10×4^5 also includes the straight flushes (ranks 1 and 2). Subtracting these gives

$$n(\text{straight}) = 10 \times 4^5 - 36 - 4) = 10{,}200.$$

EXAMPLE 2. Find the probability of drawing a bridge hand in which every card is a 9 or lower (ace is high in bridge).

Solution. There are $8 \times 4 = 32$ such cards. Hence $n(E) = C(32,13)$. Thus

$$P[E] = \frac{C(32,13)}{C(52,13)} = \frac{32 \times 31 \times \cdots \times 20}{52 \times 51 \times \cdots \times 40} \approx 0.000547.$$

Thus if one does draw such a hand, he is quite justified in complaining about his bad luck.

EXAMPLE 3. It is traditional to refer to bridge players as North, East, South, and West. In a particular game in which North won the contract and South was the dummy (his cards are face up on the table), North observed that together he and his partner (South) had 7 spades. Of the 13 spades in a standard bridge deck, the remaining 6 were held jointly by East and West. Find the probability that the 6 spades are distributed 4-2 between East and West.

Solution. North and South together hold 26 cards, which are known to North. Hence we are concerned with the distribution of the remaining 26 cards of which 6 are spades and 20 are other suits. Suppose that East has 4 of the spades. This can happen in $C(6,4)$ different ways. The remaining 9 cards in the East hand can be selected from 20 nonspades in $C(20,9)$ different ways. We must also consider the symmetrical situation in which East has 2 spades and West has 4 spades. Thus we find that

$$P[\text{4-2 split}] = \frac{C(6,4)C(20,9) + C(6,2)C(20,11)}{C(26,13)}.$$

But $C(6,4) = C(6,2)$, and $C(20,9) = C(20,11)$. Hence

$P[\text{4-2 split}]$

$$= \frac{2\dfrac{6 \times 5}{2 \times 1} \times \dfrac{20 \times 19 \times 18 \times 17 \times 16 \times 15 \times 14 \times 13 \times 12}{9 \times 8 \times 7 \times 6 \times 5 \times 4 \times 3 \times 2 \times 1}}{\dfrac{26!}{13!13!}}$$

$$= 30\frac{13 \times 12 \times 13 \times 12 \times 11 \times 10}{26 \times 25 = 24 \times 23 \times 22 \times 21}$$

$$= \frac{78}{161} \approx 0.4845.$$

Other results of the same type are exhibited in Table 3.

TABLE 3

Distribution of Cards in a Single Suit

Held jointly by N and S	Held jointly by E and W	Distribution between E and W	Probability
7	6	3-3	$286/805 \approx 0.3553$
		4-2	$78/161 \approx 0.4845$
		5-1	$117/805 \approx 0.1453$
		6-0	$12/805 \approx 0.0149$
8	5	3-2	$78/115 \approx 0.6783$
		4-1	$13/46 \approx 0.2826$
		5-0	$9/230 \approx 0.0391$
9	4	2-2	$234/575 \approx 0.4070$
		3-1	$286/575 \approx 0.4974$
		4-0	$11/115 \approx 0.0957$
10	3	2-1	$39/50 \approx 0.78$
		3-0	$11/50 \approx 0.22$
11	2	1-1	$13/25 \approx 0.52$
		2-0	$12/25 \approx 0.48$

When one speaks of the odds in favor of an event he always quotes a ratio of two integers. If a gambler states that he will give 7 to 5 odds that E will occur, he usually means that if E occurs he will receive 5 dollars and that if E does not occur he will pay 7 dollars.

To arrive at a fair bet let us suppose that $p = P[E]$ and that the gambler is giving odds of x to y that E happens. Suppose further that after N repeated trials the event E does occur pN times and fails to occur $(1 - p)N$ times. If the bet is made with the same odds x to y on each trial, the gambler will pay out $x(1 - p)N$ dollars and receive ypN dollars. For a fair bet we must have

$$(11) \qquad\qquad x(1 - p)N = ypN$$

and hence

$$(12) \qquad\qquad \frac{x}{y} = \frac{p}{1 - p} = \frac{p[E]}{p[E']}.$$

Definition 5. Odds. Let $p = P[E]$. The odds in favor of the event E are x to y if x and y are any pair of numbers (usually integers) that satisfy equation (12). A bet on E is called a fair bet if the bettor gives x dollars if E does not occur and receives y dollars if E occurs.

EXAMPLE 4. A gambler bets that he will draw a diamond from a standard deck of cards. What odds should he give for a fair bet?

Solution. $P[E] = 13/52 = 1/4$ and $P[E'] = 3/4$. Hence

$$\frac{x}{y} = \frac{1/4}{3/4} = \frac{1}{3}.$$

The gambler should give odds of 1 to 3.

Exercise 3

1. Check at least four of the entries in Table 2.
2. Check at least five of the entries in Table 3.

In problems 3 through 7, use the symbols $C(n,k)$ to give the probability of obtaining the hand described.

3. A poker hand with two aces, one king, and the remaining cards different.
4. A poker hand with four cards of one suit and one card of another suit.
5. A bridge hand in which every card is a 9 or above.
6. A bridge hand in which every card is a 10 or above.
7. A bridge hand with a distribution of cards in the various suits as follows:

    (a) 4, 3, 3, 3.   **(b)** 5, 5, 2, 1.
    (c) 6, 4, 2, 1.   **(d)** 7, 5, 1, 0.

8. Two players each draw a single card from a deck. If the man with the high card wins (ace is highest), find the probability that the second player will **(i)** win, **(ii)** tie, if the first player has already drawn a

    (a) Jack.   **(b)** Ten.
    (c) Seven.   **(d)** Three.

9. Balls numbered 1 to 10 are placed in an urn, and two balls are drawn at random without replacement. Find the probability that of the two numbers on the balls one is a divisor (factor) of the other.
10. In a simplified version of poker, each hand consists of only three cards. For this game find the probability of drawing **(a)** a pair, **(b)** a three-card straight flush, and **(c)** a flush that is not a straight.

In problems 11 through 15, determine the odds for a fair bet on the event described.

11. The roll of a number $S \geq 7$ with a pair of dice.
12. The roll of an odd number with a pair of dice.
13. The draw of a 7 or better from a standard deck of cards.
14. The draw of at least one card that is higher than an 8 when two cards are selected from a standard deck.

15. The throw of two or more heads when five coins are tossed.

16. Ajax, Big Boy, Crawling Carcass, and Dubious Dolt are entered in a horse race. Just prior to the race, the odds posted on the board for a win are A, 3 to 2; B, 1 to 4; C, 1 to 9; and D, 1 to 24. Find the probability for each horse to win. Do these probabilities sum to 1?

17. Prove that if the odds are x to y that E occurs, then $P[E] = x/(x + y)$.

18. Experience shows that in 75 percent of the fatal accidents involving two cars, at least one of the drivers is drunk. If you witness such an accident, what odds should you give your companion that at least one of the drivers is drunk?

19. In Uproar University, the calculus course is given in three semesters. It is customary to pass 3/4 of the class in each of the first two semesters, and 4/5 of the class in the third semester. If the grades are assigned in a random fashion, what are the odds that a student will get through the calculus without flunking once?

★4. the birthday paradox

Fifty people are gathered for a party. What is the probability that at least two of them have the same birthday; i.e., the same day and the same month of birth, but not necessarily the same year? Before reading the solution, try to estimate the answer, keeping in mind that there are 365 possible days but only 50 people at the party. If someone offered to give odds of 3 to 1 that two of the people have the same birthday, should you accept the bet?

   Records show that not all days are equally likely, but the variation is so slight that we may assume the days form an equiprobable space. We shall also ignore February 29 as a possibility, since the error involved is very small.

   Let p_n be the probability that in a random selection of n people at least two have the same birthday. Rather than compute p_n directly, it is simpler to find q_n, the probability that no two have the same birthday. Then it is easy to find p_n from the relation $p_n = 1 - q_n$.

   Since the selection of birthdays of n people can be regarded as a selection of n things from 365 things (days), the number of such selections with no two alike is $P(365,n)$, the number of permutations of 365 things taken n at a time. When repetitions are allowed, the number of such selections is 365^n. Consequently

$$(13) \quad q_n = \frac{P(365,n)}{365^n} = \frac{365 \times 364 \times 363 \times \cdots \times (365 - n + 1)}{365^n}.$$

The computation of q_n from (13) is a nuisance even for small values of n.

TABLE 4

Number of people	E, at least two with the same birthday		Approximate odds for a fair bet that E occurs, p/q
	q = P[E']	p = P[E]	
5	0.973	0.027	1 to 36
10	0.883	0.117	3 to 23
15	0.747	0.253	1 to 3
18	0.653	0.347	53 to 100
20	0.589	0.411	70 to 100
21	0.556	0.444	80 to 100
22	0.524	0.476	91 to 100
23	0.493	0.507	103 to 100
24	0.462	0.538	116 to 100
25	0.431	0.569	132 to 100
27	0.373	0.627	168 to 100
30	0.294	0.706	240 to 100
35	0.186	0.814	438 to 100
40	0.109	0.891	817 to 100
50	0.030	0.970	32 to 1
60	0.0059	0.9951	169 to 1
70	0.00084	0.99916	1190 to 1
80	8.57×10^{-5}	0.99991	11,500 to 1
90	6.15×10^{-6}	0.99999	160,000 to 1
100	3.07×10^{-7}	—	3,300,000 to 1
125	3.19×10^{-11}	—	31,000,000,000 to 1
150	2.45×10^{-16}	—	4,100,000,000,000,000 to 1

However, with the aid of a suitable computing machine, equation (13) gives the numerical values listed in Table 4.

As the table indicates, if there are 23 people or more at the party, then $p > 1/2$. At a gathering of 50 people or more, the bettor can make money by offering 5 to 1 odds. If he offers greater odds, as he can safely do, people may become suspicious, and they will be reluctant to bet. The teacher may test the theory in class by having each student write his birthday (or any date selected at random) on a piece of paper. A neutral official can then check the papers for duplications.

★5. the Kaos problem

Suppose that 10 Control agents, numbered 1 to 10, are assigned the task of capturing 10 Kaos agents, numbered 1 to 10, where each Control agent is supposed to capture the Kaos agent with the same number. The specified 10 Kaos agents are indeed captured, but in the usual confusion no Control agent caught the right Kaos agent. Assuming that the captures were completely at random, what is the probability of this event?

The same problem can be posed in other (and less romantic) terms:

(1) Ten letters are placed at random in 10 properly addressed envelopes. What is the probability that every letter is in the wrong envelope?

(2) Ten guests at a party in Alaska are so inebriated that when they leave each one grabs a coat at random. What is the probability that no one gets his own coat?

In abstract terms, we are concerned with matching elements from two sets, A and B, where each set consists of n elements. We number the elements in each set from 1 to n. If the elements in A are considered in numerical order, then any pairing with the elements in B can be regarded as a permutation of the elements of B, or a permutation of the numbers 1 to n. In the permutation, there is a match if the number k is in the kth place.

Now the number of permutations of the numbers 1 to n in which each number is in the wrong place was determined in Chapter 4, and by equation (48) [see page 120] we find that

$$(14) \qquad n[C] = n!\left[1 - 1 + \frac{1}{2!} - \frac{1}{3!} + \frac{1}{4!} - \cdots + (-1)^n\frac{1}{n!}\right].$$

Since the total number of permutations is $n!$, equation (14) gives

$$(15) \qquad P[C] = \frac{1}{2!} - \frac{1}{3!} + \frac{1}{4!} - \cdots + (-1)^n\frac{1}{n!} = \sum_{k=2}^{n}\frac{(-1)^k}{k!}$$

for the probability in question: the probability that if n things are ordered at random (when there is a correct order), then none of the things will be in the correct place.

How does $P[C]$ change as the number of things gets large? Here intuition may fail us completely. One person might argue that if $n = 100$, then surely one item will fall in the right place and hence $P[C]$ is very close to 0. Another might contend that, for large n, there are more ways of mixing the items and hence $P[C]$ is very close to 1.

It is an interesting fact that both would be wrong. Indeed, equation (15) gives the numerical values listed in Table 5. The reader will note that there is no change in the first six significant figures of $P[C]$ as n goes from 10 to 20. In fact, for $n \geq 10$, we have $P[C] = 0.367879$ to six decimal places. It turns out that as n gets larger and larger, $P[C]$ approaches more and more closely to a certain fixed number. This limiting value for $P[C]$ is $1/e$, where $e = 2.71828\ldots$ is a special transcendental number known as Euler's number. This constant plays a fundamental role in many branches of mathematics.

Returning now to the problem posed at the beginning of this section, the

TABLE 5

n Number of things to be ordered	$P[C]$ Probability that none of the things are in the right place	$P[C']$ Probability that at least one thing is in the right place
2	0.500000	0.500000
3	0.333333	0.666667
4	0.375000	0.625000
5	0.366667	0.633333
6	0.368056	0.631944
7	0.367857	0.632143
8	0.367882	0.632118
10	0.367879	0.632121
20	0.367879	0.632121

probability that each Control agent will capture the wrong Kaos agent is 0.367879 . . . .

Exercise 4

1. Suppose E is the event that two people have birthdays in the same month (disregarding the day or year). Make a table for $P[E]$ when $n = 2, 3, 4,$ and 5.
2. Two decks of cards are well shuffled and placed on the table, and the top card from each deck is turned face up simultaneously. If both cards are the same (for example, 7 of diamonds), we call this a *match*. We continue to turn up pairs of cards from each deck until both decks are exhausted. Find the probability that there will be at least one match. Find an approximation for the odds for a fair bet. If a gambler bets even money (odds 1 to 1) consistently on a match, will he make money?
★3. Find a formula for the probability that *exactly one number* will be in the "right place" when the numbers from 1 to n are ordered at random.

6. conditional probability

The probability of a certain event may change as we obtain more information about the situation. For example, if we select a person at random, the probability that he is a college graduate is rather low. But if after selecting the person, we find out that his yearly income is over $10,000, we would feel that it is much more likely that he is a college graduate. Our aim is to discover a reasonable and systematic method of revising our original probability function to obtain a new probability function which takes into account any additional information we may have.

Continuing with the example, if U is the set of people under consideration we partition the sample space U into three mutually disjoint subsets D, H, and C, where D is the set of people who did not finish high school (dropouts), H is the set of those who finished high school but not college, and C is the set of college graduates. By the definition of a probability function,

(16) $$P[D] + P[H] + P[C] = 1.$$

Now let T be the set of people with yearly income over \$10,000. Having selected a person at random and having discovered that he belongs to T, we are no longer concerned with the three sets D, H, and C but with the three new sets $D \cap T$, $H \cap T$, and $C \cap T$ because our sample space has shrunk from U to T. Because of this decrease in the size of the space, equation (16) is automatically replaced by the inequality

(17) $$P[D \cap T] + P[H \cap T] + P[C \cap T] \leqq 1.$$

In fact, the union of these three sets must be T:

(18) $$(D \cap T) \cup (H \cap T) \cup (C \cap T) = T.$$

Therefore (17) can be made more precise, namely

(19) $$P[D \cap T] + P[H \cap T] + P[C \cap T] = P[T].$$

If we divide both sides of (19) by $P[T]$, we obtain

(20) $$\frac{P[D \cap T]}{P[T]} + \frac{P[H \cap T]}{P[T]} + \frac{P[C \cap T]}{P[T]} = 1,$$

and this gives us a new set of numbers that has 1 for its sum. This new set of three numbers is proportional to the set of numbers $P[D \cap T]$, $P[H \cap T]$, and $P[C \cap T]$. Hence the ratios in (20) are the revised probabilities that we are seeking. We use the symbol $P[A \mid B]$ to denote the probability of the event A, given that the event B has occurred. The symbol $P[A \mid B]$ can be read as "the conditional probability of A on the hypothesis B" or more briefly "the probability of A given B." Using this new notation we have the following:

If x is selected at random and we find that his income is over \$10,000, then

(21)   Probability that $x \in D = P[D \mid T] = \dfrac{P[D \cap T]}{P[T]}.$

(22)   Probability that $x \in H = P[H \mid T] = \dfrac{P[H \cap T]}{P[T]}$.

(23)   Probability that $x \in C = P[C \mid T] = \dfrac{P[C \cap T]}{P[T]}$.

We have arrived at (21), (22), and (23) in an intuitive manner, and one might now expect a theorem. However, a deeper investigation reveals that one cannot prove that a certain event has a certain probability. Instead, we have the reasonable

Definition 6. Conditional Probability. Let A and B be subsets of a sample space U and suppose that $P[B] \neq 0$. Then

(24) $P[A \mid B] = \dfrac{P[A \cap B]}{P[B]}$.

EXAMPLE 1. Let S be the sum obtained when two dice are thrown in sequence. Find $P[5 \leq S \leq 9]$ if it is known that the first die shows a 4.

Solution. In Section 1 we found that $P[5 \leq S \leq 9] = 2/3$, when we had no further information about the dice. Let A be the event $5 \leq S \leq 9$, and let B be the event "the first die shows a 4." Then the set $A \cap B$ is the set of pairs (4,1), (4,2), (4,3), (4,4), and (4,5), and $P[A \cap B] = 5/36$. Further, $P[B] = 1/6$. Then Definition 6 gives

$$P[5 \leq S \leq 9 \mid \text{first die is } 4] = \frac{5/36}{1/6} = 5/6.$$

EXAMPLE 2. Consider the sample space of all families with exactly two children, and assume that we select a family at random from U. Given that one child in the family is a boy, find the probability that the other child is also a boy. Assume that the probability of a child being a boy is $1/2$ (actually at birth it is slightly higher).

Solution. We represent the points of U by the pairs (B,B), $(B,G,)$, (G,B), and (G,G), where the first letter in each pair refers to the sex of the older child. By hypothesis the probability for each of these pairs is $1/4$. If A denotes the event "the other child is a boy" and B denotes the event "the family has at least one boy," then we are asked to compute $P[A \mid B]$. By Definition 6,

$$P[A \mid B] = \frac{P[A \cap B]}{P[B]} = \frac{1/4}{3/4} = \frac{1}{3}.$$

This result is interesting because one might guess that the probability of the second child being a boy is 1/2.

A good way to test this result experimentally is to take any stag gathering and ask each person from a family of two children to state whether the other child is a brother or sister. Of course the fraction of those eligible that have a brother need not be exactly 1/3, but it should be considerably less than 1/2.

Exercise 5

In problems 1 through 3 a person draws two cards from a well-shuffled deck.

1. **(a)** Find the probability of drawing two aces. **(b)** Find the probability of drawing two aces given that the first card is an ace. **(c)** Find the probability of drawing two aces, given that the first card is a queen.
2. **(a)** Find the probability of drawing two cards of the same suit. **(b)** Find the probability of drawing two cards of the same suit if the first card is the ace of spades.
3. **(a)** Find the probability of drawing a pair. **(b)** Find the probability of drawing a pair if the first card is the queen of hearts.

4. It is known that among the families in Euphoria, 30 percent have no children, 20 percent have one child, 20 percent have two children, 14 percent have three children, 9 percent have four children, and 7 percent have five or more children. If a Euphorian family is selected at random, what is the probability that they have more than two children? What is the probability that they have more than two children, given that they have at least one child?
5. About 80 percent of the senators in Zampolia are opinionated and are willing to lecture on the slightest provocation. The judges there are a little more reticent and only 30 percent suffer from this fault. At a recent gathering of 30 senators and 10 judges from Zampolia, a man selected at random to win the door prize was very talkative. What is the probability that the man is a senator? If the last man to leave the party is very quiet, what is the probability that he is a senator?
6. In a certain bridge deal, North has a hand with no aces. What is the probability that his partner South also has no aces?
7. Records show that about 5 percent of the men are color-blind but that only 0.25 percent of the women are color-blind. If J. Doe is color-blind, what is the probability that it is John and not Jane? *Hint:* Assume half of the population is male.
8. In one of the southern states, registered Democrats outnumber registered Republicans 4 to 1. In the latest election all Republican voters remained loyal, and enough Democrats crossed over to elect a Republican gover-

nor by a ratio of 7 to 6. If X is selected at random from the voters, what is the probability that he is a Democrat? What is the probability if it is known that he voted for the Republican governor?

9. In Definition 6, we must assume that $P[B] \neq 0$ because it appears in the denominator in equation (24). Explain why $P[B] = 0$ never occurs in a "practical" problem.

10. Interpret equation (24) when $A \cap B = \varnothing$.

11. Let U be an equiprobable space. Prove that $P[A \mid B] = n(A \cap B)/n(B)$.

12. Let $A_1, A_2, \ldots, A_r$ be a partition of U. Thus $A_1 \cup A_2 \cup \cdots \cup A_r = U$, and any two of the sets have empty intersection. Prove that if $P[B] \neq 0$, then

$$P[A_1 \mid B] + P[A_2 \mid B] + \cdots + P[A_r \mid B] = 1.$$

7. independent events

The concept of independent events is usually associated with a sequence of experiments. The question involved is simply this: Does the outcome of the first experiment have any effect on the probabilities of the various outcomes of the subsequent experiments? When the answer is no, we regard the experiments as independent. For simplicity we consider only two experiments in

EXAMPLE 1.   A standard die is rolled, and then a card is drawn at random from a well-shuffled deck. Let E be the event that the die shows either a 2 or a 5, and the card drawn is a heart. Find $P[E]$.

Solution.   The sample space U consists of $6 \times 52 = 312$ points. The set E consists of $2 \times 13 = 26$ points. Since this U is an equiprobable space, we have

(25) $$P[E] = \frac{26}{312} = \frac{1}{12}.$$

There is an alternative approach. Let U_1 be the sample space for rolling a die, and let U_2 be the sample space for drawing a card. Let E_1 and E_2 be the events already described for these two experiments. Then the space U can be regarded as the Cartesian product $U_1 \times U_2$, and similarly we have $E = E_1 \times E_2$. Clearly

(26) $$P[E_1] = \frac{2}{6}, \qquad P[E_2] = \frac{13}{52},$$

and we observe that

(27) $$P[E] = \frac{1}{12} = \frac{1}{3} \cdot \frac{1}{4} = \frac{2}{6} \cdot \frac{13}{52} = P[E_1]P[E_2].$$

The crux of the matter is that we can obtain $P[E]$ by multiplying the probabilities of E_1 and E_2, the components of E. This multiplication rule does not hold for all sequences of experiments, but it does indeed hold if they are *independent*. In the classical theory of probability it was regarded as intuitively obvious that

(28) $$P[E] = P[E_1]P[E_2]$$

and equation (28) was regarded as a consequence of the independent nature of the two experiments. Although there is strong experimental evidence that under the circumstances described in Example 1, equation (28) does hold, we now recognize that a (mathematical) proof is impossible. Instead we make equation (28) the definition of the term "independent events."

Definition 7. Independent Events. Let U_1 and U_2 be sample spaces corresponding to two experiments, and let $E_1 \subset U_1$ and $E_2 \subset U_2$. Further, let $E = E_1 \times E_2$ and $U = U_1 \times U_2$. If equation (28) holds, then E_1 and E_2 are called independent events. If equation (28) holds for every pair of subsets $E_1 \subset U_1$ and $E_2 \subset U_2$, then the experiments are said to be independent experiments.

This definition of independence generalizes to any sequence of n experiments, and we leave the labor of stating this generalization for the energetic reader.

EXAMPLE 2. A pair of dice is rolled four times in a row. Assuming that the experiments are independent, find the probability that $S \leq 7$ on every roll.

Solution. For the first roll $P[S \leq 7] = 21/36 = 7/12$. By the multiplication rule for independent events we find that

$$P[S \leq 7 \text{ on four successive rolls}] = \left(\frac{7}{12}\right)^4 = \frac{2401}{20{,}736} \approx 0.116.$$

The idea of independent events also occurs when one considers the outcome of a single experiment (rather than a sequence of experiments). We have

Definition 8. Let A and B be events in U. If

(29) $$P[A \cap B] = P[A]P[B],$$

then A and B are said to be independent events.

One should observe the relation of this concept to that of conditional probability. Does event B have any influence on event A? If it does not, then we must have

(30) $$P[A] = P[A \mid B].$$

But we recall (Definition 6) that

(31) $$P[A \mid B] = \frac{P[A \cap B]}{P[B]}.$$

If we use (30) in (31), then we deduce (29). Conversely, if (29) holds, we can prove that (30) does. This gives

Theorem 2. The events A and B are independent if and only if

(30) $$P[A] = P[A \mid B].$$

EXAMPLE 3. Let $A = \{1,2,3\}$ and $B = \{3,4,5\}$. Show that for one roll of a standard die, these two events are not independent.

Solution. Clearly $P[A] = 1/2$, $P[B] = 1/2$, and

(32) $$P[A \cap B] = P[3] = \frac{1}{6} \neq \frac{1}{4} = P[A]P[B].$$

EXAMPLE 4. Suppose that the die used in Example 3 is loaded in such a way that

$$P[1] = P[2] = P[4] = P[5] = \frac{1}{8}, \quad \text{and} \quad P[3] = P[6] = \frac{1}{4}.$$

Prove that for this loaded die the events A and B of Example 3 are independent.

$$P[A] = P[1,2,3] = \frac{1}{8} + \frac{1}{8} + \frac{1}{4} = \frac{1}{2},$$

and

$$P[B] = P[3,4,5] = \frac{1}{4} + \frac{1}{8} + \frac{1}{8} = \frac{1}{2}.$$

Further,

$$P[A \cap B] = P[3] = \frac{1}{4} = \frac{1}{2} \cdot \frac{1}{2} = P[A]P[B].$$

Hence A and B are independent.

EXAMPLE 5. Three identical urns A, B, and C contain red and green balls as follows: A, four red and one green; B, two red and three green; and C, three red and seven green. An urn is selected at random, and then a ball is drawn from that urn. Find the probability that the ball is green.

Solution. At first glance the events do not seem to be independent because the probability of drawing a green ball depends on which urn is selected. But if $E_1 = (A,G_A)$ denotes the compound event of selecting the urn A and drawing a green ball from urn A, then the events A and G_A are independent and we can write

(33) $$P[E_1] = P[(A,G_A)] = P[A]P[G_A] = \frac{1}{3} \cdot \frac{1}{5} = \frac{1}{15}.$$

Similarly

(34) $$P[E_2] = P[(B,G_B)] = P[B]P[G_B] = \frac{1}{3} \cdot \frac{3}{5} = \frac{3}{15},$$

and

(35) $$P[E_3] = P[(C,G_C)] = P[C]P[G_C] = \frac{1}{3} \cdot \frac{7}{10} = \frac{7}{30}.$$

On the other hand, E_1, E_2, and E_3 are pairwise disjoint and give all ways of drawing a green ball. Hence

(36) $$P[G] = P[E_1] + P[E_2] + P[E_3] = \frac{2}{30} + \frac{6}{30} + \frac{7}{30} = \frac{1}{2}.$$

Observe that if all the balls were in one urn, then we would have $P[G] = 11/20 \neq 1/2$.

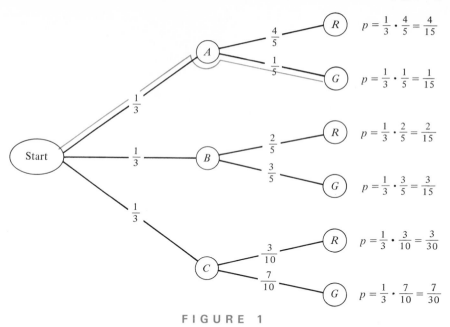

FIGURE 1

One can use a tree diagram to help visualize the above computations. Such a diagram is given in Fig. 1. Each branch is labeled with the probability of the event that it represents, and the probability of a certain path is exactly the product of the probabilities associated with the branches of that path. For example, the computation in equation (33) is represented in Fig. 1 by the path indicated with color.

Exercise 6

1. There are three urns A, B, and C each containing red and green balls as follows: A, 3 red and 1 green; B, 2 red and 3 green; and C, 731 red and 0 green. A ball is drawn at random from each of urns A, B, and C. Describe the sample space for this compound experiment and determine the probability for each point.

2. A good die is rolled four times. What is the probability that it shows a 6 at least once? Is this greater or less than $1/2$?

3. A good coin is tossed four times. What is the probability of getting the same number of heads and tails?

4. An urn contains 10 balls numbered from 1 to 10. Two balls are drawn at random without replacement. Let A_1 be the event that the first ball drawn is odd, and let A_2 be the same event for the second ball. Find

$P[A_1]$, $P[A_2]$, and $P[A_1 \cap A_2]$. Are A_1 and A_2 independent events?

5. An urn contains seven red balls and three green balls. Two balls are drawn at random in sequence. Find (a) the probability that the second ball is green, (b) the probability that the second ball is green given that the first ball was red, and (c) the probability that the second ball is green given that the first one was green.

6. Generalize the result of problem 5 to an urn with r red balls and g green balls, assuming that $r \geq 1$ and $g \geq 1$.

7. Referring to problems 5 and 6, let A, B, and C be the events described in (a), (b), and (c), respectively. Prove that

$$P[B] > P[A] > P[C].$$

Do these inequalities seem reasonable?

8. One of two urns is selected at random, and from the urn selected a ball is drawn at random. It is observed that the ball is red. We know that the first urn contains five red balls and two green balls and the second urn contains one red ball and six green balls. Find the probability that the first urn was selected.

9. Suppose that in problem 8 there is a third urn containing 10 red balls and 0 green balls. An urn is selected at random from among the three, and a ball is drawn at random from the urn selected. If the ball drawn is red, find the probability that the third urn was selected.

10. Suppose that in problem 9 all conditions are the same except that the third urn contains n red balls and 0 green balls, with $n \geq 1$. Given that a red ball is drawn, find the probability that the third urn was selected.

11. Suppose that for the urns of problem 10 the ball drawn turns out to be green. Find the probability that the third urn was selected. Find the probability that it was the second urn.

12. A true coin is flipped 10 times in a row. Find the probability of obtaining 10 heads. Given that the first 9 throws are heads, find the probability that the tenth throw will also be heads.

13. Suppose a die is loaded so that $P[1] = P[3] = P[5] = 1/4$ and $P[2] = P[4] = P[6] = 1/12$. Let $A = \{1,3\}$, $B = \{3,5\}$, and $C = \{1,5\}$. Prove that any pair of events A, B, and C are independent. Prove that the three events are not independent.

14. A student taking mathematics, science, English, and philosophy estimates that his probability of receiving an A is 1/10, 3/10, 5/10, and 7/10, respectively. Further, since the four professors involved never talk to each other, the student assumes that the grades can be regarded as independent events. Based on these assumptions, compute the probability that he receives (a) no A's and (b) exactly one A.

★8. Bayes' theorem

This theorem is merely a restatement of the equation for the conditional probability. However, its importance lies in the manner in which it is applied. Briefly, we have a number of hypotheses, $H_1, H_2, \ldots, H_n$, each of which predicts a certain event E with a certain probability.[1] Now suppose that E does in fact occur. We would like to deduce from this that one of the hypotheses is correct—but this is too much. We can, however, revise the probabilities for the various hypotheses in view of the occurrence of E. Before giving the theorem, we illustrate a typical application in

EXAMPLE 1. Three distinct economic theories H_1, H_2, and H_3 are proposed by their supporters. We are not certain which theory really represents the behavior of our economic system, but on the basis of the arguments put forth and the reputation of the proponents, we assign the following probabilities for being correct:

$$(37) \qquad P[H_1] = \frac{1}{2}, \qquad P[H_2] = \frac{1}{3}, \qquad P[H_3] = \frac{1}{6}.$$

None of the theories can predict with certainty, but each theory produces a certain probability that the cost of living will behave in a certain way during the coming year. These are displayed in Table 6, where, for example, the theory H_1 gives a probability of $1/5$ that the cost will go up; i.e., $P[A \mid H_1] = 1/5$. A year later it is observed that the cost of living has indeed gone up. How does this affect the probability that the theory H_1 is the correct theory?

TABLE 6

Economic theory	P[E], according to the theory		
	A: Up	B: Unchanged	C: Down
H_1	$\frac{1}{5}$	$\frac{3}{5}$	$\frac{1}{5}$
H_2	$\frac{1}{5}$	$\frac{1}{5}$	$\frac{3}{5}$
H_3	$\frac{4}{5}$	$\frac{1}{10}$	$\frac{1}{10}$

We pause for a proof of Bayes' theorem because it is the key to the solution of this problem. Rather than refer to economic theories, we consider a

---

[1] The sets $H_1, H_2, \ldots, H_n$ are events in the sample space, but it is desirable to give them distinctive symbols H_k in order to remind us that they may be regarded as hypotheses.

partition of U into a set of events $H_1, H_2, \ldots, H_n$. By the definition of a
partition $H_i \cap H_j = \emptyset$ whenever $i \neq j$, and moreover,

(38) $$H_1 \cup H_2 \cup \cdots \cup H_n = U.$$

If we intersect both sides of (38) with an arbitrary set A, we obtain

(39) $$(A \cap H_1) \cup (A \cap H_2) \cup \cdots \cup (A \cap H_n) = A,$$

and hence

(40) $$P[A \cap H_1] + P[A \cap H_2] + \cdots + P[A \cap H_n] = P[A].$$

For each index k, we have

(41) $$P[A \mid H_k] = \frac{P[A \cap H_k]}{P[H_k]},$$

and consequently

(42) $$P[A \cap H_k] = P[A \mid H_k]P[H_k].$$

Using this result in (40), we can rewrite (40) in the form

(43) $$P[A] = \sum_{i=1}^{n} P[A \mid H_i]P[H_i].$$

We interchange the role of A and H_k in (41) and obtain

(44) $$P[H_k \mid A] = \frac{P[A \cap H_k]}{P[A]}.$$

In (44) we replace the numerator by the right-hand side of (42). For the
denominator of (44) we use (43). These replacements give

(45) $$P[H_k \mid A] = \frac{P[A \mid H_k]P[H_k]}{\sum_{i=1}^{n} P[A \mid H_i]P[H_i]}.$$

Theorem 3. Bayes' Theorem. If $H_1, H_2, \ldots, H_n$ form a partition of
U, then equation (45) holds.

In many cases, the hypotheses H_i are initially all equally likely, and then
the terms $P[H_i]$ cancel in (45). This gives a simpler form of Bayes' theorem.

Theorem 4. If $P[H_i] = 1/n$ for $i = 1, 2, \ldots, n$, then

$$(46) \qquad P[H_k \mid A] = \frac{P[A \mid H_k]}{\displaystyle\sum_{i=1}^{n} P[A \mid H_i]}.$$

Solution for Example 1. Since $P[H_1] \neq P[H_2]$, we must use equation (45), and not (46). We are asked to find the probability of H_1 given that the cost of living did indeed go up. Using equation (45) and the entries in Table 6 we find that

$$(47) \qquad P[H_1 \mid A] = \frac{\dfrac{1}{5} \cdot \dfrac{1}{2}}{\dfrac{1}{5} \cdot \dfrac{1}{2} + \dfrac{1}{5} \cdot \dfrac{1}{3} + \dfrac{4}{5} \cdot \dfrac{1}{6}} = \frac{1}{3}.$$

Similarly

$$(48) \qquad P[H_2 \mid A] = \frac{\dfrac{1}{5} \cdot \dfrac{1}{3}}{\dfrac{1}{5} \cdot \dfrac{1}{2} + \dfrac{1}{5} \cdot \dfrac{1}{3} + \dfrac{4}{5} \cdot \dfrac{1}{6}} = \frac{2}{9},$$

and

$$(49) \qquad P[H_3 \mid A] = \frac{\dfrac{4}{5} \cdot \dfrac{1}{6}}{\dfrac{1}{5} \cdot \dfrac{1}{2} + \dfrac{1}{5} \cdot \dfrac{1}{3} + \dfrac{4}{5} \cdot \dfrac{1}{6}} = \frac{4}{9}.$$

Observe that before the event A, the theory H_3 seemed least likely to be correct, but that after A, the theory H_3 became most likely. Does this seem reasonable?

The numbers in Table 6 are often presented as an array, free from any indication of their source. Following this custom we may write

$$M = \begin{bmatrix} \dfrac{1}{5} & \dfrac{3}{5} & \dfrac{1}{5} \\[2mm] \dfrac{1}{5} & \dfrac{1}{5} & \dfrac{3}{5} \\[2mm] \dfrac{4}{5} & \dfrac{1}{10} & \dfrac{1}{10} \end{bmatrix}.$$

An array such as M is called a *matrix*. If a matrix M does arise from a probability problem of the type considered in Example 1, then the entries

(numbers) must all be nonnegative and the sum of the entries in each row $\qquad$
must be 1. Such a matrix is called a *stochastic matrix* (a *probability matrix*). The reader should check that M is a stochastic matrix.

Matrices and their applications will be studied extensively in Chapters 7 and 8.

Exercise 7

1. Suppose that in Example 1 the cost of living was unchanged. Find the probability for each economic theory in this event.
2. In Example 1 find $P[H_k \mid C]$ for $k = 1, 2, 3$.
3. Make a tree diagram representing the data in Example 1. Interpret equations (47), (48), and (49) in terms of the tree diagram.
4. Mr. X has one night out a week, and he always returns home sometime between 11:00 P.M. and 3:00 A.M. either sourly sober, somewhat happy, or disgustingly drunk. His arrival time is a probability function of his condition, and these probabilities are given in Table 7.

TABLE 7

Condition	Arrival time		
	11:00–12:00	12:00–1:00	1:00–3:00
Sober	0.80	0.20	0.00
Happy	0.15	0.50	0.35
Drunk	0.05	0.30	0.65

Through the years Mrs. X (whose hobby is statistics) observes that her husband is sober just as often as he is drunk and that he is happy twice as often as he is sober. One night Mrs. X hears the key turning in the lock and observes that the time is 12:34 A.M. Find the probability that Mr. X is happy.

5. A week later Mrs. X (from problem 4) hears the key turn at 1:56 A.M. Find the probability that on this night Mr. X is drunk.
★6. Standards are rather high for the master's Degree at Uplift University and the committee is resolved to flunk any C student. The trouble is that students become frightened on their final oral exams and occasionally do not perform as well as they should. Sometimes a student is lucky and is questioned on the few things that he happens to know. The probabilities for performances on the final exam are given in Table 8.

TABLE 8

True ability of student	Actual grade on final exam		
	A	B	C
A	0.80	0.15	0.05
B	0.20	0.70	0.10
C	0.00	0.15	0.85

Suppose now that Mr. M.T. takes the exam and scores a C. The committee is in doubt about the validity of the exam and passes him anyway. Find the probability that the committee made a mistake and passed a man who should have been flunked, given that 60 percent of the students are A students, 30 percent are B students, and 10 percent are C students.

7. Machines A, B, and C produce 20, 30, and 50 percent, respectively, of the day's production of superwidgets. The number of defective superwidgets turned out by machines A, B, and C is 1, 2, and 3 percent, respectively. A checker selects a superwidget at random and finds that it is defective. Find the probability that it came from C.

8. Usually a standard box of subgudgets contains 1 defective one and 99 good ones. An inspector faced with 10 such boxes suddenly receives word that one of the boxes contains 10 bad subgudgets and 90 good ones. He selects a box at random and tests 2 subgudgets, finding that one is defective and one is satisfactory. Find the probability that the inspector picked the bad box.

9. Automobile drivers can be classified as G (good risks), M (medium risks), and B (bad risks). According to data collected by the Instant Aid Insurance Company, 20 percent of their policyholders are class G, 30 percent are class M, and 50 percent are class B. The probability of having at least one accident in a year is 0.001 for G, 0.005 for M, and 0.010 for B. Mr. Nuro Pridemore buys a policy from the company and in 6 months has an accident. What is the probability that he is a class B driver?

10. Miss Kautious Feersum also buys a policy from the company described in problem 9 and goes 2 years without an accident. Find the probability that she is a class G driver.

11. A test for a disease Z may be considered good if it works 95 percent of the time. This means that if X has Z the test is positive 95 percent of the time, and if X does not have Z the test is negative 95 percent of the time. The disease is very rare among the general population, but of those who go to the hospital and take the test, 10 percent have Z. Find the probability that a patient has Z given that the test is negative. Find the probability that the patient does not have Z given that the test is positive.

12. Among male students from middle- and upper-class families, 75 percent reject their parents' standards and wear long hair and go without shoes. Among the lower-class families, 90 percent of the male students are striving to better themselves and are careful to cut their hair and wear shoes. At Uplift U. 60 percent of the students are upper-middle class and 40 percent are from the lower classes. A student selected at random obviously needs a haircut and a pair of shoes. Find the probability that he can afford them.

★**13.** According to the Mendelian theory of genetics the presence of certain measurable characteristics (such as the color of eyes) is governed by genes. These genes, which are present in pairs in both parents, are customarily denoted by letters such as AA, Aa, and aa. An individual is either dominant (AA), heterozygous (Aa), or recessive (aa). According to the theory each offspring inherits one gene from each parent, and either one of the two genes present in the parent may be passed to the offspring with equal probability. If the offspring is AA or Aa, then it always exhibits the dominant characteristic. If the offspring is aa, it will exhibit the recessive characteristic. Suppose that a female rabbit, known to be recessive, is bred to a male chosen at random and that each of the three offspring exhibit the dominant characteristic. Find the probability that the male is (1) AA, (2) Aa, and (3) aa. Assume that among rabbits 25 percent are dominant, 50 percent are heterozygous, and 25 percent are recessive.

14. Referring to Bayes' theorem, prove that

$$\sum_{k=1}^{n} P[H_k \mid A] = 1.$$

15. Given a partition $\{H_1, H_2, \cdots, H_n\}$ of U as in Bayes' theorem and a specific event A, is it true that

$$\sum_{k=1}^{n} P[A \mid H_k] = 1?$$

9. Bernoulli trials

We now consider a sequence of experiments in which each experiment is the same and is run under the same conditions. Consequently the probability function is the same for each experiment. Further, we assume that the experiments are independent (see Definition 7). Under these conditions the sequence is called a *sequence of independent repeated trials under identical conditions* or, more briefly, *repeated trials*.

Now repeated trials are very common in gambling. Rolling dice, dealing cards, and tossing coins are the common examples, but there are also repeated trials apart from games. The manufacture of any product can be thought of as repeated trials in which the result is frequently good (the product passes certain minimum specifications) and occasionally bad (fails to pass). The application of probability theory to the control of the quality of manufactured articles forms a vast and very important field.

Whenever we have a sequence of repeated trials, each trial may have a number of possible results. For example, in the throw of a pair of dice there are 11 possible sums. But in most applications these results can be divided into two disjoint subsets labeled S (success) and F (failure). In such a case the trials are called *Bernoulli trials*.

> **Definition 9. Bernoulli Trials.** A sequence of Bernoulli trials is a sequence of independent repeated trials under identical conditions in which on each trial there are only two possible outcomes, S and F.

It is customary to set $P[S] = p \geq 0$, and $P[F] = q \geq 0$, where, of course, $p + q = 1$. Obviously any sequence of Bernoulli trials can be simulated by repeatedly flipping a coin, loaded so that $P[H] = p$ and $P[T] = q$. The central result for Bernoulli trials is given in

> **Theorem 5.** The probability of exactly k successes in a sequence of n Bernoulli trials with $P[S] = p$, is denoted by $b(k,n,p)$ and is given by

$$(50) \qquad b(k,n,p) = C(n,k)p^k q^{n-k}.$$

Proof. Let us look at a particular sequence of n outcomes, $X = (F,S,S,F,F,S, \ldots, S)$ in which there are exactly k successes and, consequently, $n - k$ failures. Since the experiments are independent, the probability of this particular sequence is the product of p^k for k successes and $(1 - p)^{n-k}$ for $n - k$ failures. Hence

$$(51) \qquad P[X] = p^k(1 - p)^{n-k} = p^k q^{n-k}.$$

How many such sequences X are there with exactly k successes? We can think of this as the number of ways of selecting k of the blanks in $(—,—,—, \ldots, —)$ for the letter S and filling the remaining blanks with F. This is exactly the number of combinations of n things taken k at a time. Hence we have $C(n,k)$ different sequences X with exactly k successes. Since each such X has the probability given by (51), the sum gives (50). ∎

EXAMPLE 1. A student is faced with a true-false examination consisting of 10 questions. He plans on guessing, but he estimates that with his knowledge of the material he has a probability of 2/3 of guessing correctly on each question. Find the probability of his guessing at least 7 correct answers. Find the probability for the same result if the guessing is truly random, $P[S] = 1/2$.

Solution. We regard the 10 consecutive guesses as Bernoulli trials. The probability of exactly k correct answers is

(52) $$b(k,10, 2/3) = C(10,k)\left(\frac{2}{3}\right)^k \left(\frac{1}{3}\right)^{10-k} = C(10,k)\frac{2^k}{3^{10}}.$$

Using equation (52) with $k = 7, 8, 9,$ and 10 we find that

$$P[k \geq 7] = C(10,7)\frac{2^7}{3^{10}} + C(10,8)\frac{2^8}{3^{10}} + C(10,9)\frac{2^9}{3^{10}} + C(10,10)\frac{2^{10}}{3^{10}}$$

$$= \frac{1}{3^{10}}(120 \cdot 2^7 + 45 \cdot 2^8 + 10 \cdot 2^9 + 2^{10}) = \frac{33024}{59049} \approx 0.559.$$

If $p = 1/2$, the same type of computation gives

$$P[k \geq 7] = \frac{1}{2^{10}}(120 + 45 + 10 + 1) = \frac{176}{2^{10}} = \frac{11}{64} \approx 0.172.$$

EXAMPLE 2. Most historians give 1654 as the birth year for probability theory. At that time a French nobleman Chevalier de Méré was puzzled about a certain dice problem and presented this problem to Blaise Pascal (the originator of Pascal's triangle). Pascal in turn wrote to Pierre Fermat and, in the exchange of letters that followed, these two men worked out in great detail much of the classical theory of probability.

It was known to the gamblers of that day that the probability of obtaining at least one 6 in 4 throws of a single die is greater than 1/2. Accordingly it seemed reasonable to the gamblers that if two dice were used, there would be 6 times as many possibilities, and as a result the probability of obtaining at least one double 6 in $6 \times 4 = 24$ throws should also be greater than 1/2. Yet de Méré thought otherwise, and this is the problem he presented to Fermat. Discuss this problem.

Solution. Let A_1 denote the event that at least one 6 is thrown in 4 throws of a single die. Let A_2 denote the event that at least one double 6 is thrown in 24 throws of two dice. In both cases we have a sequence of Bernoulli trials.

For computation, it is easier to find the probability of the complementary event and subtract the result from 1. Now A_1' occurs if, on each of 4 throws, we obtain any number other than 6. For one throw, $p = 5/6$. By Theorem 5 with $k = 4$ and $n = 4$, we obtain

$$P[A_1'] = \left(\frac{5}{6}\right)^4 = \frac{625}{1296}.$$

Consequently

(53) $$P[A_1] = 1 - P[A_1'] = 1 - \frac{625}{1296} = \frac{671}{1296} > \frac{1}{2}.$$

When we consider A_2, the method is the same, but the computations become laborious. The probability for not obtaining a double 6 on 1 throw of two dice is $35/36$. Hence, by Theorem 5,

$$P[A_2'] = \left(\frac{35}{36}\right)^{24}.$$

The reader who is familiar with logarithms can easily check that

$$\log P[A_2'] = 24(\log 35 - \log 36) \approx 24(1.54407 - 1.55630)$$
$$\approx -0.29352 = 0.70648 - 1.$$

Consequently, we find that $P[A_2'] \approx 0.509$, and

(54) $$P[A_2] = 1 - P[A_2'] \approx 0.491 < \frac{1}{2}.$$

The reader who is not familiar with logarithms may accept this computation or check the result on a suitable machine.

Thus we see that de Méré was correct in questioning the old gambler's rule, because equation (53) gives $P[A_1] > 1/2$, and equation (54) gives $P[A_2] < 1/2$.

Exercise 8

1. Without reference to probability theory prove that if $p + q = 1$, then

$$\sum_{k=0}^{n} b(k,n,p) = 1.$$

2. A good coin is tossed five times in a row. Compute the probability of obtaining exactly k heads, for $k = 0, 1, 2, 3, 4, 5$. Use this to check the formula in problem 1.

3. Suppose that the coin in problem 2 is loaded so that $P[H] = 2/3$. Compute the new probabilities for obtaining exactly k heads.

4. The probability of hitting a target with one shot is $1/4$, and five shots are fired. What is the probability of scoring at least two hits?

5. Two evenly matched baseball teams play in the World Series, so that for any one game either team has probability $1/2$ of winning. The series is over as soon as one team wins four games. Find the probability that the series is over in just four games.

6. For the teams of problem 5, find the probability that all seven games are necessary to decide the world championship.

7. Find the probability of throwing a 7 at least twice in five throws of a standard pair of dice.

8. A true coin is tossed n times. Find the probability of obtaining heads one-half of the time if (a) $n = 2$, (b) $n = 6$, and (c) $n = 10$.

9. A coin is weighted so that $P[H] = 1/3$. Find the probability of obtaining $n/3$ heads on n throws if (a) $n = 3$, (b) $n = 6$, and (c) $n = 9$.

10. A batting average for a baseball player is merely a statistical approximation to his probability of getting a hit. It is customary to multiply $P[\text{Hit}]$ by 1000, so that a 300 hitter is one for which $P[\text{Hit}] = 0.3$. For such a hitter find the probability of getting three hits or more in four times at bat.

11. Opinion polls can often give erroneous results when a small sample is taken. Suppose that 60 percent of the population of a town are opposed to a certain action and that the others are for it. If five people are asked for their opinion, what is the probability that this small poll will show the majority favor the action (contrary to fact).

12. Suppose that an airplane dropping a bomb has probability $1/3$ of hitting the target. Find the smallest number of (identical) airplanes making the same run that are necessary to be at least 95 percent certain of hitting the target at least once.

13. Suppose that the probability that a couple will get a divorce within the next 10 years is $1/5$. Given eight couples, selected at random, what is the probability that they will all have the same mates after 10 years?

14. Assuming that in birth either sex is equally probable, find the probability that a family of six children will consist of three boys and three girls.

15. Suppose that security conditions in a jail are so good that the probability of a jail break in any one year is 0.01. Suppose that there are n identically run jails in the country. Prove that the probability of at least one jail break in the country is $1 - (0.99)^n$. Note that if $n = 300$, then $1 - (0.99)^n \approx 0.95$. Hence if there are 300 jails, it is almost certain that there will be at least one jail break during the year.

16. One unit of Smarts Rent a Car Company has 300 cars at its disposal. By carefully selecting its customers they figure the probability of a

serious accident with any one car for a two-year period is only 0.01. Find the probability that during the next two years at least one car will be involved in a serious accident.

17. The President of Carcinoma College figures that for the next semester the probability of militant students seizing the administration building is only 0.01. Suppose now that throughout the country there are 300 liberal arts colleges in the same situation (same rules, same type of student body, etc.). Find the probability that in at least one of the colleges the militant students will seize the administration building during the next semester. The 1969 edition of the World Almanac lists over 1200 senior colleges in the United States.

★18. Let n and p be fixed with $0 < p < 1$. Let k and n be positive integers with $0 \leq k \leq n$. Find the value (or values) of k that maximize $b(k,n,p)$. *Hint:* Consider the condition $b(k,n,p)/b(k-1, n, p) \geq 1$.

10.   expected value

When we consider a single experiment such as throwing a pair of dice, it must be admitted that the probability function is not very useful. One would hesitate to risk his entire fortune on a single throw of dice or a single game of cards. If the experiment is repeated frequently and under identical conditions (Bernoulli trials), then the probability function is physically meaningful, for in this situation we expect $P[A]$ to be a good approximation to the fraction of times that event A actually occurs. This is expressed accurately in

Pseudo-Theorem 1.   Let $P[A]$ be the probability of the event A and let $s(n)$ be the number of times that A actually occurs on n repeated trials. Then

(55) $$\lim_{n \to \infty} \frac{s(n)}{n} = P[A].$$

The unusual looking collection of symbols on the left side of (55) is read "the limit of $s(n)/n$ as n approaches infinity." The idea expressed is quite simple: There is a number $P[A]$, and the fraction $s(n)/n$ gets closer and closer to $P[A]$ as n gets larger and larger. Clearly, equation (55) involves some type of infinite process that is undesirable in a finite mathematics text, but this cannot be avoided. However, the concept of a limit (as just described) should be intuitively clear without any elaborate discussion.

Why do we label the assertion Pseudo-Theorem 1 instead of Theorem 6? Simply because equation (55) really states something about the outcome of an infinite sequence of physical experiments. Certainly we cannot check it experimentally, because we cannot run any experiment an infinite number of times. If we perform the experiment a large number of times, we might hope to discern a pattern and use equation (55) to guess at $P[A]$. But in any long sequence there might be unusual runs that would lead us to the "wrong" value, or worse still the left-hand side of (55) may not have a limit.

There is a theorem closely related to equation (55) that can be proved. This theorem is rather complicated and will be presented as Theorem 2 in the next chapter. In the meantime, however, we can use our pseudo-theorem as a guide to further topics of interest. One of these topics is the expected value of a function associated with an experiment.

To introduce E, the expected value, we adopt the point of view of a gambler playing against the house. The experiment may be regarded as any gambling game (dice, roulette, poker, etc.) in which there are a finite number of mutually exclusive and exhaustive events, $A_1, A_2, \ldots, A_m$. We suppose that for each event A_k there is an exchange of $|V_k|$ dollars (or any other convenient monetary unit), where V_k is positive if the house pays the gambler and V_k is negative if the gambler pays the house.

Suppose that in n plays of the game the event A_k actually occurs in $s_k(n)$ times, $k = 1, 2, \ldots, m$. Then M, the total amount of money exchanged, is given by the sum

$$(56) \qquad M = \sum_{k=1}^{m} V_k s_k(n) = V_1 s_1(n) + V_2 s_2(n) + \cdots + V_m s_m(n).$$

If $M > 0$, then the gambler has won a total of M dollars after playing the game n times. If $M < 0$, then the gambler has lost $|M|$ dollars. Dividing by n gives the average gain (or loss) per game:

$$(57) \qquad \frac{M}{n} = V_1 \frac{s_1(n)}{n} + V_2 \frac{s_2(n)}{n} + \cdots + V_m \frac{s_m(n)}{n}.$$

We now let n approach infinity in equation (57) and we set $E = \lim_{n \to \infty} M/n$, assuming of course that this limit exists. Using Pseudo-Theorem 1 on each term on the right-hand side of (57), we see that[1]

$$E = \lim_{n \to \infty} \frac{M}{n} = \lim_{n \to \infty} \left(V_1 \frac{s_1(n)}{n} + V_2 \frac{s_2(n)}{n} + \cdots + V_m \frac{s_m(n)}{n} \right)$$

$$= V_1 \lim_{n \to \infty} \frac{s_1(n)}{n} + V_2 \lim_{n \to \infty} \frac{s_2(n)}{n} + \cdots + V_m \lim_{n \to \infty} \frac{s_m(n)}{n}.$$

$$(58) \qquad E = V_1 P[A_1] + V_2 P[A_2] + \cdots + V_m P[A_m].$$

[1] Here we use a reasonably obvious theorem on limits:

$$\lim_{n \to \infty} \left(V_1 \frac{s_1(n)}{n} + V_2 \frac{s_2(n)}{n} \right) = V_1 \lim_{n \to \infty} \frac{s_1(n)}{n} + V_2 \lim_{n \to \infty} \frac{s_2(n)}{n}.$$

It is clear that the right-hand side of (58) represents in some way the value of the game to the gambler. If $E > 0$, the game is advantageous to the gambler and he may expect to win on the average E dollars per game. If $E < 0$, the gambler may expect to lose if he plays often enough. If $E = 0$, we say that the game is a *fair game*.

One can give another interpretation to E by altering the background slightly. Suppose that in equation (58) we have $V_k \geq 0$ for $k = 1, 2, \ldots, m$. Then certainly $E \geq 0$, so the gambler cannot lose. In this situation, the gambler is expected to pay the house for the privilege of playing the game. If he pays E dollars to play each game, then the game is *fair*. If he pays less than E dollars, the game is advantageous to the gambler, or the game is in his favor. If he is foolish enough to pay more than E dollars each time for the privilege of playing, then in the long run he must lose.

We have given a detailed discussion of the origin of E, the expected value. But the definition of E stands on its own and is independent of any gambling interpretation. The experiment need not be a game, and the numbers V_1, $V_2, \ldots, V_m$ need not refer to money. Since to each set A_i in the partition there is associated a real number V_i, this association defines a real-valued function V.

> **Definition 10. Expected Value.** Let $A_1, A_2, \ldots, A_m$ be a partition of the sample space U, and let V be the real-valued function defined above. Then E, given by equation (58), is called the expected value of V. The quantity E is also called the expectation.

If we set $p_k = P[A_k]$, we can shorten (58) to

$$(59) \qquad E = V_1 p_1 + V_2 p_2 + \cdots + V_m p_m = \sum_{k=1}^{m} V_k p_k.$$

EXAMPLE 1. The game of roulette is played with a wheel containing 37 slots. The slots are numbered with the integers from 0 to 36 inclusive. A player may place a $1 bet (or any other amount) on any number. If the ball falls in the slot bearing that number, the player recovers his bet and receives in addition 35 times the amount wagered. Otherwise he loses his bet. Using a $1 bet, find the expectation for this game.

Solution. The probability of the ball rolling into any given slot is $1/37$, because (presumably) all of the slots are the same size. Hence from equation (59), with $m = 2$,

$$(60) \qquad E = 35 \cdot \frac{1}{37} + (-1) \cdot \frac{36}{37} = -\frac{1}{37} \approx -0.027.$$

This can be interpreted to mean that if a gambler plays a long time, then E will be close to his average loss. For example, if he plays 370 games, he may expect to lose $370 \times 1/37 = 10$ dollars.

To show that E does not necessarily refer to money we consider

EXAMPLE 2. An automobile dealer receives nine new cars for sale. The next day he receives a telegram from the factory saying that two of the cars are from a defective lot and somehow give very low gas mileage. Find the expected value of the number of cars the dealer will have to test in order to locate both defective cars.

Solution. When we arrange the cars in order for testing, the only important item is the position of the defective ones. Suppose that they occupy the jth and kth positions, where $1 \leq j < k$. These positions can be selected in $C(9,2) = 36$ different ways. Let A_k be the event that the second defective car is in the kth position. The other car must be in any one of the $k - 1$ preceding places. Hence $P[A_k] = (k - 1)/36$, and the number of cars to be tested in event A_k is $V_k = k$. Then

$$E = \sum_{k=1}^{9} V_k p_k = 1 \cdot \frac{0}{36} + 2 \cdot \frac{1}{36} + 3 \cdot \frac{2}{36} + \cdots + 9 \cdot \frac{8}{36} = \frac{240}{36} = 6\frac{2}{3}.$$

EXAMPLE 3. The standard game of dice is usually played according to the following rules. The player throws a pair of dice. If he rolls a 7 or 11 (known as *naturals*) on the first throw, he wins. If on the first throw, he rolls a 2, 3, or 12 he loses. If he rolls any other number, 4, 5, 6, 8, 9, or 10, this number becomes his "*point*," and he must continue to roll until he rolls his "point" again or rolls a 7, whichever occurs first. If his point occurs first, he wins. If the 7 occurs first, he loses. Find the probability that the player wins. Find the expected value of this game.

Solution. We assume that the dice are not loaded. Hence the probability of each individual event is easy to determine (see Section 1, Example 2). Before launching into the computations let us discuss a logical difficulty. To be specific, suppose that the player first rolls a 9. This is now his point, and on continued rolls suppose he throws a 5, 6, 2, 11, 5, 8, 3, 11, 2, ... He must continue to roll until he throws 9, and wins, or throws a 7, and loses. Now all of the various sequences of throws can be taken into consideration, but considerable technique and labor are required. If we use our intuition here we can argue that once the point 9 has been established, all rolls other than 7 or 9 "do not count" and can be ignored. Since 7 can be rolled in six different ways and 9 can be rolled in four different ways, we have $P[W] = 4/10$, and $P[L] = 6/10$, given that 9 is his point. With this

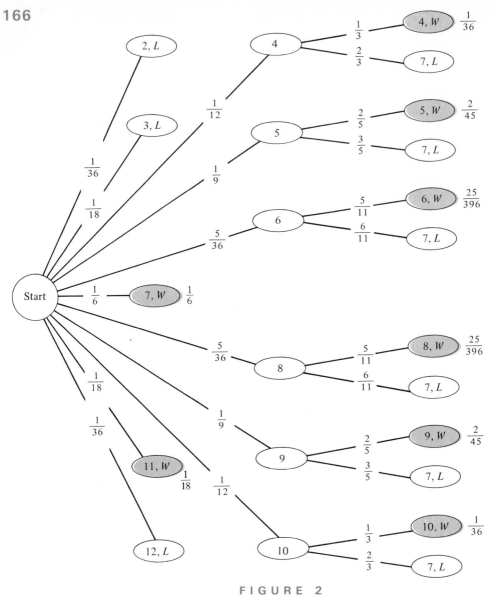

FIGURE 2

understanding, the tree diagram for the game is highly simplified (see Fig. 2). The individual probabilities are placed on the branches, and the probability of each winning path is placed at the end of the path.

Since the various paths are mutually disjoint,

$$P[W] = \frac{2}{36} + \frac{4}{45} + \frac{50}{396} + \frac{1}{6} + \frac{1}{18} = \frac{244}{495}.$$

To find the expected value we assume that a player receives \$1 if he wins and pays \$1 if he loses. Then

(61) $$E = 1 \cdot \frac{244}{495} + (-1)\frac{251}{495} = -\frac{7}{495} \approx -0.0141.$$

Therefore the player who continues to throw will lose in the long run, unless he can "control" the dice.

EXAMPLE 4. The Petersburg Paradox.   Suppose that a player flips a good coin until heads first appears. If heads appears on the first throw, the house pays the player x dollars. If heads first appears on the second throw, the house pays the player x^2 dollars. In general if the sequence of throws is $(T,T,T,\ldots,T,H)$ with H in the nth place, then the house pays the player x^n dollars. Find the expected value of this game if **(a)** $x = \$1.50$, **(b)** $x = \$1.95$, and **(c)** $x = \$2.00$.

Solution.   We first note that the sample space is infinite because it is conceivable that the player might throw an infinite sequence of tails. Hence the expected value involves the sum of an infinite number of terms (an infinite series). The probability of the sequence $(T,T,T,\ldots,T,H)$ on n throws is $1/2^n$, and for this event $V = x^n$. Hence

(62) $$E = \frac{x}{2} + \frac{x^2}{4} + \frac{x^3}{8} + \cdots + \frac{x^n}{2^n} + \cdots,$$

an infinite series. It turns out that if $-2 < x < 2$, then this series is easy to sum (see problem 14 in Exercise 9) and we find that

(63) $$E = \frac{x}{2 - x}.$$

(a) If $x = \$1.50$, then equation (63) gives $E = \$3$. For a fair game the player should pay $3 each time for the privilege of playing.
(b) If $x = \$1.95$, then equation (63) gives $E = \$39$.
(c) The paradox arises when $x = \$2.00$. In this case formula (63) is no longer applicable. Returning to (62) we see that if $x = 2$, then

(64) $$E = 1 + 1 + 1 + \cdots + 1 + \cdots,$$

or the sum can be made as large as we please by taking as many terms on the right as we need. We abbreviate this result by writing $E = \infty$ (the expected value is infinite). This result is certainly paradoxical because it seems to say that if a person paid $1,000,000 each time he played the game, he would eventually win. This startling result bothered mathematicians of the eighteenth and nineteenth century, because the error (if there is one) was difficult to spot, and yet the conclusion, $E = \infty$, could not be accepted. The curious reader should consult *Lady Luck* by Warren Weaver, Anchor

Books, New York, 1963, or the article by Karl Menger, "The Role of Uncertainty in Economics," pp. 211–232, in *Essays in Honor of Oskar Morganstern,* Princeton University Press, Princeton, N.J., 1967.

We close this chapter with a very important theorem. If p is the probability of an event E and the experiment is repeated n times, we would expect E to occur roughly np times. This is expressed accurately in

Theorem 6. In a sequence of n Bernoulli trials the expected value of the number of successes is np, where p is the probability of a success on each trial.

Proof. We use equation (58) with $V_k = k$, the number of successes, and $p_k = b(k,n,p)$. Then

$$(65) \qquad E = \sum_{k=0}^{n} k b(k,n,p) = \sum_{k=0}^{n} k \frac{n!}{k!(n-k)!} p^k q^{n-k}.$$

Since the first term in the sum is zero ($k = 0$), we can drop it. Then n and p both occur as factors in each of the other terms. Finally, we note that $k/(k!) = 1/(k-1)!$, and the right-hand side of (65) yields

$$(66) \qquad E = np \sum_{k=1}^{n} \frac{(n-1)!}{(k-1)!(n-k)!} p^{k-1} q^{n-k}$$

$$= np \sum_{k=1}^{n} \binom{n-1}{k-1} p^{k-1} q^{(n-1)-(k-1)}$$

$$= np \sum_{k=0}^{n-1} \binom{n-1}{k} p^k q^{n-1-k}$$

$$= np(p+q)^{n-1} = np \cdot 1 = np. \quad \blacksquare$$

Exercise 9

1. A player rolls a die and receives a number of dollars equal to the number of dots showing on the face of the die. What should the player pay each time to play this game in order for the game to be a fair one?
2. Suppose that in problem 1 the player rolls a pair of dice. What is the expected value of the game?
3. In roulette a player may bet on red or black. The numbers from 1 to 36 are evenly divided between these two colors. If a player bets $1 on red and the ball comes to rest in a red slot, he recovers his bet and receives an additional $1. If a black number comes up, he loses his bet.

If the ball comes to rest on 0, then the player also loses. Find the expected value of the game.

4. A good coin is tossed six times in a row. Find the expected value of the number of heads.

5. A dealer offers to pay $25 to any player who draws the ace of spades from a well-shuffled standard deck of cards. The dealer is willing to pay $4 for any other spade drawn provided the player will pay the dealer $2 if he fails to draw a spade. Find the expectation for this game. Who has the advantage in this game?

6. One popular game played with three dice allows the player to select as his point any number x from 1 to 6. If on the roll of three dice he gets three x's he wins $4. For a pair of x's he receives $3, for one x he receives $2, and otherwise he receives nothing. If he pays $1 entrance fee, find his average gain (or loss) per game.

7. A drunk has six keys. He tries them at random on his front door, but if a key does not work, he is sober enough not to try it again. Find the expected value of the number of keys he must try before the door will open (assuming that only one key works).

8. Find the expected value of a lottery ticket if the prizes are 1 prize of $50,000, 2 prizes of $25,000, 6 prizes of $5000, 100 prizes of $100, and 1000 prizes of $10; 1 million tickets are sold. Would it be to your advantage to buy tickets at $1 apiece?

9. The Instant Aid Insurance Company carries insurance for 10,000 cars. They estimate that in any one year, $P[X]$, the probability of paying a liability claim for X dollars is as follows: $P[50,000] = 0.001$, $P[25,000] = 0.002$, $P[5000] = 0.006$, and $P[100] = 0.05$. Find the expected amount of money to be payed per car. What premium should the drivers pay Instant Aid for a liability policy?

10. Suppose that in Example 4 the house pays 3^n dollars for a sequence $(T,T,T, \ldots, T,H)$ of n throws, but it is agreed that the game never proceeds beyond four throws. If the house charges $10 to play the game, is this advantageous to the player?

11. Suppose that on the draw of a card, the dealer will pay $10 for an ace, $5 for a king or queen, and $3 for a jack or a ten. What is the expected value of this game.

12. Clever Clem agrees to a game of dice with Honest John, whom he just met in the school library. Without further knowledge Clem assumes the probability $1/2$ that Honest John is honest, but if the contrary is true, he assumes that John can throw a 7 with probability $3/4$. Honest John proceeds to win the first three games by throwing 7 each time. Find the probability that John is crooked. Suppose that before the game starts Clem assigns probability $5/6$ that Honest John is honest. After the first three plays, what is the probability that John is crooked?

13. **Mathematicians dice.** A mathematician rolling a pair of dice is willing to pay n dollars if the number thrown is the product of n primes (not necessarily distinct), where $n = 2$ or 3. If, however, he rolls a prime number, then he receives \$3. What is the expected value of this game?

14. **Derivation of equation (63).** Multiply both sides of equation (62) by $x/2$ (assuming that this is permitted for an infinite series). Subtract the new equation from (62) and show that this gives $E - Ex/2 = x/2$. Show that this gives equation (63).

15. Check that Theorem 6 is true by direct evaluation of the left-hand side of (65) in each of the following cases: **(a)** $n = 4$, $p = 1/2$; **(b)** $n = 6$, $p = 1/3$; **(c)** $n = 6$, $p = 2/3$; and **(d)** $n = 5$, $p = 3/5$.

16. Find the expected value of the number of 7's in a sequence of 1200 throws of a pair of standard dice. If you throw a pair of dice 1200 times, would you expect to roll exactly this many 7's?

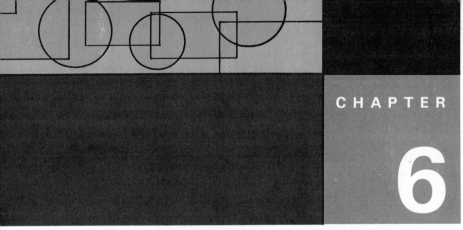

6

further topics in probability

In Chapter 5 we touched on the salient features of probability theory. In this chapter we dig a little deeper.

1. stochastic processes

Any sequence of experiments in which the outcomes depend on some chance element is called a *stochastic process*. A stochastic process is also called a *random process*, or a *chance process*. The reader will realize that Chapter 5 was devoted to stochastic processes, although they were not given that name. The processes studied in Chapter 5 were usually of a simplified nature: either (a) the experiments were independent, or (b) the probability function associated with the kth experiment was independent of the outcomes of the preceding experiments. In a general stochastic process, neither (a) nor (b) is necessarily true.

Consider, for example, the probability of a win between two tennis players A and B who are presumably of the same ability at the outset. It may be that player A tires more readily so that as the match progresses the probability of his winning a given game grows smaller. On the other hand, A may also be stimulated by earlier successes so that in the event of a previous sequence of earlier victories his probability of winning the game increases.

A more serious example may be the spread of a contagious disease. In this case the probability that a certain individual may contract the disease during a certain time interval (say, a month) is clearly dependent on the number of people who have previously contracted the disease and the number of people who have been cured and who presumably are no longer carriers (contagious). Consequently this probability will change from month to month, in a complicated manner.

To handle such problems we need to introduce a suitable notation. Suppose that on each trial there are three outcomes of interest, A, B, and C. (One may imagine a game such as chess where the player may win, lose, or draw.) We let p_a, p_b, and p_c denote probability of A, B, and C, respectively, on the first trial. On the second trial, the outcome now depends on the results of the first trial. We let $p_{a,b}$ denote the probability that B occurs on the second trial given that A occurred on the first trial. In this case the second trial gives rise to nine numbers $p_{x,y}$, where x and y may each be a, b, or c. These numbers can be arranged in an array (a matrix), as indicated in Table 1.

TABLE 1

Outcome of first trial	Probability of indicated outcome on second trial		
	A	B	C
A	$p_{a,a}$	$p_{a,b}$	$p_{a,c}$
B	$p_{b,a}$	$p_{b,b}$	$p_{b,c}$
C	$p_{c,a}$	$p_{c,b}$	$p_{c,c}$

When we proceed to the third trial, we use the notation $p_{ba,c}$ to indicate the probability that C occurs on the third trial given that B occurred on the first trial and A occurred on the second trial. We have, then, 27 numbers $p_{xy,z}$ since each of x, y, and z may be a, b, or c. These cannot be arranged as in Table 1; they require a three-dimensional array for a systematic presentation. As the reader may anticipate, the situation is still more complex for the fourth trial where symbols such as $p_{aaa,a}$ and $p_{bca,b}$ make their appearance. The only restriction on these numbers is that at each trial the sum of the probabilities must be 1. For example, if there are three possible outcomes, then for each fixed set of indices, x, y, z, we must have

$$p_{xyz,a} + p_{xyz,b} + p_{xyz,c} = 1.$$

Thus in Table 1 the sum of the numbers in each row must be 1.

A tree diagram may be helpful in understanding a stochastic process, but the diagram will soon become unwieldy if the number of trials is very large. In the tree diagram we place on each branch the probability of the event it represents, given the events that must have occurred to arrive at that branch.

EXAMPLE 1. Players A and B are equally matched. Each time player A loses a game he becomes tired and discouraged and his probability of winning the next game decreases by 2/10. However, if A wins, then his elation overcomes his fatigue and his probability of winning the next game increases by 1/10. Make a tree diagram for a three-game sequence. Find the probability that A wins either two games or three games.

Solution. The tree diagram is shown in Fig. 1. The probability of each path through the tree is placed at the end of the path and is obtained by multiplying the probabilities on the various branches that compose the path. The reader should observe that the sum of the probabilities for all paths is 1.

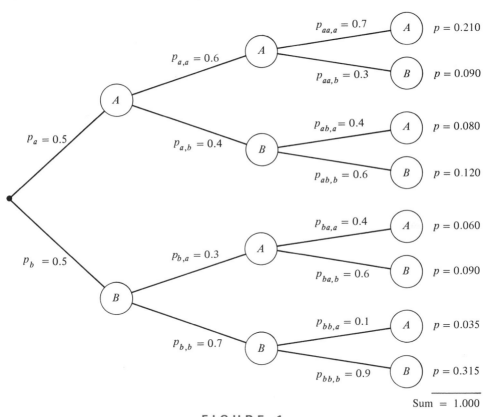

FIGURE 1

The probability that A wins either two games or three games is

$$0.210 + 0.090 + 0.080 + 0.060 = 0.440.$$

EXAMPLE 2. An Urn Representation for the Spread of a Disease. Suppose that an urn contains N balls: x_0 are green (immune people), y_0 are red (sick people), and the rest are white (people who have not had the disease). A ball is drawn at random and then placed back in the urn. If the ball drawn is red, then r of the white balls are replaced by red balls. If the ball drawn is green, then g of the red balls are replaced by green balls. If the ball is white, no change is made. Find the probability of drawing the sequence R, G, W, R.

Solution. We assume that the balls are well mixed before each drawing, so that the probability of drawing any particular color is proportional to the number of balls present of that color. Then

$$P_{rgwr} = \frac{y_0}{N} \cdot \frac{x_0}{N} \cdot \frac{N - r - x_0 - y_0}{N} \cdot \frac{y_0 + r - g}{N}.$$

In general, after a sequence in which a green balls, b red balls, and c white balls have been drawn, the composition of the urn is

$$
\begin{aligned}
x_0 + ag & \qquad \text{green balls,} \\
y_0 + br - ag & \qquad \text{red balls,} \\
N - x_0 - y_0 - br & \qquad \text{white balls.}
\end{aligned}
$$

Of course there are limitations. As soon as $y_0 + br - ag = 0$, it becomes impossible to draw a red ball (the disease has been eradicated). If $N - x_0 - y_0 - br = 0$, then everyone has already had the disease or is immune to it.

Exercise 1

1. Suppose that in Example 1, A has won m games and B has won n games. Find the probability for a victory for A on the next game. Show that under certain conditions the restrictions for a probability function are no longer satisfied.

2. **Odd man out.** A, B, and C enter a tournament in which A plays B and in the second round the winner plays C. In each round the winner of the previous round has for his opponent the man who did not play in the previous round. The tournament is over as soon as one player (the victor) wins two consecutive games. Draw a tree for the first four rounds. Show that the tree for this tournament may have paths of infinite length.

3. Assuming that the players in problem 2 are evenly matched, find the probability for each player to win the tournament in four rounds. Find the probability that the tournament requires more than four rounds.

4. Suppose that in problem 2, A and B are equally skillful but that when C plays either A or B the probability that C wins is 3/4. Find the probabilities requested in problem 3.

5. The game of odd man out (problem 2) is revised so that play continues for four rounds, no matter who wins in each round. The winner of the tournament is the player who has won the most games. Find the probabilities for a victory **(a)** assuming the players are equally matched and **(b)** assuming the players are as described in problem 4.

6. In a chess tournament, player A is dashing but erratic and his probability of winning the next game is $(n + 1)/(n + 2)$, where n is the number of games previously won. However, he is easily depressed and this figure must be multiplied by $(2/3)^k$ if he has previously lost k games. **(a)** Find the probability that A will win a three-game match with B, who is unimaginative but very steady. Assume that none of the games are drawn. **(b)** Find the probability that A will win a four-game match.

7. The players in problem 6 play five games. Let W denote a win, and let L denote a loss. Find the probability for A to obtain the sequence **(a)** (W,W,W,W,L), **(b)** (W,W,L,W,W), and **(c)** (L,W,W,W,W).

8. Jones and Smith play a sequence of games. In each game Jones has probability 1/2 of winning \$1 from Smith and probability 1/2 of losing \$1 to Smith. Smith is willing to play as long as Jones wishes, but Jones decides to play until he has won \$2 and then quit. Show by a tree diagram that the game might continue indefinitely.

9. Suppose that in the game described in problem 8, Smith decides to quit after four games if Jones has not already stopped playing. Find the expected value of this game for Jones.

10. Because of patronage and publicity, the party in power, the Conserverals, has probability 5/7 of winning the next election. In fact, with each successive victory their ability to influence public opinion increases and their probability of winning the next election is $(5 + k)/(7 + k)$, given that they have won k elections in a row after coming to power. Find the probability that once in power, the Conserverals will win three more elections in a row. Show that despite their power, this probability is less than 1/2. Find the probability for five more victories after coming to power.

11. An urn contains one green ball and one red ball. A ball is drawn at random and replaced. After each drawing two more balls of the same color as the one drawn are added to the urn. Find the probability for each possible content of the urn after three draws and replacement. Find the probability of drawing six green balls in a row.

12. **A growing confidence model.** In various fields of human endeavors, a success greatly increases the probability of the next success, perhaps due to the increased skill or the increased confidence of the experi-

menter. For example, after each successful recording of a certain popular singer, the probability that the next recording will be a success is increased. To simulate this situation, suppose that an urn originally has two green balls and one red ball and that a ball is drawn at random and replaced together with four more balls of the same color as the one drawn. After the nth drawing, 2^{n+1} more balls of the same color are added, where $n = 1, 2, 3, \ldots$. Find the probability of drawing (a) two green balls in a row, (b) four green balls in a row, and (c) six green balls in a row.

★13. Show that under the conditions of problem 12, the probability of drawing N green balls in a row is always greater than $1/2$, no matter how large N may be.

2.   Markov chains

We now specialize the general type of stochastic process considered in Section 1. We again have a sequence of experiments (trials), but it is convenient to think of the experiment as being performed on a certain system S and that the outcome leaves the system in a certain state. For example, the system may be a certain country, the experiment may be an election, and the outcome leaves the country in a certain state, namely in control of the winning party. In general the outcome on the kth trial may be influenced by the outcomes of all or many of the outcomes of the previous $k - 1$ trials. In a Markov process the probability of the outcome E_j on the kth trial depends only on E_j and on the outcome E_i of the previous experiment. In formalizing this concept we insist that the outcome of the experiment is a state of the system.

> **Definition 1. Markov Chain.**   A Markov chain is a sequence of experiments performed on a system S with the following properties:
>
> (I) At any given time the system is in one of the states $E_1, E_2, \ldots,$ E_n and the outcome of each experiment is that S either is unchanged or is changed to a new state from the set $\{E_1, E_2, \ldots, E_n\}$.
>
> (II) The system can change from one state E_i to another state E_j only as a result of the experiment.
>
> (III) The probability of changing the system from the state E_i to the state E_j depends only on E_j and on E_i, the state of the system at the time of the experiment. This probability is denoted by p_{ij}.

It is convenient to arrange the various probabilities in a matrix. Thus if there are four different states, the Markov chain can be represented by the matrix

(1)
$$P = \begin{bmatrix} p_{11} & p_{12} & p_{13} & p_{14} \\ p_{21} & p_{22} & p_{23} & p_{24} \\ p_{31} & p_{32} & p_{33} & p_{34} \\ p_{41} & p_{42} & p_{43} & p_{44} \end{bmatrix}.$$

If the system is in the state E_1, we look in the first row of P to find the probability that it will stay in E_1 or move to E_2, E_3, or E_4. Similarly if the system is in E_2, we look in the second row of P, etc.

EXAMPLE 1. In Neurosisburg the Conservatives, Democrats, Republicans, and Socialists always nominate candidates for mayor. The probability of winning depends on the party in control and is given by the matrix

(2)
$$P = \begin{bmatrix} 0.25 & 0.30 & 0.40 & 0.05 \\ 0.05 & 0.40 & 0.30 & 0.25 \\ 0.10 & 0.30 & 0.60 & 0.00 \\ 0.00 & 0.45 & 0.30 & 0.25 \end{bmatrix},$$

where the states are E_1, Conservative; E_2, Democrat; E_3, Republican; and E_4, Socialist. Given that a Republican is mayor, find the probability that after two elections a Republican will be mayor.

Solution. Once we have learned to multiply matrices (in the next chapter), the solution to this problem will be purely mechanical. For the present, we resort to a tree diagram. However, we are interested only in those paths which end in a Republican mayor, and hence we need only that portion of the tree shown in Fig. 2 (see next page). The probability of each path is shown at the end of the path. Hence

$$P[(R,X,R)] = 0.04 + 0.09 + 0.36 + 0.00 = 0.49.$$

We observe that because the matrix P represents probabilities, the elements p_{ij} are all nonnegative. Further, for each row of the matrix the sum of the elements in that row must be 1. Any matrix that satisfies these two conditions is called a *stochastic matrix*.

Since p_{ij} gives the probability for the transition of S from the state E_i to the state E_j, it is called the *transition probability*.

The transition probabilities can be represented geometrically as follows. Each state is represented by a point in the plane. The transition from E_i to E_j is indicated by an arrow (curved) from E_i to E_j, and this arrow is

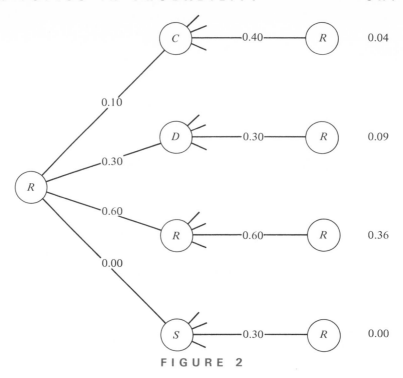

FIGURE 2

labeled with the probability p_{ij}. Such a diagram for a Markov chain is called a *transition diagram*. Of course if the number of states is large, then some of the arrows will have to cross others, but in many simple processes the transition diagram is very helpful.

EXAMPLE 2.   Make a transition diagram for the matrix

(3) $$P = \begin{bmatrix} \dfrac{1}{3} & \dfrac{1}{3} & \dfrac{1}{3} \\[2mm] \dfrac{1}{4} & \dfrac{1}{4} & \dfrac{1}{2} \\[2mm] \dfrac{1}{2} & \dfrac{1}{2} & 0 \end{bmatrix}.$$

Given that the system starts in E_1, find the probability that it will be in state E_3 after two steps (experiments, trials).

Solution.   The required diagram is given in Fig. 3. We leave it for the reader to make a suitable portion of the tree diagram and show that

$$P[(E_1, X, E_3)] = \frac{1}{3} \cdot \frac{1}{3} + \frac{1}{3} \cdot \frac{1}{2} + \frac{1}{3} \cdot 0 = \frac{5}{18}.$$

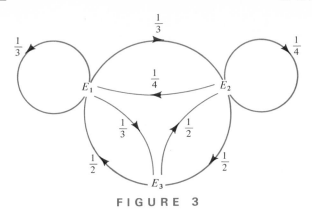

FIGURE 3

We can generalize the type of question asked in Example 2. If the system starts in state E_i, what is the probability that after k steps it will be in state E_j? We denote this probability by $p_{ij}^{(k)}$ (read "p upper k sub ij"). These numbers can also be arranged in a matrix, which we denote by $P^{(k)}$. For example, if there are only three states, then

(4)
$$P^{(k)} = \begin{bmatrix} p_{11}^{(k)} & p_{12}^{(k)} & p_{13}^{(k)} \\ p_{21}^{(k)} & p_{22}^{(k)} & p_{23}^{(k)} \\ p_{31}^{(k)} & p_{32}^{(k)} & p_{33}^{(k)} \end{bmatrix}.$$

In Example 2 we proved that $p_{13}^{(2)} = 5/18$ for the Markov chain defined by equation (3).

EXAMPLE 3. The Random Walk Problem. In studying the motion of a particle it is sometimes convenient to think of the motion as resulting from a sequence at random steps taken in a random direction. Even this idealization is too complicated for our present study, because the particle may move in any one of an infinite number of directions and the steps may be of various lengths. For simplicity we restrict the motion to a line segment of length N and we assume that motion takes place in steps of unit length, either to the right with probability p or to the left with probability q. Give the matrix for this Markov process.

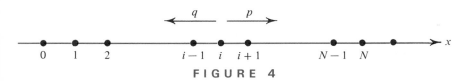

FIGURE 4

Solution. We introduce a coordinate system on a line segment, as indicated in Fig. 4. We let E_i denote the state that the particle is located at the point $x = i$ on the line segment. Then by definition $p_{i,i+1} = p$, and $p_{i,i-1} = q$ for

$i = 1, 2, \ldots, N - 1$. Suppose now that the particle arrives at an end point of the segment ($x = 0$ or $x = N$). Here different possibilities present themselves. If the particle must bounce back, we say that we have *reflecting barriers*. In this case we have $p_{01} = 1$ and $p_{N,N-1} = 1$. If we agree that the particle remains at an end point of the segment once it arrives there, then we say that we have *absorbing barriers*. In this case $p_{00} = p_{NN} = 1$. For the case of absorbing barriers the matrix is

$$
(5) \qquad G = \begin{bmatrix}
1 & 0 & 0 & 0 & \cdots & 0 & 0 & 0 \\
q & 0 & p & 0 & \cdots & 0 & 0 & 0 \\
0 & q & 0 & p & \cdots & 0 & 0 & 0 \\
0 & 0 & q & 0 & \cdots & 0 & 0 & 0 \\
\vdots & \vdots & \vdots & \vdots & \cdots & \vdots & \vdots & \vdots \\
0 & 0 & 0 & 0 & \cdots & q & 0 & p \\
0 & 0 & 0 & 0 & \cdots & 0 & 0 & 1
\end{bmatrix},
$$

where G has $N + 1$ rows and $N + 1$ columns. We shall return to this Markov chain and make an important application in Section 3. Observe that if there are no barriers or only one barrier, the corresponding matrix for this Markov process will have infinitely many rows and columns.

Exercise 2

1. Each of the transition diagrams in Fig. 5 (see next page) represents a Markov chain. Give the associated matrix.

2. Show that each matrix that you obtained in problem 1 is a stochastic matrix.

3. Make a transition diagram for the Markov chain associated with each of the matrices

$$
A = \begin{bmatrix}
0 & \frac{1}{2} & \frac{1}{2} \\
\frac{1}{3} & 0 & \frac{2}{3} \\
\frac{1}{4} & \frac{3}{4} & 0
\end{bmatrix}, \quad
B = \begin{bmatrix}
0 & \frac{1}{2} & \frac{1}{2} & 0 \\
0 & \frac{1}{3} & \frac{1}{3} & \frac{1}{3} \\
\frac{1}{4} & \frac{1}{4} & \frac{1}{4} & \frac{1}{4} \\
\frac{1}{2} & 0 & \frac{1}{3} & \frac{1}{6}
\end{bmatrix}.
$$

4. For the process described by the matrix in Example 2, find a few of the entries for $P^{(2)}$.

5. Repeat problem 4 for the matrix given in Example 1.

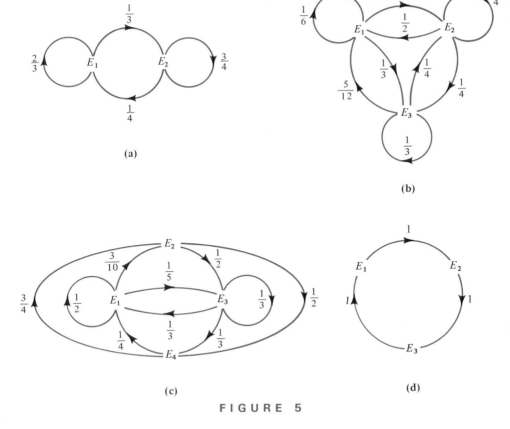

FIGURE 5

6. Find $P^{(2)}$ for the matrix associated with the transition diagram in problem 1(a).

7. Find $P^{(k)}$ for the matrix associated with the transition diagram in problem 1(d). *Hint:* No tree diagram is needed because the problem can be solved purely on a logical basis. There are three cases: $k = 3m$, $k = 3m + 1$, and $k = 3m + 2$, where m is an integer.

8. Find $P^{(k)}$ for a Markov chain determined by each of the following matrices:

(a) $P = \begin{bmatrix} 1 & 0 \\ 0 & 1 \end{bmatrix}$,

(b) $P = \begin{bmatrix} 0 & 1 \\ 1 & 0 \end{bmatrix}$,

(c) $P = \begin{bmatrix} 1 & 0 & 0 \\ 1 & 0 & 0 \\ 0 & 0 & 1 \end{bmatrix}$,

(d) $P = \begin{bmatrix} 0 & 1 & 0 & 0 \\ 1 & 0 & 0 & 0 \\ 0 & 0 & 0 & 1 \\ 0 & 0 & 1 & 0 \end{bmatrix}$.

9. Why is the matrix $P^{(k)}$ a stochastic matrix?

181

10. In 80 percent of the families where the fathers are staunch Republicans, the sons are also. The same fraction holds for sons of Democrats. Among the remaining sons, one-half switch parties, and one-half become discouraged and apathetic and do not vote. Among the sons of the nonvoters, one-half follow their fathers' example, while the remaining are divided equally between the major parties. Describe an appropriate Markov process and give the matrix. What is the probability that the grandson of a Republican will be **(a)** a Republican and **(b)** a nonvoter?

11. A study of the weather in Miami, Fla., shows that if it is sunny on one day, then 90 percent of the time the next day will also be sunny. Fog or rain are equally likely to follow a sunny day (it never snows in Miami). A foggy day is followed 50 percent of the time by a sunny day and 30 percent of the time by a foggy day. A rainy day is followed 90 percent of the time by a sunny day and 10 percent of the time by more rain. Give the matrix that represents this situation.

★12. A gas will slowly seep through a membrane. Suppose that two different gases are separated by a membrane; we may simulate this seepage (diffusion process) by the following urn model. Two urns each have N balls, and at the start of the process there are N green balls in urn A and N red balls in urn B. The experiment consists of selecting a ball at random from each urn, placing the ball from urn A in urn B, and placing the ball from urn B in urn A. Let E_i denote the state in which urn B has i green balls. Give the matrix for this Markov chain.

★3.　gambler's ruin

So far we have ignored the practical aspects of gambling. Gamblers cannot play forever, nor do they enter a game with an infinite amount of money. We now examine the situation in which the two players involved have only a finite amount of money and when either player is broke, the game stops.

To be specific, suppose that a gambler G is playing a game against a house H. G enters the game with A dollars, the house has B dollars, and the game terminates when G has either won B dollars or lost A dollars. When G has lost A dollars, he is naturally ruined, and our problem is to compute $P(A,B)$, the probability that he is ruined, given the initial conditions already described. We assume that on each play of the game G wins \$1 with probability p or loses \$1 with probability q, where $p + q = 1$. We let $N = A + B$ be the total amount of money, and we set up a Markov chain which is identical with the one used for a random walk in Example 3 of the preceding section.

We let the states be the amount of money that G has at any time during the play and we denote this by x. We let p_x be the probability that G will be ruined, given that he is in the state x; that is, G has x dollars. If on the next play he wins, then he is in the state $x + 1$ and has probability p_{x+1} of being ruined. If instead he loses, then he is in the state $x - 1$ and has probability p_{x-1} of being ruined. Since these are the only two possibilities, starting from the state x we have

(6) $$p_x = pp_{x+1} + qp_{x-1}.$$

This equation is valid provided that the game is still in progress, i.e., if $x = 1$, $2, \ldots, N - 1$. If $x = 0$, the gambler is already ruined and no play takes place. If $x = N$, the gambler has broken the house (also called the bank) and play again stops. Thus the special values 0 and N for x yield

(7) $$p_0 = 1 \quad \text{and} \quad p_N = 0,$$

which merely state that if G is broke, he is certainly ruined, and that if the house is broke, then G cannot be ruined. Equations (7) are called *boundary conditions*, because the values $x = 0$ and $x = N$ form "boundaries" for the set of possible values of x.

To solve the set of $N - 1$ equations given by (6), it is convenient to introduce the ratio of probabilities. Thus we set

(8) $$r = \frac{p}{q},$$

and we rearrange (6) as follows. Since $1 = p + q$, we have

$$(p + q)p_x = pp_{x+1} + qp_{x-1},$$
$$qp_x - qp_{x-1} = pp_{x+1} - pp_x,$$
(9) $$q(p_x - p_{x-1}) = p(p_{x+1} - p_x).$$

We divide both sides of (9) by q and use (8). This gives

(10) $$p_x - p_{x-1} = r(p_{x+1} - p_x), \quad x = 1, 2, \cdots, N - 1.$$

A direct attack upon the system of equations (10) will yield the solution, but there is a shorter way. Suppose that we are handed a formula and told that it is the solution. Naturally we would be suspicious, but if we check and find that it does solve the given system of equations, and if further we know there is only one solution, then we must accept the formula, although we may be curious as to the source of the formula. We adopt this easier approach and state the results in the following theorem.

Theorem 1. If $r = p/q \neq 1$, then the formula

$$(11) \qquad p_x = \frac{1 - r^{N-x}}{1 - r^N},$$

$x = 0, 1, 2, \ldots, N$, gives the solution of the system of equations

$$(10) \qquad p_x - p_{x-1} = r(p_{x+1} - p_x), \qquad x = 1, 2, \ldots, N-1,$$

with the boundary conditions

$$(7) \qquad\qquad p_0 = 1, \quad \text{and} \quad p_N = 0.$$

If $r = 1$, then the solution is given by

$$(12) \qquad\qquad p_x = \frac{N - x}{N}.$$

Proof. Since the probability that G is ruined is uniquely determined by p, N, and the amount of money G has, it follows that the system of equations has only one solution.[1]

Suppose now that $r \neq 1$. We set $x = 0$ and $x = N$ in equation (11) and we find that

$$(13) \quad p_0 = \frac{1 - r^{N-0}}{1 - r^N} = 1 \quad \text{and} \quad p_N = \frac{1 - r^{N-N}}{1 - r^N} = \frac{1 - 1}{1 - r^N} = 0.$$

Hence p_x as given by (11) satisfies the boundary conditions (7).

Next we test (11) in equation (10). For the left-hand side of (10)

$$(14) \qquad p_x - p_{x-1} = \frac{1 - r^{N-x}}{1 - r^N} - \frac{1 - r^{N-(x-1)}}{1 - r^N} = \frac{r^{N-x+1} - r^{N-x}}{1 - r^N}.$$

For the right-hand side of (10), equation (11) gives

$$(15) \qquad r(p_{x+1} - p_x) = r\left(\frac{1 - r^{N-(x+1)}}{1 - r^N} - \frac{1 - r^{N-x}}{1 - r^N} \right)$$

$$= r\left(\frac{r^{N-x} - r^{N-x-1}}{1 - r^N} \right) = \frac{r^{N-x+1} - r^{N-x}}{1 - r^N}.$$

---

[1] An argument of this type, based on the physical aspects of a problem, is regarded as intuitive and is not generally acceptable as rigorous mathematics. The basis for a purely mathematical proof will be covered in Chapter 8.

Since the right-hand side of (14) is identical with the right-hand side of (15), this proves that (11) gives the solution for the system (10).

If $r = 1$, the denominator on the right-hand side of (11) is zero, and the formula is meaningless. We leave it for the student to show that when $r = 1$, equation (12) provides a solution.

To appreciate these formulas we make a table of values for various values of r, x, and N. Let us recall that in solving our problem, it was convenient to regard x as a variable, the amount of money that G has at a particular stage of the game. If he enters the game with A dollars, then we replace x by A and N by $A + B$ to obtain the formulas

(16)
$$P(A,B) = \frac{1 - r^B}{1 - r^{A+B}}, \quad \text{if } r \neq 1,$$

and

(17)
$$P(A,B) = \frac{B}{A + B}, \quad \text{if } r = 1,$$

for the probability that G is ruined if he starts with A dollars. In Table 2 we have tabulated $P(A,B)$ for selected values of A and B using $r = 244/251 \approx 0.972$. This is the ratio that applies when G is throwing a pair of dice. In Table 3 we find the values, using $r = 18/19 \approx 0.947$. This is the ratio that applies in one form of roulette where G bets on either red or black but loses if the ball rolls into the zero slot (see Chapter 5, pages 164, 168).

TABLE 2. The Probability of Losing A Dollars Before Winning B Dollars at Dice ($p \approx 0.493$).

A \ B	2	4	8	16	32
2	0.514	0.685	0.822	0.912	0.964
4	0.352	0.528	0.704	0.843	0.932
8	0.223	0.372	0.556	0.739	0.879
16	0.138	0.248	0.411	0.611	0.802
32	0.089	0.167	0.299	0.490	0.712

TABLE 3. The Probability of Losing A Dollars Before Winning B Dollars in Playing Red or Black at Roulette ($p \approx 0.486$).

A \ B	2	4	8	16	32
2	0.527	0.702	0.841	0.931	0.978
4	0.370	0.554	0.736	0.876	0.960
8	0.245	0.407	0.606	0.797	0.930
16	0.165	0.294	0.483	0.704	0.889
32	0.122	0.227	0.397	0.626	0.849

When the game is even, such as matching dollars, formula (17) yields useful information. Suppose G enters such a game with $10, while H has $90 that he is able to risk. Then, according to equation (17), we find that $P(A,B) = 90/100 = 0.9$. Thus it is 90 percent certain that G will lose all his money (be ruined).

Although the tables are computed on the basis of the gambler's ruin, they need not be used in this manner. Let us suppose that a player G who has $100 decides to play at dice until he loses $32 and then he intends to quit (reserving the remaining $68 for other purposes). In this case $A = 32$, and not 100. Further, suppose his opponent H has $10,000. If G (who has perfect self-command) resolves to quit after winning $2, then $B = 2$, and not 10,000. With such a strategy Table 2 gives G a probability of 0.089 of losing, and hence a probability of 0.911 of winning. Although this sounds wonderful for G, we must realize that he is balancing a high probability of winning a small amount, $2, against a low probability of losing a large amount, $32. Indeed, the expected value for a sequence of such strategies is

$$E = 0.911 \times 2 - 0.089 \times 32 = -1.026.$$

Thus if G plays dice each night until he either wins $2 or loses $32, in the long run his losses will average a little over $1 per night.

The reader naturally understands that the amount bet is arbitrary. Thus G might bet $100 on each play. If he resolves to quit throwing the dice as soon as he either wins $200 or loses $3200, whichever is first, then Table 2 still applies. He is almost certain of winning $200 because $q = 1 - 0.089 = 0.911$. But he does run a slight risk ($p = 0.089$) of losing $3200.

The above example illustrates the fallacy in various conservative systems of gambling. One approach used in horse racing, dog racing, etc., is for the gambler always to bet on the favorite to show (come in third). His gain will be small on each success, but the probability of success will be close to 1. However, if the expected value is negative (as it surely must be), this system will also lose in the long run.

Exercise 3

1. If G, who always throws the dice, has $16 and is playing H who has $32 and $1 is bet on each game, what is the probability that G wins $32 before going broke?
2. Solve problem 1 if (a) G bets $2 on each game, (b) G bets $4 on each game, and (c) G bets $8 on each game.
3. The Apex Co. and the Bildfine Co. both manufacture and sell very good widgets. Each year the market will absorb exactly 100,000 widgets. The Apex Co. currently gets 20 percent of the sales, while

Bildfine Co. handles the rest of the business. Both companies secretly plan aggressive advertising campaigns, and in the absence of other information each company assumes that each year it has probability 1/2 of increasing its sale of widgets by 1000. Find the probability that the Apex Co. will lose all its customers and go broke.

4. Suppose that in problem 3, the factory can no longer be operated at a profit when sales fall to 10,000 widgets a year and that when this happens, the company stops operating. Find the probability that in due time the Apex Co. will cease to make widgets.

5. Complete the proof of Theorem 1 by considering the case $r = 1$.

★6. Use equations (7) and (10) and the condition $r \neq 1$ to derive equation (11) in the special cases $N = 3$ and $N = 4$.

7. Suppose that in a game G has probability 3/5 of winning \$1 from H on each play. Show that if G starts with only \$1 and H starts with \$3, then the probability that G goes broke is greater than 1/2.

8. Suppose that the game in problem 7 is modified so that G has probability 2/3 of winning \$1 on each play. Prove that if G starts with \$1, then the probability that G goes broke is less than 1/2, no matter how much money H has at the start.

Problems 9 through 16 are concerned with the expected value of the number of games played before G or H is ruined.

9. Suppose that G has x dollars, H has $N - x$ dollars, \$1 is bet on each play, and G has probability p of winning each game. Let L_x denote the expected value of the number of games that will be played before either G or H is ruined. Show that

(18) $$L_x = pL_{x+1} + qL_{x-1} + 1.$$

10. Explain why $L_0 = 0$ and $L_N = 0$.

11. If $p \neq q$, set

(19) $$L_x = \frac{x}{q - p} - \frac{N}{q - p} \cdot \frac{r^{N-x} - r^N}{1 - r^N}, \quad \text{where } r = p/q.$$

If $p = q = 1/2$, set

(20) $$L_x = x(N - x).$$

Show that each of these formulas for L_x satisfies the boundary conditions obtained in problem 10.

★12. Show that each of the formulas for L_x given in problem 11 satisfies the equation for the expected number of games found in problem 9.

13. Mr. Skidmore is playing pool with Prof. Bankball at the club. They are equally matched ($p = 1/2$), each game takes about 10 minutes, and they bet $1 on each game. At 8:00 P.M. Skidmore calls his wife and explains that both he and the professor have $6 each and will play until one of them goes broke. He assures Mrs. Skidmore that he will certainly be home by midnight. Is his confidence justified?

14. Under the conditions of problem 4 find the expected value of the number of years for one of the companies to stop making widgets.

15. Referring again to the Apex and Bildfine companies of problem 4, suppose that because of the dominant position of the Bildfine Co. it has a probability 3/4 of increasing the sale of widgets by 1000 each year and a probability 1/4 of a corresponding decrease. Find the expected value of the number of years for one of the companies to stop making widgets. Find the probability that it will be the Apex Co.

16. A gentleman G agrees to play a hustler H, betting even money on a game in which G has probability 2/5 of winning. If G has $2 and H has $5, find (a) the probability that G will lose $2 and (b) the expected value of the number of games before either G or H is broke.

4. testing a hypothesis

In certain ideal situations it is relatively easy to assign a probability function. The reader may recall the fair dice, the well-shuffled deck of cards, etc. But in most applications of probability theory the correct value of p for a certain outcome is unknown. In such a situation one may have an approximate value of p, or one may feel that p lies in a certain interval. When one assumes a certain value for p, or that p lies in a certain interval, that assumption is a hypothesis which we may wish to test by performing the relevant experiment a given number of times. We shall develop the concept of testing a hypothesis in some detail while discussing

EXAMPLE 1. The Gurd Co. proposes a new drug ZY which it claims will cure xerrophobia in at least 70 percent of the cases. However, about 30 percent of the people recover from xerrophobia in two days without any treatment. Suppose that 10 people who have this disease are to be treated with this new drug. How can we make a decision about the reliability of the claim of the company on the basis of the experimental results?

Discussion. We regard the 10 people treated with ZY as forming a sequence of Bernoulli trials. We must recognize that even if all 10 are cured by ZY this does not prove that the drug is helpful because this may occur on 10 trials even when the probability of a cure is only 0.30, although such an event is highly unlikely.

The customary procedure in a situation of this type is to select a particular hypothesis to be tested. This particular hypothesis is called the *null hypothesis* and is denoted by H_0. The alternative to H_0 is denoted by H_1, and by definition H_0 and H_1 are both mutually exclusive and exhaustive; i.e., if H_0 is true, then H_1 is false, and if H_0 is false, then H_1 is true. Further, it is customary to select for H_0 that hypothesis that is conservative. In the example under consideration the conservative point of view is that the drug ZY is either useless, or perhaps even harmful. Consequently we arrive at:

$$H_0 \text{ is the hypothesis that } p \leq 0.30,$$

where p is the probability that a patient will recover in two days when treated with ZY. Once H_0 is selected to be $p \leq 0.30$, we automatically have for the *alternative hypothesis*

$$H_1 \text{ is the hypothesis that } p > 0.30.$$

Now that the null hypothesis has been selected we want to formulate a rule which will lead to a decision to accept or to reject H_0. Let k be the number of persons that are cured by ZY, and suppose that we adopt (arbitrarily) the following rule:

$$\text{Accept } H_0 \text{ if } k < 6.$$
$$\text{Reject } H_0 \text{ if } k \geq 7.$$

To investigate this rule, we want to compute $P[k \geq 7]$, the probability of rejecting H_0, but this is impossible because we do not know p (the probability that ZY will cure a patient). However, we can compute $P[k \geq 7 \mid p = x]$ for various values of x. Here it is helpful to have available a table of values of $b(k,n,p)$, the probability of k successes on n trials when p is the probability of a success on a single trial. Such a table for $n = 10$ is given in Table 4, and more extensive tables are given in the Appendix.

TABLE 4. $b(k,n,p) = C(10,k)p^k q^{n-k}$

k \ p	0.1	0.3	0.5	0.7	0.9
0	0.349	0.028	0.001	0.000	0.000
1	0.387	0.121	0.010	0.000	0.000
2	0.194	0.233	0.044	0.001	0.000
3	0.057	0.267	0.117	0.009	0.000
4	0.011	0.200	0.205	0.037	0.000
5	0.001	0.103	0.246	0.103	0.001
6	0.000	0.037	0.205	0.200	0.011
7	0.000	0.009	0.117	0.267	0.057
8	0.000	0.001	0.044	0.233	0.194
9	0.000	0.000	0.010	0.121	0.387
10	0.000	0.000	0.001	0.028	0.349

For example, from Table 4 we find that

$$P[k \geq 7 | p = 0.3] = 0.009 + 0.001 + 0.000 + 0.000 = 0.010,$$

and

$$P[k \geq 7 | p = 0.5] = 0.117 + 0.044 + 0.010 + 0.001 = 0.172,$$

etc. The graph, called the *power curve for the test,* is shown in Fig. 6.

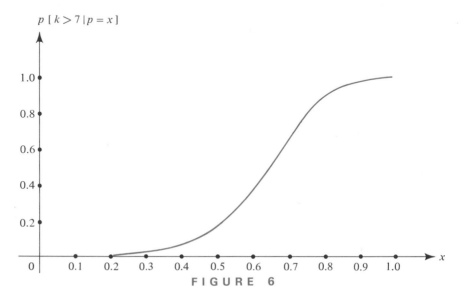

FIGURE 6

Suppose now that $p \leq 0.3$. From the graph we see immediately that $P[k \geq 7] \leq 0.010$, the value when $p = 0.30$. Consequently if $k \geq 7$ and we reject H_0, the probability is at most 0.010 that we have made an error by rejecting H_0 when in fact H_0 is true.

Before proceeding with this example we summarize the essential features and introduce some new terminology.

There are two hypotheses: the *null hypothesis*, H_0, and the *alternative hypothesis*, H_1. An event E (a rule) has been selected such that if E occurs, then H_0 is rejected; and if E does not occur, then H_0 is accepted. As a result of the experiment there are four possibilities:

(I) We may reject H_0 when it is true.
(II) We may accept H_0 when it is false.
(III) We may accept H_0 when it is true.
(IV) We may reject H_0 when it is false.

Of course if either possibility (III) or (IV) is the case, we are quite content. Our principal concern is the possibility of making an error as described in (I) or (II).

Definition 2. If we reject the null hypothesis when it is true, we say that a type I error has been committed. If we accept the null hypothesis when it is false, we say that a type II error has been committed.

The effect of a type I error is usually quite different from that of a type II error. In our example a type I error may mean that we allow a drug to come on the market when the drug is useless or perhaps harmful. A type II error may mean that research and development of ZY is stopped when in fact ZY may be just the drug needed to cure the patient.

Definition 3. A test of a statistical hypothesis is a rule which, when the experimental results have been obtained, leads to a decision to accept or reject H_0.

Definition 4. The critical region of a test is the event E that leads to the rejection of H_0.

In our example H_0 is the hypothesis that $p \leq 0.30$. The event E is the recovery (in two days) of 7 or more of the 10 patients treated with ZY. Thus the critical region is $k \geq 7$.

Definition 5. The power function of a test is the probability of the event E that leads to the rejection of H_0.

This probability $P[E]$ usually depends on one or more variables. In our example the power function is a function of p, and the graph is the one shown in Fig. 6.

Definition 6. The significance level of a test is the maximum value of the power function when H_0 is true.

This number, which we denote by α, is just the maximum value of the probability of rejecting H_0 when H_0 is true. Thus α is the maximum value of the probability of making a type I error. We can use the notation $\alpha = \max P[E|H_0]$ to indicate that $P[E|H_0]$ is a function and α is the maximum value of this function. For example, in Fig. 6, $P[E|H_0]$ is represented by that part of the graph for which $p \leq 0.30$, and $\alpha = 0.010$ is the maximum of $P[E|H_0]$ for $p \leq 0.30$.

In our example, we selected a test arbitrarily and then computed α. The usual procedure is the reverse of this. With due consideration to the serious consequences of a type I error, we select a suitable small value for α and then select our test to obtain this preassigned α. A detailed discussion of the problem of selecting α lies outside the aims of this text. However, one is frequently content to set $\alpha = 0.05$ or $\alpha = 0.01$. We then say that the *significance level* of the test is 5 percent or 1 percent, respectively.

We return now to Example 1. The wording of that example seems to suggest two different hypotheses: $p \leq 0.30$ and $p \geq 0.70$. For clarity these must be treated separately. As we have already observed, the conservative hypothesis is $p \leq 0.30$, so we treat this first in

EXAMPLE 1A. Under the conditions of Example 1, design a test for the null hypothesis $p \leq 0.30$ at the 5 percent significance level.

Solution. Let E be the event $k \geq x$. We have already computed the quantity $P[E|p = 0.3]$ when $x = 7$ and found $\alpha = 0.010$. Since this is too small, we enlarge our critical region by taking $x = 6$. The maximum of $P[E|H_0]$ occurs when $p = 0.3$, and hence we compute $P[k \geq 6|p = 0.3]$. From Table 4 we find that

$$P[k \geq 6 \mid p = 0.3] = 0.037 + 0.009 + 0.001 = 0.047,$$

and this is sufficiently close to 0.05 for our purposes. Hence the test is as follows: Accept H_0 if $k \leq 5$; reject H_0 if $k \geq 6$. The significance level of this test is 4.7 percent.

Why do we prefer the critical region $k \geq 6$ over the region $k \geq 7$? We have just seen that $P[k \geq 6|p = 0.3] = 0.047$ and we found earlier that $P[k \geq 7|p = 0.3] = 0.010$. Consequently, if we change the test from "Reject H_0 if $k \geq 6$" to "Reject H_0 if $k \geq 7$," then we decrease α from 0.047 to 0.010. Since α is the maximum of the probability of making a type I error, this decrease in α suggests that it might be a good idea to change the test as indicated. The difficulty is simply this: Any decrease in the probability of making a type I error will give a corresponding increase in the probability of making a type II error.

In our example the probability of a type II error is a function of p and p is still unknown, but we can observe the behavior of this function as we change the test by selecting a particular value for p. Suppose, indeed, that $p = 0.5$. From Table 4, we find that

$$P[k < 6|p = 0.5]$$
$$= 0.246 + 0.205 + 0.117 + 0.044 + 0.010 + 0.001 = 0.623,$$

and

$$P[k < 7|p = 0.5] = 0.623 + 0.117 = 0.740.$$

This computation shows clearly why a decrease in the probability of making a type I error gives an increase in the probability of making a type II error.

EXAMPLE 1B. Under the conditions of Example 1, let H_0 be the hypothesis that the claim of the Gurd Co. is correct. Find α if we decide to accept H_0 if $k \geq 6$ and reject H_0 if $k < 6$.

Solution. The Gurd Co. claims $p \geq 0.7$. The probability of rejecting H_0: $p \geq 0.7$ is $P[k < 6 | p \geq 0.7]$. Clearly, the maximum value of this function occurs when $p = 0.7$. Using Table 4, we find that

$$\alpha = P[k < 6 | p = 0.7] = 0.103 + 0.037 + 0.009 + 0.001 = 0.150.$$

EXAMPLE 2. As an entrance requirement, a business college gives a multiple-choice examination with 20 questions. Each question is followed by five possible answers, only one of which is correct. The questions are rather difficult and are designed to separate those students who have some knowledge about business from those who know nothing and are merely guessing. If the significance level of the test is to be not more than 5 percent, find the number of correct answers necessary for admission.

Solution. Here the null hypothesis is that the student is merely guessing (or perhaps worse, knows some "facts" that are wrong). Consequently $p \leq 1/5$ for a correct answer on any one question if H_0 is true. As usual we regard the sequence of 20 questions as a sequence of Bernoulli trials with $p \leq 1/5$. We let x be the number of correct answers obtained by a student on the examination. Now $P[x \geq n | H_0] \leq P[x \geq n | p = 1/5]$, and so to obtain the maximum for the probability of making a type I error, we use $p = 1/5$. From Table B in the Appendix we find that

$$P[x \geq 7 | p = 1/5] = 0.055 + 0.022 + 0.007 + 0.002 = 0.086,$$

and

$$P[x \geq 8 | p = 1/5] = 0.022 + 0.007 + 0.002 = 0.031.$$

If we set $x \geq 7$ for admission, then $\alpha = 0.086$ or 8.6 percent. This is too much. Hence under the conditions given, we set $x \geq 8$ for admission, and then the probability of making a type I error (admitting a student who knows nothing about business) is 0.031. The significance level of the test is 3.1 percent.

Exercise 4

1. Prove the identity $b(k, n, p) = b(n - k, n, 1 - p)$. Explain how it is used in Table B to find $b(8, 10, 0.9)$, $b(4, 20, 0.8)$, and $b(13, 15, 0.95)$.
2. The Bildfine Co. makes microwidgets and the manufacturing process is such that at least 95 percent of them are good. Each day 20 microwidgets

are carefully tested. Find the critical number of bad microwidgets that will indicate that the production process is out of control (something has gone wrong) **(a)** using a 5 percent significance level and **(b)** using a 1 percent significance level.

3. It is estimated that 90 percent or more of the adult males in a certain mining town have never had TB. To test this hypothesis a sample of 15 men is selected. We agree to accept this estimate if not more than 4 of the men have had TB. Find α for this test.

4. Clever Clem claims that at least 70 percent of the time he can distinguish the sound of a gun being fired from other similar noises (firecracker, backfire of a car, etc.). He is willing to wager a good sum and his supporters arrange a sequence of 20 tests. His opponents agree to pay Clem if he identifies 12 or more of the sounds correctly (gunshots or not). Find the significance level of this test.

5. Suppose that in problem 4, Clem is only guessing. Find the probability that Clem will win his bet anyway.

6. Dr. Jones claims he can predict the sex of a baby five months before it is born with probability 0.90 or better. Dr. Smith claims that unless $p > 0.60$ the method is of no use. When tested in 20 cases, Jones actually made 15 correct predictions. Using a 5 percent significance level, would you reject or accept **(a)** $H_0: p \leqq 0.60$, and **(b)** $H_0: p \geqq 0.90$?

7. In a recent rain-making experiment, the investigators searched for pairs of clouds that seemed to be roughly identical. The airplane then seeded one cloud with a water spray, but left the other one untouched. In many cases neither cloud produced rain (this was carefully defined) and in some cases both clouds produced rain. Here it was assumed that forces more powerful than the seeding were at work, and these data were rejected. However, in 14 cases the seeded cloud produced rain and the unseeded cloud did not. In 6 cases the unseeded cloud produced rain and the seeded cloud did not. If H_0 is the hypothesis that seeding has no effect ($p \leq 0.5$), would you reject H_0 **(a)** at the 10 percent significance level, and **(b)** at the 5 percent significance level?

8. In a random sample of 20 voters, x of the voters state that they favor an increase in the state sales tax from 3 to 4 percent. We make the hypothesis that at most 40 percent of the voters of the state favor the increase. Find the smallest x_0 such that if $x \geqq x_0$, then we reject the hypothesis at the 5 percent significance level.

9. Tom, Dick, and Harry disagree about McMillan's hamburgers. Tom claims that 90 percent of the time the meat in a hamburger weighs more than the standard 1.6 ounces. Dick claims it is just as likely to be overweight as underweight. Harry feels that 90 percent of the time it weighs less than the standard. They test 10 hamburgers purchased at random, and they agree that Tom wins if 8 or more hamburgers are overweight, Harry wins if 8 or more are underweight, and Dick wins in all other cases (the

meat never weighs exactly 1.6 ounces when an accurate scale is used).
For each man find the probability that he will win the bet, given that
he is right.

★5.   the law of large numbers

We return for a moment to the foundations of probability and examine the
statement that if k is the number of successes in n repeated trials, then

$$(21) \qquad \lim_{n \to \infty} \frac{k}{n} = p,$$

where p is the probability of a success on one trial. As explained in Chapter
5, the peculiar looking symbols in (21) mean that as n gets larger and larger,
the ratio k/n gets closer and closer to p. This assertion cannot be proved
by experiment, because one cannot repeat an experiment an infinite number
of times. We cannot give a mathematical argument for (21) because we have
at the moment no foundation on which to base the proof. But if we alter
the statement slightly, we can obtain a new statement that can be proved.

It is customary in mathematics to let ϵ (Greek letter epsilon) denote a
small positive number. The statement that k/n is close to p can be symbolized by

$$(22) \qquad \left| \frac{k}{n} - p \right| < \epsilon.$$

The inequality (22) merely states that k/n does not differ much from p and
that ϵ is an upper bound for the absolute value of this difference.

Now (22) is no easier to prove than (21) as long as we look at a *particular
sequence*. But suppose we look at *all* possible outcomes for a sequence of
n trials. For some of these k/n will be far from p; for others k/n will be
close to p. What can be proved is that those sequences for which k/n is
close to p are much more likely to occur, and indeed, as n gets larger the
probability that k/n is close to p approaches 1. This descriptive statement
is phrased accurately in

Theorem 2. The Law of Large Numbers. Let $p = P[E]$ on a single
trial. Let k be the number of times that E occurs on a sequence of
n repeated trials. Let ϵ be a fixed positive number. Then

$$(23) \qquad P\left[\left| \frac{k}{n} - p \right| < \epsilon \right]$$

approaches 1 as n approaches infinity.

This is the theorem that replaces the pseudo-theorem of Chapter 5 (see page 162). As the reader may have already suspected, the proof of Theorem 2 is difficult and we shall omit it. However, we illustrate the meaning of Theorem 2 in

EXAMPLE 1. Suppose that $p = 0.8$ and $\epsilon = 0.11$. Find the probability expressed by (23) when (a) $n = 10$ and (b) $n = 20$.

Solution. We use the values given in Table B.
(a) If $n = 10$, we find that

$$\left| \frac{k}{10} - 0.8 \right| < 0.11 \Longleftrightarrow k = 7, 8, \text{ or } 9.$$

(24) $$P[7 \leqq k \leqq 9] \approx 0.201 + 0.302 + 0.268 = 0.771.$$

(b) If $n = 20$, we find that

$$\left| \frac{k}{20} - 0.8 \right| < 0.11 \Longleftrightarrow k = 14, 15, 16, 17, \text{ or } 18.$$

(25) $$P[14 \leqq k \leqq 18] \approx 0.109 + 0.175 + 0.218 + 0.205 + 0.137 = 0.844.$$

Clearly as n increases from 10 to 20 the probability that $|k/n - 0.8| < 0.11$ increases from 0.771 to 0.844. To pursue the matter further we need more extensive tables. If $n = 30$, such tables give

(26) $$P\left[\left| \frac{k}{30} - 0.8 \right| < 0.11 \right] = 0.895,$$

and this is still closer to 1. The reader may well imagine that if $n = 10,000$, then this probability will be very close to 1, but of course we have not proved this fact.

Exercise 5

1. Suppose that $p = 0.5$. Find

$$P\left[\left| \frac{k}{n} - 0.5 \right| < 0.11 \right]$$

when (a) $n = 10$, (b) $n = 15$, and (c) $n = 20$.

6. the DeMoivre-Laplace limit theorem

In Section 4 we observed a practical need for certain sums of $b(k,n,p)$. Although tables are helpful, even the most extensive tables cannot cover all of the needs that occur. Fortunately there is a simple way to approximate these sums as areas under a certain bell-shaped curve, which we now explain.

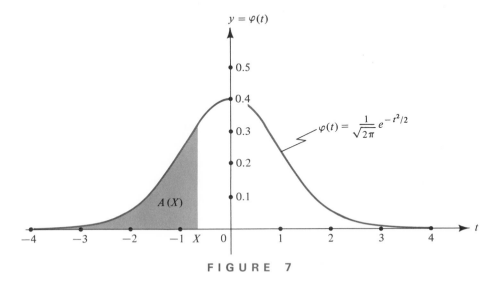

FIGURE 7

The bell-shaped curve is the graph of the function

(27)
$$\varphi(t) = \frac{1}{\sqrt{2\pi}} e^{-t^2/2},$$

and it is shown in Fig. 7. On the left-hand side, φ (Greek lowercase phi) merely names the function defined by the formula on the right. On the right-hand side, we have an exponential function (the variable t is in the exponent) and the base e is the same Euler number that we met earlier. If we must compute, then we use the approximation $e \approx 2.718$, but fortunately this will not be necessary because extensive tables have been prepared giving values for this function.

We are not concerned with the curve itself but rather with the area of the region bounded above by the curve, below by the horizontal axis, and on the right by the vertical line constructed at $t = X$. This region is of infinite extent (there is no vertical boundary line on the left), but the area, which we denote by $A(X)$, is finite[1] and in fact $A(X) \leq 1$. The region of interest

[1] It may come as a surprise to the reader that a region may have infinite extent and finite area. An example that illustrates this possibility is given in problems 17, 18, and 19 of the next exercise.

TABLE 5. Values of $A(X)$.

X	0.00	0.20	0.40	0.60	0.80
$-3.$	0.0013	0.0007	0.0003	0.0002	0.0001
$-2.$	0.0228	0.0139	0.0082	0.0047	0.0026
$-1.$	0.1587	0.1151	0.0808	0.0548	0.0359
$-0.$	0.5000	0.4207	0.3446	0.2743	0.2119
$+0.$	0.5000	0.5793	0.6554	0.7257	0.7881
$1.$	0.8413	0.8849	0.9192	0.9452	0.9641
$2.$	0.9772	0.9861	0.9918	0.9953	0.9974
$3.$	0.9987	0.9993	0.9997	0.9998	0.9999+

is shown shaded in Fig. 7. Thus

(28) $A(X) =$ area of the shaded region to the left of $t = X$.

The computation of these areas[1] is rather complicated, but again we are
fortunate because extensive tables of $A(X)$ are also available. A small table
is presented here as Table 5, and a somewhat larger one is given as Table C
in the Appendix.

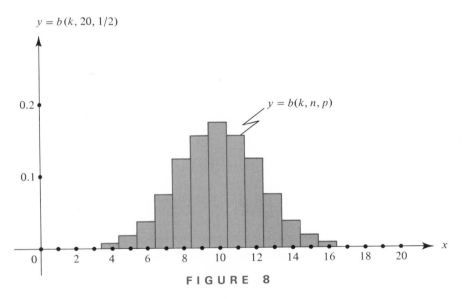

FIGURE 8

To observe the relation between the bell-shaped curve of Fig. 7 and the
probability $b(k,n,p)$ we make a graph of $b(k,n,p)$ in the special case $n = 20$,

[1] For those familiar with the calculus

$$A(X) = \frac{1}{\sqrt{2\pi}} \int_{-\infty}^{X} e^{-t^2/2}\, dt.$$

Many authors denote this function by $\Phi(X)$ (capital Greek phi).

$p = 1/2$, using the values from Table B. The result is shown in Fig. 8. Although the similarity in the two figures is reasonably clear, there are certain marked differences that must be reconciled.

In the first place the peak of the curve in Fig. 7 occurs at $t = 0$. In Fig. 8, the peak occurs at $x = 10$. This is obviously the expected value of the number of successes when $n = 20$ and $p = 1/2$. This suggests that we move the y-axis to the right in Fig. 8. This adjustment is made by considering $x - 10$ or $x - np$ rather than x.

Second, the peak values are different. In Fig. 7, the greatest height is $\varphi(0) = 1/\sqrt{2\pi} \approx 0.3989$. In Fig. 8, the greatest height is $y = b(10,20,1/2) \approx 0.176$. This difference should not be too surprising because it is the areas under the two curves that are related and not the height.

To bring the areas into agreement we introduce the *change of variable*

(29)
$$X = \frac{x - np}{\sqrt{npq}}.$$

In this fundamental equation, x represents the independent variable for the Bernoulli distribution (see Fig. 8), and X represents the independent variable for the normal distribution (see Fig. 7). When suitable adjustments are made in accordance with equation (29), and the adjusted normal curve is superimposed on the Bernoulli curve, there is indeed close agreement,

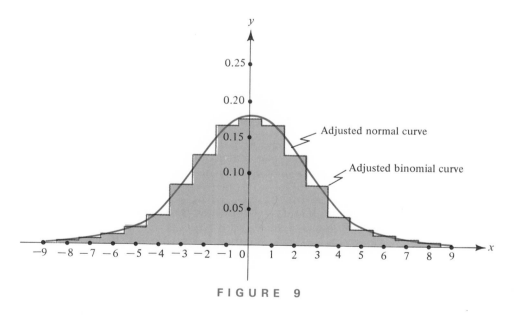

FIGURE 9

as indicated in Fig. 9. Consequently, the area under one curve can be approximated by the area under the other curve, as stated in

Theorem 3. DeMoivre-Laplace Limit Theorem. Let s be the number of successes in a sequence of n Bernoulli trials, where p is the probability of a success on each trial. Then for large n

(30) $$P[s \leqq x] \approx A(X),$$

where

(29) $$X = \frac{x - np}{\sqrt{npq}}.$$

In other words we obtain a good approximation for

(31) $$P[s \leqq x] = \sum_{k=0}^{x} b(k,n,p)$$

by using the area under the curve $y = \varphi(t)$ to the left of the line $t = X$ (see Fig. 7).

The proof of this theorem is too difficult to be included here, but the curious reader may locate it in any advanced book on probability or statistics.

We recall (Chapter 5, Theorem 6) that np is the expected value of the number of successes. The quantity $\sqrt{npq}$ also has a special title as described in

Definition 7. The quantity $\sqrt{npq}$ is called the standard deviation for a sequence of n Bernoulli trials with probability p for a success on each trial.

EXAMPLE 1. Let s be the number of times that a 2 shows in 180 rolls of a standard die. Find an approximate value for the probability that (a) $s \leqq 22$, (b) $s \leqq 26$, and (c) $s \leqq 33$.

Solution. In this problem $n = 180$ and $p = 1/6$, so the expected value of the number of 2's is $np = 180/6 = 30$.

For the standard deviation we find

$$\sqrt{npq} = \sqrt{180 \cdot \frac{1}{6} \cdot \frac{5}{6}} = \sqrt{25} = 5.$$

From equation (29) we find the following for X:

(a) If $x = 22$, then $X = \dfrac{22 - 30}{5} = -1.6$.

(b) If $x = 26$, then $X = \dfrac{26 - 30}{5} = -0.8$.

(c) If $x = 33$, then $X = \dfrac{33 - 30}{5} = +0.6$.

Selecting the proper entry for $A(X)$ from Table 5 (and invoking Theorem 3) we find that

(a) $P[s \leqq 22] \approx 0.0548$.

(b) $P[s \leqq 26] \approx 0.2119$.

(c) $P[s \leqq 33] \approx 0.7257$.

EXAMPLE 2.   Under the conditions of Example 1, find $P[27 \leqq s \leqq 33]$.

First Solution.   We can use the results of Example 1. Indeed,

$$p = P[27 \leqq s \leqq 33]$$
$$= P[s \leqq 33] - P[s \leqq 26] \approx 0.7257 - 0.2119 \approx 0.5138.$$

Second Solution.   If a and b are integers, the event $a \leqq s \leqq b$ is the same as the event $a - 1/2 \leqq s \leqq b + 1/2$, since s, the number of successes, must be an integer. Applying this remark to the example under consideration, we see that

(32) $$P[27 \leqq s \leqq 33] = P\left[26\frac{1}{2} \leqq s \leqq 33\frac{1}{2}\right].$$

If $x = 26.5$, then equation (29) gives $X = -0.7$, and if $x = 33.5$, then equation (29) gives $X = +0.7$. Using Table C in the Appendix, the right-hand side of (32) now yields

$$P\left[s \leqq 33\frac{1}{2}\right] - P\left[s \leqq 26\frac{1}{2}\right] \approx A(0.7) - A(-0.7)$$

$$\approx 0.7580 - 0.2420 = 0.5160.$$

In most cases the procedure used in the second solution will give a value closer to the correct probability than the one obtained using the procedure in the first solution. In our examples and exercises we shall use the second method.

The standard deviation $\sqrt{npq}$ has an important interpretation which we now investigate. Suppose we ask for the probability that the number of successes cluster about np, the expected value of the number of successes. We consider three events of interest:

$$E_1: |s - np| \leq \sqrt{npq}.$$
$$E_2: |s - np| \leq 2\sqrt{npq}.$$
$$E_3: |s - np| \leq 3\sqrt{npq}.$$

For these events our new-found technique gives

(33) $P[E_1] \approx A(1) - A(-1) \approx 0.8413 - 0.1587 = 0.6826.$

(34) $P[E_2] \approx A(2) - A(-2) \approx 0.9772 - 0.0228 = 0.9544.$

(35) $P[E_3] \approx A(3) - A(-3) \approx 0.9987 - 0.0013 = 0.9974.$

Equation (33) tells us that with probability (approximately) $2/3$, the difference between s and the expected number of successes does not exceed one standard deviation. Consequently, a difference that is greater than $\sqrt{npq}$ is not surprising. The numbers $2/3$ and $\sqrt{npq}$ and merely convenient guide numbers.

In contrast, equation (34) tells us that a difference that exceeds $2\sqrt{npq}$ is highly unlikely and in fact has probability 0.0456. This is close to 0.05 and it is often used to determine the critical region near the 5 percent significance level. Thus if we are testing a hypothetical value of p at the 4.56 percent significance level, we can:

Accept the value of p if $|s - np| \leq 2\sqrt{npq}$.

Reject the value of p if $|s - np| > 2\sqrt{npq}$.

Similarly equation (35) tells us that if $|s - np| > 3\sqrt{npq}$ we can reject the hypothesis at the $1/4$ percent significance level.

EXAMPLE 3.   Before nominating Xero Zorro for Mayor of Neuro City the party leaders want to test the hypothesis that he is *just as popular* as his opponent. In making a survey they test 3600 registered voters. If s is the number of voters that prefer Zorro, find the values of s for the hypothesis to be accepted at the 4.56 percent significance level.

Solution.   We are testing the hypothesis that $p = 1/2$. Then $\sqrt{npq} = \sqrt{3600/4} = 30$. At the 4.56 percent significance level we accept H_0 if $|s - np| \leq 2\sqrt{npq}$ or $|s - 1800| \leq 60$. Hence if $1740 \leq s \leq 1860$, we accept the hypothesis that Zorro is just as popular as his opponent.

The test $1740 \leq s \leq 1860$ is called a "two-tailed" test because the critical region consists of two "tails," the events $s < 1740$ and the events $s > 1860$. In the particular example at hand, it is more natural to make the hypothesis that $p \geq 1/2$. This leads to a "one-tail" test and consequently a different critical region. This is illustrated in

EXAMPLE 4.   Suppose that in Example 3 the party leaders want to test the hypothesis that Zorro is *at least as popular* as his opponent. Find the values of s for this hypothesis to be accepted at the 5 percent significance level.

Solution.   Here H_0 is the hypothesis that $p \geq 1/2$. In this case we are quite content if $s \geq np$ and disturbed only if s is considerably less than np. Thus we reject H_0 if $s \leq x$, where x is determined by the condition

(36)
$$P\left[s \leq x \mid p = \frac{1}{2}\right] = 0.05.$$

From Table C, we find that $A(X) = 0.05$ for $X = -1.645$. If we solve

(29)
$$X = \frac{x - np}{\sqrt{npq}}$$

for x, we obtain

(37)
$$x = np + X\sqrt{npq}.$$

Using $X = -1.645$, $n = 3600$, and $p = q = 1/2$ in equation (37) we find that $x = 1800 - (1.645)(30) = 1750.65$. Consequently if $s \leq 1750$, we reject the hypothesis that Zorro is at least as popular as his opponent; and if $s \geq 1751$, we accept this hypothesis. Notice that the critical number 1751 is larger than the corresponding critical number 1740 found in the solution to Example 3.

EXAMPLE 5.   We return to the problem posed in Section 1 of Chapter 5 (Example 2). In that example, p is the probability of throwing a sum S with two dice, where $5 \leq S \leq 9$, and in that section we presented two different arguments that give two values for p: $p = 13/21$ and $p = 2/3$. We run an experiment by throwing the dice 7200 times and we observe that the number of successes is 4751. Determine the critical region for the hypothesis H_0: $p = 2/3$ at the 5 percent significance level.

Solution.   If $p = 2/3$ and $n = 7200$, then $np = 4800$ and $\sqrt{npq} = 40$. The situation dictates a two-tailed test, with the area of each "tail" set at 0.025. When $A(X) = 0.025$, Table C gives $X = -1.96$, and (because of the symmetry of the normal curve) $A(X) = 0.975$ when $X = 1.96$. Further we note that $1.96\sqrt{npq} = 1.96 \times 40 = 78.4$. Hence we accept H_0 at the 5 percent significance level if $|s - 4800| \leq 78.4$, or if $4722 \leq s \leq 4878$. Since 4751 lies in this interval, we can accept the hypothesis that $p = 2/3$.

We can also test the hypothesis H_0: $p = 13/21$. In this case $np \approx 4457$ and $1.96 \sqrt{npq} \approx 81$. Now $4457 - 81 = 4376$ and $4457 + 81 = 4538$ so we may accept H_0 at the 5 percent significance level if $4376 \leq s \leq 4538$. Since 4751 is not in this interval we can reject $p = 13/21$ at the 5 percent significance level.

Exercise 6

1. A true coin is tossed 10,000 times. Find **(a)** the expected number of heads and **(b)** the standard deviation.

2. A pair of dice is rolled 720 times. Find **(a)** the expected number of 7's and **(b)** the standard deviation.

3. In problem 2 find **(a)** the expected number of occurrences of a number other than 7 and **(b)** the standard deviation.

4. A deck of cards is thoroughly shuffled 300 times and after each shuffle the top card is examined. Find **(a)** the expected number of times the card is a spade and **(b)** the standard deviation.

5. Suppose that in problem 4 we are testing the hypothesis that the deck is well shuffled. We shall accept this hypothesis at the 5 percent significance level if the number of spades is within certain limits. Find those limits.

6. Mountainview Airlines and Transvalley Airlines plan to operate competing flights between Sirloin City and Steakville. It is estimated that 100 passengers will use one of the two available airplanes on any one trip. Find the proper seating capacity of the airplanes if the customers select the airline at random and each company wishes to have enough seats 95 percent of the time. *Hint:* Each company is concerned only with the possibility that too many customers want tickets.

7. Suppose that in problem 6, Mountainview Airlines wants to have sufficient seating capacity 99 percent of the time. Find the proper seating capacity of the airplane.

8. A gambler spends a year gambling at a certain fair game. In each play of the game he either wins or loses $1 with probability $1/2$. If he plays 10,000 games, find the probability that at the end of the year his total gain or loss will be **(a)** less than $51 and **(b)** less than $201.

9. Suppose that 10 percent of the population of a nation is between 15 and 21 years of age. Find the probability that in a city of 40,000 people, the number of residents in that age group is between 3900 and 4100. *Hint:* You may assume that the residents are selected at random from the nation.

10. The Bildfine Co. estimates that the probability of manufacturing a defective transwidget is $1/50$. The company puts 100 transwidgets in each box, and to avoid testing each transwidget, they adjust the price and further offer to replace any box that contains more than 5 defective transwidgets and to pay a bonus for the troubles caused. Find the probability that a box contains 6 or more defective transwidgets. *Hint:* Use $x = 5.5$ as the critical number.

11. Suppose that we want to determine the number of Conserveral voters in a given state by a survey and that we want to be 95 percent certain that the error in our estimate is less than $d/100$ (d percent). Prove that we should include approximately $n = 40{,}000pq/d^2$ voters in the survey. It can be proved that for any probability p, we have $4pq \leq 1$. Hence a suitable upper bound for n is $10{,}000/d^2$. This formula is very useful because we do not know the correct value for p.

12. Major Motors, Inc. plans a survey to discover what fraction of its customers would like triple windshield wipers. Estimate the number of customers that should be included in the survey, if the manufacturers want a probability of 0.95 that the fraction obtained is within 2 percent of the correct value.

13. The admissions office at Utopia University has discovered through the years that among the applicants accepted for the freshman class only three-fourths actually attend (the remaining applicants go to other schools, into the army, etc.). The freshman dormitory is designed for 900 students but can be stretched to house 940 students under emergency conditions. If the admissions office sends acceptance notices to 1200 applicants, find the probability that the school will need to provide dormitory space for more than 940 freshmen.

14. In Neuro City, 10 percent of those admitted to the hospitals for treatment have lung cancer. Among those admitted 1600 were heavy smokers and among these heavy smokers 200 had lung cancer. Let H_0 be the hypothesis that smoking does not increase the incidence of lung cancer. Do we accept H_0 or reject it at the 5 percent significance level?

15. A professional baseball player, with a minor in statistics, has a lifetime batting average of 300. He complains bitterly of a slump because during the last month he got only 21 hits in 84 times at bat (average 250). Is his complaint justified?

16. There is a rather little known book that contains only random numbers. An energetic student is suspicious and decides to check that the numbers are really random. Among the first 25,600 numbers in the book he finds that 12,759 are odd (and hence 12,841 are even). Does this prove that the numbers in the book are random?

★**17.** Let f be the function defined by

$$f(x) = \frac{1}{2^n}, \qquad \text{if } n - 1 \leq x < n,$$

for $n = 1, 2, 3, \ldots$. Make a graph of $y = f(x)$ for x in the interval $0 \leq x < 4$ and indicate how the graph is extended for large values of x. A function of this type is called a *step function*.

★**18.** The graph of the function defined in problem 17 is "disconnected," but we can connect the horizontal pieces by adding suitable vertical line segments. Thus, at $x = 1$, we add the segment $1/2 \leq y \leq 1/4$; at $x = 2$, we add the segment $1/4 \leq y \leq 1/8$; etc. Indicate these additions on the graph obtained in problem 17.

★**19.** Let R be the region bounded "above" by the "connected graph" obtained in problem 18, bounded below by the x-axis, and bounded on the left by the segment of the y-axis $0 \leq y \leq 1$. Show that this region is infinite in extent. Find the area of R.

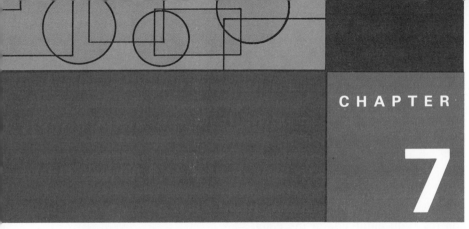

vectors and matrices

In this chapter, we shall introduce two new mathematical systems—the algebra of vectors and the algebra of matrices. Our treatment here will be purely theoretical. Various applications of these new concepts will be developed in Chapters 8 through 12.

Since the advent of computers there has been a tremendous renewed interest in vector and matrix theory. During the last twenty years, this theory has been successfully employed by the social, biological, and physical scientists.

1. vectors

Many times we express an idea by means of a set of numbers which are arranged in a specific order. For example, the number of students in a given university (8000 freshmen, 6000 sophomores, 4000 juniors, 2000 seniors) might be indicated by merely writing [8000, 6000, 4000, 2000]. Such ordered sets of numbers, which can be added together and multiplied by a number, subject to some of the laws of arithmetic are called vectors.

> **Definition 1. Row Vector.** A row vector is an ordered k-tuple of real numbers written in a row.

Examples of row vectors are

$$[2, 3], \quad [4, -1, 6], \quad [2, -5, 3.2, 2], \quad [a_1, a_2, \ldots, a_n].$$

207

The individual numbers in these vectors are called *components*. If a vector has n components, it is called a *vector of order n* or an *n-dimensional vector*. Thus, the first vector above is two-dimensional, the second is three-dimensional, the third is four-dimensional, and the last one is n-dimensional. In general we shall use letters such as A, B, ... to denote vectors, and reserve the corresponding lowercase letters for their components. Thus the equations

$$A = [a_1, a_2, \ldots, a_n] \quad \text{and} \quad X = [x_1, x_2, \ldots, x_n]$$

merely state that the vectors A and X are to have the components indicated on the right-hand side of the equal sign.

In the same way we define an n-dimensional column vector in

Definition 2. Column Vector. A column vector is an ordered k-tuple of real numbers written in a column. If a column vector has n components, it is said to be an n-dimensional column vector or a column vector of order n.

Examples of column vectors are

$$\begin{bmatrix} 1 \\ 2 \end{bmatrix}, \quad \begin{bmatrix} 2 \\ 3 \\ -4 \end{bmatrix}, \quad \begin{bmatrix} 1 \\ 2.5 \\ 7 \\ 9 \end{bmatrix}, \quad \begin{bmatrix} b_1 \\ b_2 \\ \vdots \\ b_n \end{bmatrix}.$$

Definition 3. Equal Vectors. Two row vectors of the same dimension are equal if and only if their corresponding components are equal.

Thus if $A = [a_1, a_2, \ldots, a_n]$ and $B = [b_1, b_2, \ldots, b_n]$, then $A = B$ if and only if

$$a_1 = b_1, \quad a_2 = b_2, \quad \ldots, \quad a_n = b_n.$$

Similarly, two column vectors

$$C = \begin{bmatrix} c_1 \\ c_2 \\ \vdots \\ c_n \end{bmatrix} \quad \text{and} \quad D = \begin{bmatrix} d_1 \\ d_2 \\ \vdots \\ d_n \end{bmatrix}$$

are equal, and we write $C = D$ if and only if

$$c_1 = d_1, \quad c_2 = d_2, \quad \ldots, \quad c_n = d_n.$$

EXAMPLE 1. Check the following vectors for equality:

$$A = [3, 4], \quad B = [4, 3], \quad C = \begin{bmatrix} 3 \\ 4 \end{bmatrix}, \quad D = \begin{bmatrix} 4 \\ 3 \end{bmatrix}.$$

Solution. No two of these vectors are equal to each other. (Explain!).

Definition 4. Addition. Let $A = [a_1, a_2, \ldots, a_n]$ and $B = [b_1, b_2, \ldots, b_n]$ be two row vectors of dimension n. We define their sum $A + B$ by the right-hand side of

(1) $$A + B = [a_1 + b_1, a_2 + b_2, \ldots, a_n + b_n].$$

The addition of two column vectors of the same dimension is defined analogously. Thus, two row vectors or two column vectors of the same dimension are added by adding their corresponding components. For Example,

$$\begin{bmatrix} 1 \\ 2 \\ 3 \end{bmatrix} + \begin{bmatrix} 2 \\ -4 \\ 5 \end{bmatrix} = \begin{bmatrix} 3 \\ -2 \\ 8 \end{bmatrix},$$

and

$$[2,4,7,6] + [3,5,-2,1] = [5,9,5,7].$$

The alert student will notice that we do not define addition of two vectors unless they are both row vectors or both column vectors of the same dimension.

Obviously the sum of two n-dimensional vectors is again an n-dimensional vector.

Definition 5. Multiplication of a Vector and a Number. Suppose that $A = [a_1, a_2, \ldots, a_n]$ is an n-dimensional row vector and c is a real number. We define the product cA by the right-hand side of

(2) $$cA = [ca_1, ca_2, \ldots, ca_n].$$

Thus to multiply a vector by a real number we multiply each of its components by the real number.[1] We define the multiplication of a column

---

[1] The multiplier c is often called a *scalar* to distinguish it from the vector A. Thus the word "scalar" means "real number" in this book.

vector and a real number analogously. For example,

$$5[3, -1, 2, 6] = [15, -5, 10, 30], \qquad -3\begin{bmatrix} 4 \\ -7 \\ -\pi \end{bmatrix} = \begin{bmatrix} -12 \\ 21 \\ 3\pi \end{bmatrix},$$

and

$$0\begin{bmatrix} 55 \\ 72 \\ -9 \\ 0 \end{bmatrix} = \begin{bmatrix} 0 \\ 0 \\ 0 \\ 0 \end{bmatrix}.$$

We naturally interpret $-A$, the negative of a vector, as the product $(-1)A$, and we denote $A + (-1)B$ by $A - B$, for brevity.

A vector is called the *zero vector* if all of its components are zero. If we use $\mathbf{0}$ to denote an n-dimensional zero vector, then $\mathbf{0} = [0, 0, \ldots, 0]$. We should have a different symbol for an n-dimensional zero column vector, but for convenience we choose to use the same symbol $\mathbf{0}$ for the zero (row and column) vector. The meaning will be clear in the context. For example, if we write

(3) $$\mathbf{0} + \begin{bmatrix} 2 \\ 3 \\ 5 \end{bmatrix} = \begin{bmatrix} 2 \\ 3 \\ 5 \end{bmatrix}$$

and

(4) $$\mathbf{0} + [6, 8, -2, -7] = [6, 8, -2, -7],$$

then it is clear that $\mathbf{0} = \begin{bmatrix} 0 \\ 0 \\ 0 \end{bmatrix}$ in equation (3), and $\mathbf{0} = [0,0,0,0]$ in equation (4).

It is easy to see that for any vector A,

(5) $$A + \mathbf{0} = \mathbf{0} + A = A$$

(6) $$0A = \mathbf{0}.$$

(7) $$c\,\mathbf{0} = \mathbf{0}.$$

Theorem 1. Commutative Law of Addition. If A and B are any n-dimensional row (or column) vectors, then

(8) $$A + B = B + A.$$

Proof. Let $A = [a_1, a_2, \ldots, a_n]$ and $B = [b_1, b_2, \ldots, b_n]$ be two row vectors of dimension n. Then

(9) $A + B = [a_1 + b_1, a_2 + b_2, \ldots, a_n + b_n]$

(10) $= [b_1 + a_1, b_2 + a_2, \ldots, b_n + a_n]$

(11) $= B + A.$ ∎

In the proof, equations (9) and (11) follow from the definition of vector addition, and equation (10) follows from equation (9) because the real numbers are commutative under addition.

Theorem 2. Associative Law of Addition. If A, B, and C are any n-dimensional row (or column) vectors, then

$$(A + B) + C = A + (B + C).$$

Proof. The proof is left as an exercise. ∎

Exercise 1

1. Let $A = [1, -5, 7, 2]$ and $B = [2, 2, 0, 5]$. Compute
   (a) $A + B$.   (b) $A - B$.
   (c) $3B$.   (d) $2A - 5B$.
2. If A and B are the vectors given in problem 1, find C so that $3A + C = 2B$.
3. When possible, compute the following sums; when not possible, give the reason:
   (a) $[2, 3] + [5, 7, 8]$.
   (b) $[3, 2] + \begin{bmatrix} 2 \\ 3 \end{bmatrix}$.
   (c) $2\begin{bmatrix} 1 \\ 2 \\ 3 \end{bmatrix} + 5\begin{bmatrix} 1 \\ -5 \\ 7 \end{bmatrix} + 7\begin{bmatrix} 2 \\ 3 \\ 5 \end{bmatrix}$.
   (d) $[2, 3, 4] + 0[1, 2, 3]$.
4. Prove Theorem 2.
5. Suppose that with each person who visited A & P on a given Saturday we associate a row vector whose components give the amounts of the different items that he purchased. Does it make sense to add together the vectors associated with two different persons? If your answer is "no," can we modify the problem to get a "yes" answer?

6. If A and B are any n-dimensional row (column) vectors and a,b are any real numbers, prove the following:
   (a) $A - A = 0$.
   (b) $(a + b)A = aA + bA$.
   (c) $a(A + B) = aA + aB$.
   (d) $(ab)A = a(bA)$.

7. Suppose the mean weight (in pounds) of 4 different breeds of chickens at two months of age is given by the vector $A = [1/2, 3/8, 7/16, 4/9]$ and the weights of the same breeds at seven months of age is given by the vector $B = [3/4, 5/8, 9/16, 5/9]$. Find the vector representing the average gain in weight from two to seven months of age for the four breeds of chickens.

8. In problem 7, find the weight of 100 chickens of each breed at two months of age.

9. Bookstore A has 800 books, 16 dozen pencils, and 12 dozen pens. Bookstore B has 1200 books, 10 dozen pencils, and 15 dozen pens. Write the stock in each of the bookstores in vector form and give the vector representation of the stock in both stores.

10. Do problem 9 if Bookstore B in addition has 15 dozen paper clips, while Bookstore A does not carry this item.

2.   geometric representation of a vector

Certain quantities in nature possess both a magnitude and a direction. Force is such a quantity. For if we add two forces of 10 pounds each we do not necessarily obtain a force of 20 pounds. The resulting force depends on the direction of the individual forces. Similarly, the velocity of a moving particle has a magnitude (called the *speed*) and a direction, the direction in which the particle is moving. To handle physical problems involving directed quantities, it is useful to interpret a real vector of two or three dimensions geometrically as a directed line segment. Before giving this interpretation we pause for a brief review of the rectangular coordinate system.

The old familiar rectangular coordinate system consists of two directed lines meeting at right angles (see Fig. 1). The point of intersection is called the *origin* and is usually denoted by O. It is customary to make one of these lines horizontal and to take the direction to the right of O as the positive direction on this line. The horizontal line is called the *x-axis*, or the *horizontal axis*. The other directed line which is perpendicular to the *x*-axis is called the *y-axis* or the *vertical axis*, and the positive direction on this axis is upward from O.

Once a rectangular coordinate system has been chosen, any point in the plane can be located with respect to it. Suppose P is some point in the plane. Let PQ be the line segment from P perpendicular to the *x*-axis at the point

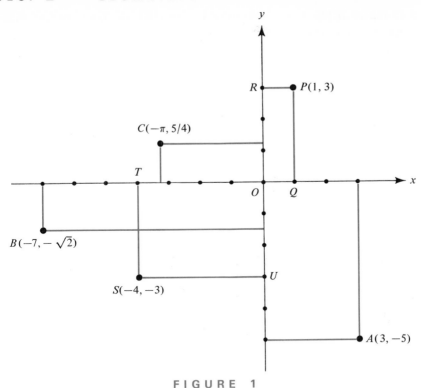

FIGURE 1

Q, and let PR be the line segment perpendicular to the y-axis at the point R (see Fig. 1). Then the directed distance $\overline{OQ}$ is called the x-coordinate of P and $\overline{OR}$ is called the y-coordinate of P. For example, in Fig. 1, $\overline{OQ} = 1$ and $\overline{OR} = 3$, so that the x-coordinate of P is 1 and the y-coordinate of P is 3. Similarly for the point S in Fig. 1 we have $\overline{OT} = -4$ and $\overline{OU} = -3$.

It is customary to enclose the coordinates of a point in parentheses, thus (x,y). In our specific case the coordinates of P and S are $(1,3)$ and $(-4,-3)$, respectively.

Of course, this procedure can be reversed. Given the coordinates $A(3,-5)$, for example, the point A can be located by moving three units to the right of O on the x-axis and then proceeding downward five units along a line parallel to the y-axis. The points $A(3, -5)$, $B(-7, -\sqrt{2})$, and $C(-\pi, 5/4)$ are shown in Fig. 1. The discussion we have just given proves

Theorem 3. With a given rectangular coordinate system each point P in the plane has a uniquely determined pair of coordinates (x,y), where x and y are real numbers. Conversely, for each ordered pair (x,y) of real numbers there is exactly one point P which has this pair for its coordinates.

Let $A = [a_1, a_2]$ be a two-dimensional vector. We associate with this vector the directed line segment $\overrightarrow{OA}$ from the origin to the point A whose coordinates[1] (a_1, a_2) are the same as the components of the vector A (see Fig. 2). This directed line segment $\overrightarrow{OA}$ is a geometric quantity that corresponds to the vector $A = [a_1, a_2]$, and because the association is one-to-one we may

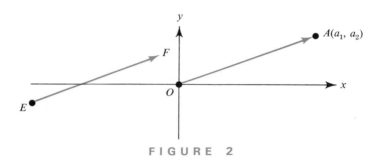

FIGURE 2

regard the directed line segment $\overrightarrow{OA}$ as a vector (the geometric representation of $[a_1, a_2]$). If (as indicated in Fig. 2) the directed line segment $\overrightarrow{EF}$ has the same length and the same direction as $\overrightarrow{OA}$, we also regard it as a geometric representation of $A = [a_1, a_2]$, and we write $\overrightarrow{OA} = \overrightarrow{EF}$ for these two (geometric) vectors. Clearly any vector in the plane is equal to a suitably chosen vector starting from the origin.

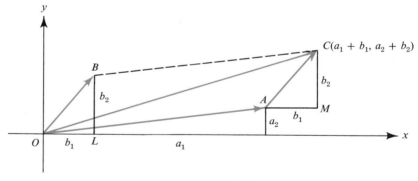

FIGURE 3

With this geometric representation of a vector, it is natural to look for a geometric representation for sum of two vectors $A = [a_1, a_2]$ and $B = [b_1, b_2]$. As indicated in Fig. 3, we move the vector B parallel to itself, so that the beginning point of B falls on the end point of A. Let C be the end point of the vector B in the new position. From the congruent triangles LOB and MAC it is clear that the coordinates of C are $(a_1 + b_1, a_2 + b_2)$.

---

[1] To distinguish between the first and second coordinate of a point one often uses distinct letters such as x and y or a and b. However, for our purposes it is better to use subscripts. Thus the notation $A(a_1, a_2)$ means that A is a point whose second coordinate, the y-coordinate, is a_2.

Consequently, in Fig. 3 the directed line segment $\overrightarrow{OC}$ is the geometric representation of the sum of $[a_1, a_2]$ and $[b_1, b_2]$. Since $\overrightarrow{OC}$ is the diagonal of the parallelogram $OACB$, this proves the familiar parallelogram law for the addition of vectors.

The multiplication of a vector by a number also has a very nice geometric interpretation. Let $A = [a_1, a_2]$ be a 2-dimensional vector and let c be a number. Consider first the case in which $c > 1$, shown in Fig. 4.

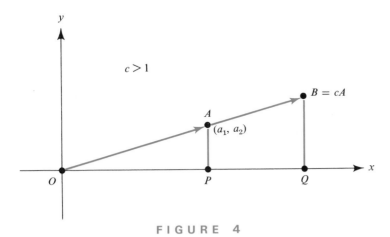

FIGURE 4

Clearly, if O, A, and B are collinear points and the length of $\overrightarrow{OB}$ is c times the length of $\overrightarrow{OA}$, then from the similarity of the triangles OPA and OQB it is easy to see that the coordinates of the point B are (ca_1, ca_2). Hence the directed line segment $\overrightarrow{OB}$ represents the vector cA.

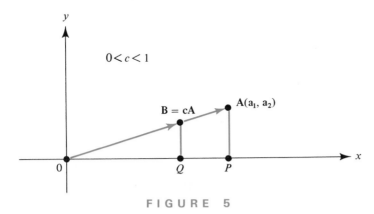

FIGURE 5

Similarly in Fig. 5 the directed line segment $\overrightarrow{OB}$ represents the vector cA when $0 < c < 1$.

When $c < 0$, the directed line segment $\overrightarrow{OB}$ that represents the vector cA is shown in Fig. 6. In this figure the points B, O, and C are still collinear but the direction of the vector $B = cA$ is just the opposite of the direction of the vector A. The length of cA is $|c|$ times the length of A.

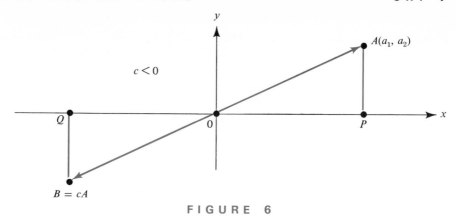

FIGURE 6

Since $A - B$ is $A + (-1)B$, the geometric rule for adding two vectors also gives a geometric rule for finding the difference of two vectors. This rule is illustrated in Fig. 7.

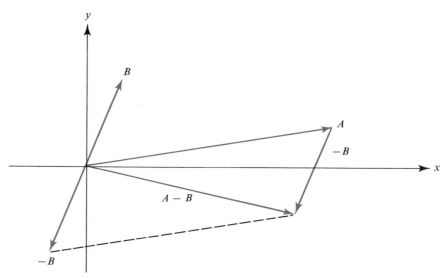

FIGURE 7

Another vector quantity which has a very useful geometric representation is the vector $C = (1 - t)A + tB$, where t is any number between 0 and 1. We assert that the points $C = ((1 - t)a_1 + tb_1, (1 - t)a_2 + tb_2)$ lie on the line segment between the two points $A(a_1, a_2)$ and $B(b_1, b_2)$.

To prove this, we observe that $B - A$ is the vector from the point A to the point B as indicated in Fig. 8. Hence for the vector from the origin to the point C, we can write

$$(12) \qquad C = A + t(B - A) = A + tB - tA = (1 - t)A + tB.$$

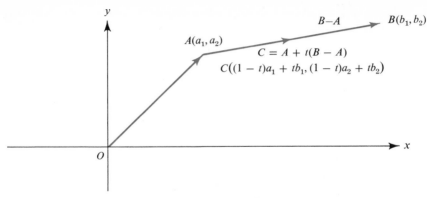

FIGURE 8

It is worth noting that if $t = 0$, equation (12) gives the vector to the point A, and if $t = 1$, equation (12) gives the vector to the point B. As t increases from 0 to 1, the point C moves steadily along the line segment from the point A to the point B.

The *magnitude* of a vector $A = [a_1, a_2]$ is denoted by $|A|$ and we define it to be $\sqrt{a_1^2 + a_2^2}$. Geometrically $|A|$ is the length of the line segment joining the origin to the point whose coordinates are (a_1, a_2).

In the same way, the *magnitude* of an n-dimensional row vector $A = [a_1, a_2, \ldots, a_n]$ is defined by the right-hand side of

(13) $|A| = \sqrt{a_1^2 + a_2^2 + \cdots + a_n^2}.$

Certain vectors of primary importance deserve a special name and symbol. If $|A| = 1$, we call A a *unit vector*. In three-dimensional space the unit vectors $[1, 0, 0]$, $[0, 1, 0]$, and $[0, 0, 1]$ are given the special symbols $\mathbf{e}_1$, $\mathbf{e}_2$, and $\mathbf{e}_3$, respectively. With these symbols, it is often convenient to express a vector $A = [a_1, a_2, a_3]$ as

$$A = [a_1, a_2, a_3] = [a_1, 0, 0] + [0, a_2, 0] + [0, 0, a_3]$$
$$= a_1[1, 0, 0] + a_2[0, 1, 0] + a_3[0, 0, 1]$$
$$= a_1\mathbf{e}_1 + a_2\mathbf{e}_2 + a_3\mathbf{e}_3.$$

Exercise 2

1. Represent the following vectors geometrically:
   (a) $A = [-3, 7]$.   (b) $B = [-3, -5]$.
   (c) $C = [2, 3]$.   (d) $D = [2, 3, 4]$.
2. Find the magnitude of the vectors in problem 1.

3. If A, B, and C are the vectors of problem 1, perform the following operations graphically:

    (a) $(A + B) + C$.   **(b)** $A + 2C$.

    (c) $C - A$.   **(d)** $(A - C) + B$.

4. If A is a nonzero vector in three-dimensional space, prove that $\dfrac{1}{|A|}A$ is a unit vector in the direction of A.

5. For each of the following vectors find a unit vector with the same direction:

    (a) $A = 3\mathbf{e}_1 - 4\mathbf{e}_2 + 5\mathbf{e}_3$.

    (b) $B = \mathbf{e}_1 + 2\mathbf{e}_2 - 2\mathbf{e}_3$.

6. From equation (12) derive a formula for the coordinates of the midpoint of the line segment AB.

7. For the points $A(-3, 7)$, $B(-5, 11)$, and $C(3, 1)$, find the midpoints of each of the line segments AB, BC, and AC. Then check your computations by making a suitable drawing.

8. Give an interpretation for equation (12) when **(a)** $t = 1/3$, **(b)** $t = 2$, and **(c)** $t = -1$. In each case make a suitable drawing.

9. How would you define a line segment in n-dimensional space when $n > 3$?

10. Using a suitable drawing, give a geometric proof of the associative law of vector addition,

$$(A + B) + C = A + (B + C),$$

for two-dimensional vectors.

11. Let ABC be a triangle in the plane. Prove geometrically that $\overrightarrow{AB} + \overrightarrow{BC} + \overrightarrow{CA} = \mathbf{0}$.

12. Generalize problem 11 to an n-sided polygon.

13. Let O be the center of an n-sided regular polygon, where n is even. Prove geometrically that the sum of the vectors from O to be the vertices is zero.

3.   the multiplication of vectors

This rule is given in

Definition 6. Let $A = [a_1, a_2, \ldots, a_n]$ be a row vector of dimension n and let

$$B = \begin{bmatrix} b_1 \\ b_2 \\ \vdots \\ b_n \end{bmatrix}$$

be a column vector of dimension n. We define the product $A \cdot B$ by the right-hand side of

(14) $$A \cdot B = a_1 b_1 + a_2 b_2 + \cdots + a_n b_n.$$

Notice that in this definition we write the row vector first and the column vector second. For the vectors A and B, the product $A \cdot B$ is a real number.

EXAMPLE 1.   Major Motors, Inc. produces the following cars per year: 1000 luxury cars (L), 3000 medium-priced cars (M), and 2000 compact cars (C). The L sells for $5000 each, M sells for $3000 each, and C sells for $2000 each. Find the total amount of money received by the company from the sale of these cars.

Solution.   We can represent the number of cars of each type sold in one year by the row vector

$$N = [L, M, C] = [1000, 3000, 2000].$$

Using the price of each type of car we can construct the corresponding price (column) vector:

$$P = \begin{bmatrix} 5000 \\ 3000 \\ 2000 \end{bmatrix}.$$

Then the total amount of money (in dollars) received by the company from the sale of these cars for one year is given by the product

$$N \cdot P = (1000)(5000) + (3000)(3000) + (2000)(2000) = 18,000,000.$$

EXAMPLE 2.   Let $A = [1, -2, 3]$ and $B = \begin{bmatrix} 2 \\ -4 \\ -1 \end{bmatrix}$. Find $A \cdot B$.

Solution. $A \cdot B = (1)(2) + (-2)(-4) + (3)(-1)$
$$= 2 + 8 - 3 = 7.$$

Theorem 4. The Distributive Law.   If A is a row vector of dimension n and B and C are column vectors of dimension n, then

$$A \cdot (B + C) = A \cdot B + A \cdot C.$$

Proof. Let

$$A = [a_1, a_2, \ldots, a_n], \quad B = \begin{bmatrix} b_1 \\ b_2 \\ \vdots \\ b_n \end{bmatrix}, \quad \text{and} \quad C = \begin{bmatrix} c_1 \\ c_2 \\ \vdots \\ c_n \end{bmatrix}.$$

Then

$$A \cdot (B + C) = [a_1, a_2, \ldots, a_n] \begin{bmatrix} b_1 + c_1 \\ b_2 + c_2 \\ \vdots \\ b_n + c_n \end{bmatrix}$$

$$= a_1(b_1 + c_1) + a_2(b_2 + c_2) + \cdots + a_n(b_n + c_n)$$

$$= (a_1 b_1 + a_2 b_2 + \cdots + a_n b_n) + (a_1 c_1 + a_2 c_2 + \cdots + a_n c_n)$$

$$= A \cdot B + A \cdot C. \quad \blacksquare$$

Exercise 3

1. Let $A = [1, -2, 3]$, $B = [2, 1, -1]$, $C = \begin{bmatrix} -3 \\ 4 \\ -5 \end{bmatrix}$, and $D = \begin{bmatrix} 2 \\ -3 \\ 4 \end{bmatrix}$.

   Find (if possible):

   (a) $A \cdot C + B \cdot D$.

   (b) $(A + B) \cdot (C + D)$.

   (c) $(A + C) \cdot D$.

   (d) $\left(\frac{1}{2}A + 3B\right) \cdot (-2C + 4D)$.

   (e) $A \cdot D - B \cdot C$.

   (f) $(A + D) \cdot (B + C)$.

   (g) $A \cdot C - D \cdot B$.

   (h) $(A + 2B) \cdot (C - D)$.

2. Let A be an n-dimensional row vector, B an n-dimensional column vector, and c a real number. Prove that

$$c(A \cdot B) = (cA) \cdot B = A \cdot (cB).$$

3. Are the following always true for vectors?

   (a) If $A \cdot B = A \cdot C$, then $B = C$.

   (b) If $A \cdot B = 0$, then either $A = \mathbf{0}$ or $B = \mathbf{0}$.

4. Solve for x:

   (a) $[x, 1, 0, 2] \cdot \begin{bmatrix} -1 \\ 3 \\ 2 \\ 7 \end{bmatrix} = 6$.

   (b) $[-1, 3x, 2x - 1] \cdot \begin{bmatrix} x \\ 1 \\ 2 \end{bmatrix} = 16$.

5. Expand $(A + B) \cdot (C + D)$. Check the formula you obtain using the vectors given in problem 1.

6. Let $A = [x_1, x_2]$, $B = \begin{bmatrix} 2 \\ 3 \end{bmatrix}$, and $C = \begin{bmatrix} -1 \\ 2 \end{bmatrix}$. If $A \cdot B = 12$ and $A \cdot C = 1$, find x_1 and x_2.

7. A store manager has three types of shirts. He has one dozen, two dozen, and three dozen shirts of types 1, 2, and 3, respectively. Write a three-component row vector A whose components give the numbers of each type of shirts he has. The shirts of types 1, 2, and 3 sell for $6, $5, and $7 apiece, respectively. Write a three-component column vector B whose components give the selling price of each type of shirt. Compute $A \cdot B$ and state what it means.

8. An investment company sells 500, 400, 600, and 700 shares of stock of companies X, Y, Z, and W, respectively. The selling prices are $30, $20, $40, and $50 per share of the companies X, Y, Z, and W, respectively. Use a product of vectors to find the total receipt from this stock sale.

9. Suppose that in an oversimplified economy there are three industries 1, 2, and 3 producing products X, Y, and Z, respectively. Suppose further that there are five types of consumers—the government, the general public, and the three industries. In each case we list the goods demanded in the form $A = $ [product X, product Y, product Z]. With suitable units of measure, the demand vectors are

By the government:	$A_g = [10, 8, 7]$
By the general public:	$A_p = [5, 6, 6]$
By industry 1:	$A_1 = [0, 3, 2]$
By industry 2:	$A_2 = [3, 0, 1]$
By industry 3:	$A_3 = [4, 2, 0].$

Let the price per unit of products X, Y, and Z be $5, $4, and $6, respectively. Assume that each industry produces exactly enough to satisfy the demand. Find

   (a) The total demand on each industry.

   (b) The profit (loss) of each industry.

10. Show that the expected value defined in Chapter 5 (page 164) can be regarded as the product of two vectors.

4.   matrices

We have already discussed matrices in an informal way. They appeared quite naturally when we studied Bayes' theorem in Chapter 5, Section 8. Further, we found in Chapter 6, Section 2, that a matrix was quite helpful in presenting the probabilities that occur in a Markov process. It is now time for a treatment of matrices in a purely theoretical manner. In Chapters 8 through

12 we shall meet further applications of the theory of matrices in a wide variety of situations.

> **Definition 7. Matrix.** A matrix A is a rectangular array of real numbers, denoted by
>
> (15)
> $$A = \begin{bmatrix} a_{11} & a_{12} & \cdots & a_{1n} \\ a_{21} & a_{22} & \cdots & a_{2n} \\ \vdots & \vdots & \vdots & \vdots \\ a_{m1} & a_{m2} & \cdots & a_{mn} \end{bmatrix}.$$

The horizontal lines of the array are called *rows* and the vertical lines are called *columns*. A matrix having m rows and n columns is referred to as an $m \times n$ (m by n) matrix. The entry in the ith row and the jth column is designated by a_{ij}; thus a_{35} is the entry in the third row and fifth column. The double subscript gives us the location or *address* of the entry.

> **Definition 8. Square Matrix.** An $m \times n$ matrix is called a square matrix if $m = n$.

> **Definition 9. Zero Matrix.** If all the entries of a matrix are zero, the matrix is called the zero matrix or the null matrix and is denoted by the symbol **0**.

> **Definition 10. Column Matrix.** When a matrix consists of a single column (an $m \times 1$ matrix) it is called a column matrix.

Notice that a column matrix is also a column vector.

> **Definition 11. Row Matrix.** When a matrix consists of a single row (a $1 \times n$ matrix) it is called a row matrix.

The following are examples of matrices:

$$A = [3, 2, 5], \quad B = \begin{bmatrix} 3 \\ 4 \\ 5 \end{bmatrix}, \quad C = \begin{bmatrix} 1 & 2 \\ 3 & 4 \end{bmatrix},$$

$$D = \begin{bmatrix} 1 & 2 \\ 3 & 4 \\ 5 & 6 \\ 7 & 8 \end{bmatrix}, \quad E = \begin{bmatrix} 1 & 2 & -3 & 4 \\ -2 & 0 & 4 & 5 \end{bmatrix}.$$

A, B, C, D, and E are 1×3, 3×1, 2×2, 4×2, and 2×4 matrices, respectively.

Definition 12. Main Diagonal. For a square matrix A, the entries $a_{11}, a_{22}, \ldots, a_{nn}$ are said to form the main diagonal of A.

Definition 13. Equality. Two matrices

(15)
$$A = \begin{bmatrix} a_{11} & a_{12} & \cdots & a_{1n} \\ a_{21} & a_{22} & \cdots & a_{2n} \\ \vdots & \vdots & \vdots & \vdots \\ a_{m1} & a_{m2} & \cdots & a_{mn} \end{bmatrix}$$

and

(16)
$$B = \begin{bmatrix} b_{11} & b_{12} & \cdots & b_{1q} \\ b_{21} & b_{22} & \cdots & b_{2q} \\ \vdots & \vdots & \vdots & \vdots \\ b_{p1} & b_{p2} & \cdots & b_{pq} \end{bmatrix}$$

are said to be equal, and we write $A = B$, if

(1) A and B have the same number of rows; $m = p$.
(2) A and B have the same number of columns; $n = q$.
(3) Their corresponding entries are equal; i.e., $a_{ij} = b_{ij}$ for all i and j.

For example

$$\begin{bmatrix} 1 & 2 \\ 3 & 4 \end{bmatrix} \neq \begin{bmatrix} 1 & 3 \\ 2 & 4 \end{bmatrix},$$

but if $x = 2$, then

$$\begin{bmatrix} x^0 & x \\ x^2 & x^3 \end{bmatrix} = \begin{bmatrix} 1 & 2 \\ 4 & 8 \end{bmatrix}.$$

Exercise 4

1. Let

$$A = \begin{bmatrix} 1 & 2 & 3 & 4 \\ 5 & 6 & 7 & 8 \\ 9 & 10 & 11 & 12 \end{bmatrix}.$$

Find a_{13}, a_{31}, a_{33}, and a_{24}.

2. In problem 1, find the address of each of the following entries: **(a)** 7, **(b)** 10, **(c)** 4, and **(d)** 12.

3. Is the following array a matrix:

$$\begin{bmatrix} 2 & 3 & 4 \\ 1 & 2 & 5 \\ 3 & 1 \end{bmatrix}?$$

4. Display the matrix A with the following entries: $a_{13} = 5$, $a_{22} = 2$, $a_{11} = 6$, $a_{12} = -5$, $a_{23} = 4$, and $a_{21} = 7$.

In problems 5 through 10, find all values of the unknowns for which the given equation is true.

5. $\begin{bmatrix} 1 & 2 \\ 3 & x \end{bmatrix} = \begin{bmatrix} 1 & 2 \\ 3 & 4 \end{bmatrix}.$

6. $\begin{bmatrix} 1 & 3 & 5 \\ 2 & 7 & 6 \end{bmatrix} = \begin{bmatrix} 1 & 2 \\ x & 3 \end{bmatrix}.$

7. $\begin{bmatrix} 3 & x \\ 4 & 7 \end{bmatrix} = \begin{bmatrix} 3 & x \\ 4 & 7 \end{bmatrix}.$

8. $\begin{bmatrix} 1 & 2 & 3 \\ x & y & 7 \\ 2 & 3 & 4 \end{bmatrix} = \begin{bmatrix} 1 & 2 & 3 \\ 3x+1 & y & 7 \\ 2 & 3 & 4 \end{bmatrix}.$

9. $\begin{bmatrix} 3 & 2 \\ x & 1 \end{bmatrix} = \begin{bmatrix} 3 & 4 \\ x & 1 \end{bmatrix}.$

10. $\begin{bmatrix} 1 & x \\ 2x & 4 \end{bmatrix} = \begin{bmatrix} 1 & 2 \\ 6 & 4 \end{bmatrix}.$

11. Suppose an arrow from point A_i to point A_j denotes the *dominance* of A_i over A_j. Translate the dominance relation given in Fig. 9 (see next page) into matrix form by writing $a_{ij} = 1$ if A_i dominates A_j, and $a_{ij} = 0$ otherwise. Assume that no point dominates itself.

12. Do problem 11 if in addition every point dominates itself.

13. Let the matrix representation of a dominance relation be

$$\begin{array}{c} \\ 1 \\ 2 \\ 3 \\ 4 \\ 5 \end{array} \begin{array}{ccccc} 1 & 2 & 3 & 4 & 5 \\ \begin{bmatrix} 0 & 1 & 0 & 1 & 1 \\ 1 & 0 & 0 & 1 & 0 \\ 0 & 1 & 0 & 1 & 0 \\ 0 & 1 & 1 & 0 & 0 \\ 0 & 0 & 1 & 0 & 0 \end{bmatrix} \end{array}$$

Draw a diagram representing this dominance.

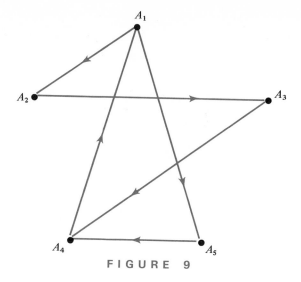

A_1

A_2

A_3

A_4

A_5

FIGURE 9

5. addition of matrices

To avoid excessive writing we shall abbreviate the matrix

(15)
$$A = \begin{bmatrix} a_{11} & a_{12} & \cdots & a_{1n} \\ a_{21} & a_{22} & \cdots & a_{2n} \\ \vdots & \vdots & \vdots & \vdots \\ a_{m1} & a_{m2} & \cdots & a_{mn} \end{bmatrix}$$

by merely writing $A = [a_{ij}]$, and if we wish to indicate the number of rows and columns, we shall use the notation $[a_{ij}]_{m,n}$. For example, the notation $[b_{ij}]_{p,q}$ for the matrix B means that

(16)
$$B = \begin{bmatrix} b_{11} & b_{12} & \cdots & b_{1q} \\ b_{21} & b_{22} & \cdots & b_{2q} \\ \vdots & \vdots & \vdots & \vdots \\ b_{p1} & b_{p2} & \cdots & P_{pq} \end{bmatrix}.$$

Definition 14. Two matrices $A = [a_{ij}]_{m,n}$ and $B = [b_{ij}]_{p,q}$ are said to be of the same size (or order) if and only if $m = p$ and $n = q$. When two matrices are of the same size, they are said to be conformable for addition.

Definition 15. Addition of Two Matrices. Let $A = [a_{ij}]_{m,n}$ and $B = [b_{ij}]_{m,n}$ be two matrices of the same size. Then the sum $A + B$ is the matrix $C = [c_{ij}]_{m,n}$, where

(17) $$c_{ij} = a_{ij} + b_{ij}$$

for every pair i,j with $i = 1, 2, \ldots, m$ and $j = 1, 2, \ldots, n$.

Theorem 5. Associative Law of Addition. If three matrices A, B, and C are conformable for addition, then

(18) $$(A + B) + C = A + (B + C).$$

Proof. Let $A = [a_{ij}]$, $B = [b_{ij}]$, and $C = [c_{ij}]$. Then

$$(A + B) + C = ([a_{ij}] + [b_{ij}]) + [c_{ij}]$$
(1) $$= [a_{ij} + b_{ij}] + [c_{ij}]$$
(2) $$= [(a_{ij} + b_{ij}) + c_{ij}]$$
(3) $$= [a_{ij} + (b_{ij} + c_{ij})]$$
(4) $$= [a_{ij}] + [b_{ij} + c_{ij}]$$
(5) $$= [a_{ij}] + ([b_{ij}] + [c_{ij}])$$
$$= A + (B + C). \quad \blacksquare$$

The steps 1, 2, 4, and 5 follow from the definition of matrix addition, while step 3 is valid since the real numbers are associative under addition.

Theorem 6. Commutative Law of Addition. If two matrices A and B are conformable for addition, then

(19) $$A + B = B + A$$

Proof. As in the proof of Theorem 1, the proof of this theorem follows from the commutative law of addition for real numbers. The details are left for the reader.

Definition 16. Multiplication of a Matrix by a Real Number. Let $A = [a_{ij}]$ and let c be a real number. Then the product of c and A is denoted by either cA or Ac and is defined by the right-hand side of

(20) $$cA = Ac = [ca_{ij}].$$

Thus, to multiply a matrix A by c, we multiply each entry of A by c. When $c = -1$, we frequently write $-A$ in place of $(-1)A$. Further, the difference of two matrices, $A - B$, is defined to be the sum $A + (-1)B$.

EXAMPLE 1. Given

$$A = \begin{bmatrix} -2 & 3 \\ 4 & -1 \\ 0 & 2 \end{bmatrix}, \quad B = \begin{bmatrix} 6 & 4 \\ -7 & 9 \\ 8 & -5 \end{bmatrix}, \quad \text{and} \quad C = \begin{bmatrix} 1 & 2 \\ 3 & 4 \end{bmatrix},$$

find $A + B$, $(-1)B$, $5A - 3B$, and $A + C$.

Solution. We naturally interpret $5A - 3B$ as $5A + (-3)B$. Then inspection gives

$$A + B = \begin{bmatrix} 4 & 7 \\ -3 & 8 \\ 8 & -3 \end{bmatrix}, \quad -B = \begin{bmatrix} -6 & -4 \\ 7 & -9 \\ -8 & 5 \end{bmatrix}, \quad 5A - 3B = \begin{bmatrix} -28 & 3 \\ 41 & -32 \\ -24 & 25 \end{bmatrix}.$$

However, $A + C$ does not exist because A and C are not of the same size.

EXAMPLE 2. A building contractor wishes to build a house. The cost of purchasing and transporting specific amounts of steel, glass, and wood from two different locations is given by the following matrices:

$$\begin{matrix} & \text{Steel} & \text{Glass} & \text{Wood} \\ A = & \begin{bmatrix} 15 & 10 & 10 \\ 7 & 5 & 4 \end{bmatrix} & & \begin{matrix} \text{Cost of material} \\ \text{Transportation cost.} \end{matrix} \end{matrix}$$

$$B = \begin{bmatrix} 14 & 12 & 2 \\ 6 & 4 & 5 \end{bmatrix} \quad \begin{matrix} \text{Cost of material} \\ \text{Transportation cost.} \end{matrix}$$

Find the matrix representing the cost of material and transportation for steel, glass, and wood from both the locations.

Solution. The required matrix is just the sum

$$A + B = \begin{bmatrix} 29 & 22 & 12 \\ 13 & 9 & 9 \end{bmatrix}.$$

Exercise 5

1. Given

$$A = \begin{bmatrix} 1 & 2 & -3 \\ 3 & 4 & 5 \\ 2 & -1 & 7 \end{bmatrix}, \quad B = \begin{bmatrix} 3 & 1 & 5 \\ 1 & -4 & 7 \\ 2 & 1 & 3 \end{bmatrix}, \quad \text{and} \quad C = \begin{bmatrix} 1 & 0 & 2 \\ 2 & 1 & 0 \\ 0 & 0 & 1 \end{bmatrix},$$

228   calculate **(a)** $3A$, **(b)** $A + 2B$, **(c)** $A - 3C$, **(d)** $2A + 2B - 3C$, and
(e) $A - A$.

2. Find a matrix D such that $A + D = B$, where A and B are the matrices given in problem 1.

3. Find a matrix E such that

$$\frac{1}{2}(A - 2C) + 3E = 5B,$$

where A, B, and C are the matrices given in problem 1.

4. Let

$$A = \begin{bmatrix} 1 & 7 & 6 \\ 4 & 9 & 6 \end{bmatrix}.$$

Find a scalar c and a matrix B such that $cB = A$ and B has 3 in the second row and first column.

5. Prove Theorem 6.

6. Let A, B, and C be three matrices of the same size.
    (a) If $A + B = A + C$, prove that $B = C$.
    (b) If $A + B = C + B$, prove that $A = C$.

6.   multiplication of matrices

We have observed that a vector may be regarded as a special kind of matrix. For example, an n-dimensional row vector is simply a $1 \times n$ matrix and an n-dimensional column vector is an $n \times 1$ matrix. In Section 3 of this chapter, we defined the multiplication of a row vector and a column vector of dimension n. In this section, we generalize that definition to the multiplication of two matrices. Before we formally give the definition of multiplication of two matrices, let us consider

EXAMPLE 1. A manufacturer makes two types of products I and II at each of the two different locations X and Y. The materials used to make each of the products are steel, glass, and rubber. Suppose it takes three units of steel, one unit of glass, and two units of rubber to make one unit of product I and four units of steel, one-half unit of glass, and three units of rubber to make one unit of product II. Suppose further that steel, glass, and rubber cost $10, $2, and $3 per unit, respectively, at the location X and that at the location Y steel, glass, and rubber cost $9, $3, and $4 per unit, respectively. Find the material cost of making one unit of each product at each of the two locations.

Solution. We may indicate the plant consumption by the matrix

$$A = \begin{bmatrix} \overset{\text{Steel}}{3} & \overset{\text{Glass}}{1} & \overset{\text{Rubber}}{2} \\ 4 & 1/2 & 3 \end{bmatrix} \begin{matrix} \text{Product I} \\ \text{Product II.} \end{matrix}$$

The cost for each unit at each location may be written as the cost matrix

$$P = \begin{bmatrix} \overset{\text{Location X}}{10} & \overset{\text{Location Y}}{9} \\ 2 & 3 \\ 3 & 4 \end{bmatrix} \begin{matrix} \text{Steel} \\ \text{Glass} \\ \text{Rubber.} \end{matrix}$$

If we use the row vectors $A_1 = [3, 1, 2]$ and $A_2 = [4, 1/2, 3]$, then the matrix A can be composed of row vectors: thus

$$A = \begin{bmatrix} A_1 \\ A_2 \end{bmatrix}.$$

Similarly we may put the matrix P in the form

$$P = [P_1 \, P_2],$$

where P_1 and P_2 are the column vectors:

$$P_1 = \begin{bmatrix} 10 \\ 2 \\ 3 \end{bmatrix} \quad \text{and} \quad P_2 = \begin{bmatrix} 9 \\ 3 \\ 4 \end{bmatrix}.$$

It is easy to see that the cost of making product I at the location X is given by the product

$$A_1 \cdot P_1 = (3)(10) + (1)(2) + (2)(3) = 38,$$

the cost of making product I at the location Y is

$$A_1 \cdot P_2 = (3)(9) + (1)(3) + (2)(4) = 38,$$

the cost of making product II at the location X is

$$A_2 \cdot P_1 = (4)(10) + \left(\frac{1}{2}\right)(2) + (3)(3) = 50,$$

and the cost of making product II at the location Y is

$$A_2 \cdot P_2 = (4)(9) + \left(\frac{1}{2}\right)(3) + (3)(4) = 49.5.$$

We let C be the 2×2 matrix given by

$$C = \begin{bmatrix} A_1 \cdot P_1 & A_1 \cdot P_2 \\ A_2 \cdot P_1 & A_2 \cdot P_2 \end{bmatrix} = \begin{bmatrix} 38 & 38 \\ 50 & 49.5 \end{bmatrix}.$$

Then the matrix C is composed from the two matrices A and P. This suggests that we write $C = AP$ and call C the product of the matrices A and P. The discussion in this example leads to

Definition 17. Multiplication of Two Matrices. Let $A = [a_{ij}]_{m,n}$ and let $B = [b_{ij}]_{n,p}$. Then the product AB is the matrix $C = [c_{ij}]_{m,p}$, where the entry c_{ij} of C is obtained by multiplying the ith row vector of A by the jth column vector of B.

When the matrix product $AB = C$ is written in full we have

(21)
$$\begin{bmatrix} a_{11} & a_{12} & \cdots & a_{1n} \\ \vdots & \vdots & \cdots & \vdots \\ a_{i1} & a_{i2} & \cdots & a_{in} \\ \vdots & \vdots & \cdots & \vdots \\ a_{m1} & a_{m2} & \cdots & a_{mn} \end{bmatrix} \begin{bmatrix} b_{11} & \cdots & b_{1j} & \cdots & b_{1p} \\ b_{21} & \cdots & b_{2j} & \cdots & b_{2p} \\ \vdots & \cdots & \vdots & \cdots & \vdots \\ b_{n1} & \cdots & b_{nj} & \cdots & b_{np} \end{bmatrix}$$
$$= \begin{bmatrix} c_{11} & \cdots & c_{1j} & \cdots & c_{1p} \\ \vdots & & \vdots & & \vdots \\ c_{i1} & \cdots & c_{ij} & \cdots & c_{ip} \\ \vdots & & \vdots & & \vdots \\ c_{m1} & \cdots & c_{mj} & \cdots & c_{mp} \end{bmatrix}.$$

Then by the definition of the product AB, we have

(22)
$$c_{ij} = a_{i1}b_{1j} + a_{i2}b_{2j} + \cdots + a_{in}b_{nj}.$$

The elements that are used in equation (22) are indicated in color in equation (21).

It should be noted that to define the product AB of two matrices A and B, it is necessary that the number of columns of the matrix A be equal to the number of rows of the matrix B.

EXAMPLE 2. Compute the products AB and BA, when

$$A = \begin{bmatrix} 1 & 2 \\ 3 & 4 \end{bmatrix} \quad \text{and} \quad B = \begin{bmatrix} 2 & -1 & 3 \\ 1 & 5 & 4 \end{bmatrix}.$$

Solution. Since A is a 2×2 matrix and B is a 2×3 matrix, we can form the product AB. We find that

$$AB = \begin{bmatrix} 1 & 2 \\ 3 & 4 \end{bmatrix} \begin{bmatrix} 2 & -1 & 3 \\ 1 & 5 & 4 \end{bmatrix}$$

$$= \begin{bmatrix} 1 \cdot 2 + 2 \cdot 1 & 1 \cdot (-1) + 2 \cdot 5 & 1 \cdot 3 + 2 \cdot 4 \\ 3 \cdot 2 + 4 \cdot 1 & 3 \cdot (-1) + 4 \cdot 5 & 3 \cdot 3 + 4 \cdot 4 \end{bmatrix} = \begin{bmatrix} 4 & 9 & 11 \\ 10 & 17 & 25 \end{bmatrix}.$$

However, BA is not defined, because the number of columns, 3, of B is not equal to the number of rows, 2, of A.

Even if both AB and BA are defined, in general AB may not be equal to BA as the next example illustrates. Thus the set of $n \times n$ matrices is noncommutative under multiplication.

EXAMPLE 3. Let $A = \begin{bmatrix} 1 & 2 \\ 3 & 4 \end{bmatrix}$ and $B = \begin{bmatrix} -1 & 0 \\ 2 & 3 \end{bmatrix}$. Compute AB and BA.

Solution.

(23) $$AB = \begin{bmatrix} 1 & 2 \\ 3 & 4 \end{bmatrix} \begin{bmatrix} -1 & 0 \\ 2 & 3 \end{bmatrix} = \begin{bmatrix} 3 & 6 \\ 5 & 12 \end{bmatrix}.$$

(24) $$BA = \begin{bmatrix} -1 & 0 \\ 2 & 3 \end{bmatrix} \begin{bmatrix} 1 & 2 \\ 3 & 4 \end{bmatrix} = \begin{bmatrix} -1 & -2 \\ 11 & 16 \end{bmatrix}.$$

It is obvious that $AB \neq BA$.

For the real numbers, we use the notation a^2 to mean $a \cdot a$ and a^3 to mean $a \cdot a \cdot a$, etc. Similarly for an $n \times n$ matrix A we shall write A^2 to mean AA and A^3 to mean AAA, etc. We define A^0 by

(25) $$A^0 = I_n = \begin{bmatrix} 1 & 0 & 0 & \cdots & 0 \\ 0 & 1 & 0 & \cdots & 0 \\ 0 & 0 & 1 & \cdots & 0 \\ \vdots & \vdots & \vdots & \vdots & \vdots \\ 0 & 0 & 0 & \cdots & 1 \end{bmatrix}.$$

The matrix I_n is called the *identity matrix.*

When the order of the matrix is clear from the context, we can drop the subscript n and write I for I_n. The title "identity matrix" for I is justified in

Theorem 7. Let A be an arbitrary square matrix. Then

(26) $$IA = A \quad \text{and} \quad AI = A.$$

Thus in matrix multiplication the matrix I plays a role similar to that of the number 1 in the multiplication of real numbers. The student will find it easy to prove this theorem.

The distributive law for matrix multiplication is also rather easy and this is reserved for problems 9 and 10 in the next exercise.

Theorem 8. Associative Law. Let A be an $m \times n$ matrix, let B be an $n \times p$ matrix, and let C be a $p \times q$ matrix. Then

(27) $$(AB)C = A(BC).$$

The proof of Theorem 8 is rather difficult with the limited material developed so far. Once it is clear that an $m \times n$ matrix represents a certain transformation from n-dimensional space to m-dimensional space, then the proof of (27) becomes rather easy. But these ideas lie outside of the scope of the text and hence we omit the proof. However, we shall use this theorem whenever the need arises.

Exercise 6

1. Find the products AB and BA for the matrices

$$A = \begin{bmatrix} 1 & 2 & 3 \\ 3 & 2 & 1 \end{bmatrix} \quad \text{and} \quad B = \begin{bmatrix} 5 & 1 \\ 6 & -2 \\ -3 & 1 \end{bmatrix}.$$

2. Compute the indicated quantities using

$$A = \begin{bmatrix} 1 & 2 \\ 3 & 4 \end{bmatrix}, \quad B = \begin{bmatrix} 2 & -3 \\ -1 & 5 \end{bmatrix}, \quad C = \begin{bmatrix} 1 & 4 \\ 5 & 1 \end{bmatrix}.$$

(a) AB.

(b) BA.

(c) BC.

(d) $(AB)C$.

(e) $A(BC)$.

(f) A^2.

(g) A^3.

(h) $A^2 - 5A + B$.

(i) $A(B + C)$.

(j) $AB + AC$.

(k) $(A + B)^2$.

(l) $A^2 + 2AB + B^2$.

3. Find A^n for every integer $n \geq 2$ if

(a) $A = \begin{bmatrix} 0 & 1 \\ 1 & 0 \end{bmatrix}$ and   (b) $A = \begin{bmatrix} 0 & 1 \\ t & 0 \end{bmatrix}$.

4. Let $A = [a_{ij}]_{3,x}$ and $B = [b_{ij}]_{5,7}$.
   (a) Under what conditions does AB exist?
   (b) Under what conditions does BA exist?
5. Let $A = [a_{ij}]_{m,p}$. Under what conditions does A^n exist?
6. Express the system of equations

$$3x + 4y = 2$$
$$2x + 3y = 7$$

in matrix notation.
7. Express the system of equations

$$a_{11}x_1 + a_{12}x_2 + a_{13}x_3 = b_1$$
$$a_{21}x_1 + a_{22}x_2 + a_{23}x_3 = b_2$$
$$a_{31}x_1 + a_{32}x_2 + a_{33}x_3 = b_3$$

in matrix notation.
8. Generalize problems 6 and 7 to find a matrix representation for a system of equations with n unknowns $x_1, x_2, \ldots, x_n$ and m equations.
9. Let A, B, and C be three matrices. Prove that

$$A(B + C) = AB + AC \quad \text{and} \quad (A + B)C = AC + BC.$$

10. In problem 9, do we need any restrictions on the order of the matrices involved?
11. If A, B, and C are the matrices given in problem 2, show that $(AB)C = A(BC)$.
12. Show that in general $(A + B)^2 \neq A^2 + 2AB + B^2$.

13. Let $A = \begin{bmatrix} 0 & 3 \\ 0 & 0 \end{bmatrix}$, $B = \begin{bmatrix} 2 & 1 \\ 3 & 0 \end{bmatrix}$, and $C = \begin{bmatrix} 5 & 4 \\ 3 & 0 \end{bmatrix}$.

Prove that $AB = AC$. This example shows that in general $AB = AC$ does not imply $B = C$.
14. Prove by means of an example that for matrices, $AB = 0$ does not imply $A = 0$ or $B = 0$. If A and B are n-dimensional vectors, does $AB = 0$ imply $A = 0$ or $B = 0$?
15. Show that in general $A^2 - B^2 \neq (A - B)(A + B)$, where A and B are square matrices of the same size.

16. Find x_1 and x_2 if

(a) $[3, -1] \begin{bmatrix} x_1 & -x_2 \\ x_2 & -x_1 \end{bmatrix} = [7, 5]$.

(b) $[x_1, 5] \begin{bmatrix} 0 & 4 \\ -3 & 2 \end{bmatrix} = [x_2, -2]$.

(c) $\begin{bmatrix} 3 & 7 \\ 4 & -10 \end{bmatrix} \begin{bmatrix} x_1 \\ x_2 \end{bmatrix} = \begin{bmatrix} -2 \\ 7 \end{bmatrix}$.

17. Let $A = \begin{bmatrix} 6 & 24 & -35 \\ 5 & 2 & 0 \end{bmatrix}$, $B = \begin{bmatrix} 2 \\ 3 \\ -4 \end{bmatrix}$, and $C = [12, 5]$. In which order

should these matrices be multiplied to produce a number?

18. Prove Theorem 7.

★**19.** Find (if possible) a matrix B such that

$$\begin{bmatrix} 1 & -2 \\ -2 & 4 \end{bmatrix} B = \begin{bmatrix} 1 & 0 \\ 0 & 1 \end{bmatrix}.$$

★**20.** Prove Theorem 8 in the special case that A, B, and C are all 2×2 matrices.

21. Under the conditions of Theorem 8, find the number of rows and columns in the product $(AB)C$.

★**7.   the transpose of a matrix**

Although the concept of the transpose of a matrix is important in the general theory of matrices, it will be used in this text mainly in Section 3 of Chapter 11. The related idea of a symmetric matrix is used quite frequently in Chapter 12.

> **Definition 18.** The transpose of a matrix A is the matrix obtained by interchanging the rows and columns of A.

We shall denote the transpose of a matrix A by A^T. By definition, the element in the ith row and jth column of A is the element in the jth row and ith column of A^T. Consequently the ith row of A is the ith column of A^T, and if A is an $m \times n$ matrix, then A^T is an $n \times m$ matrix.

EXAMPLE 1. Find the transpose of

$$A = \begin{bmatrix} 1 & 2 & 3 & 4 \\ 2 & 3 & 4 & 5 \\ 3 & 4 & 5 & 6 \end{bmatrix}.$$

Solution

$$A^T = \begin{bmatrix} 1 & 2 & 3 \\ 2 & 3 & 4 \\ 3 & 4 & 5 \\ 4 & 5 & 6 \end{bmatrix}.$$

We list here as theorems some of the important properties of this new operation. The proofs are reserved for the next exercise.

Theorem 9. $(A^T)^T = A$.

Theorem 10. $(A + B)^T = A^T + B^T$.

Theorem 11. $(AB)^T = B^T A^T$.

We illustrate Theorem 11 when A and B are 2×2 matrices. Let

$$A = \begin{bmatrix} a_{11} & a_{12} \\ a_{21} & a_{22} \end{bmatrix} \quad \text{and} \quad B = \begin{bmatrix} b_{11} & b_{12} \\ b_{21} & b_{22} \end{bmatrix}.$$

Then

(28)
$$AB = \begin{bmatrix} a_{11}b_{11} + a_{12}b_{21} & a_{11}b_{12} + a_{12}b_{22} \\ a_{21}b_{11} + a_{22}b_{21} & a_{21}b_{12} + a_{22}b_{22} \end{bmatrix}$$

and

$$B^T = \begin{bmatrix} b_{11} & b_{21} \\ b_{12} & b_{22} \end{bmatrix} \quad \text{and} \quad A^T = \begin{bmatrix} a_{11} & a_{21} \\ a_{12} & a_{22} \end{bmatrix}.$$

Thus, by direct computation,

(29)
$$B^T A^T = \begin{bmatrix} b_{11}a_{11} + b_{21}a_{12} & b_{11}a_{21} + b_{21}a_{22} \\ b_{12}a_{11} + b_{22}a_{12} & b_{12}a_{21} + b_{22}a_{22} \end{bmatrix}.$$

Clearly $B^T A^T$ is the transpose of AB.

Definition 19. A matrix A is said to be symmetric if $A = A^T$.

Note that for a symmetric matrix $a_{ij} = a_{ji}$ and the matrix must be a square matrix.

Definition 20. A matrix A is said to be skew-symmetric if $A = -A^T$.

Notice that if A is skew-symmetric, then: (I) A is a square matrix, (II) $a_{ij} = -a_{ji}$, and (III) each entry on the main diagonal is zero.

Theorem 12. Every square matrix can be expressed as the sum of a symmetric matrix and a skew-symmetric matrix.

Proof. Let A be a square matrix. Set

$$S = \frac{A + A^T}{2} \quad \text{and} \quad S\star = \frac{A - A^T}{2}.$$

It is obvious that $S + S\star = A$. The proof of the theorem will be complete if we show that S is symmetric, i.e., $S = S^T$, and $S\star$ is skew-symmetric, i.e., $S\star = -S\star^T$. We ask the reader to verify this in the next exercise.

Exercise 7

1. Find the transpose of the following matrices:

   (a) $[2 \quad 5 \quad 7]$.   **(b)** $\begin{bmatrix} 2 & -1 & 5 \\ 3 & 5 & -2 \end{bmatrix}$.

2. Prove Theorems 9, 10, and 11.
3. Let A be a matrix and c a real number. Prove that $(cA)^T = cA^T$.
4. Is $(AB)^T = A^T B^T$? Why?
5. Prove that $(ABC)^T = C^T B^T A^T$.
6. Verify Theorem 11 for the special matrices

$$A = \begin{bmatrix} 2 & 3 & 4 \\ 1 & 0 & 2 \\ 0 & 1 & 2 \end{bmatrix} \quad \text{and} \quad B = \begin{bmatrix} 2 & -1 & 3 & 1 \\ 0 & 1 & -1 & 3 \\ 1 & 0 & 0 & 2 \end{bmatrix}.$$

7. Prove that AA^T is symmetric.
8. Check that AA^T is symmetric using the matrix A given in problem 6.
★9. Give an example to show that if A and B are symmetric, then AB need not be symmetric.

10. Prove that if A is skew-symmetric, then A^2 is symmetric.

11. Prove that $(A^n)^T = (A^T)^n$ for $n \geq 2$.

12. Let A be a skew-symmetric matrix. Find the entries of the main diagonal of $A^T A$.

13. Complete the proof of Theorem 12.

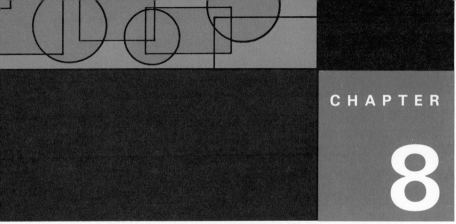

applications of matrices

The many applications of matrix theory are so deep and varied that one can touch only on a few. In this chapter we use matrix theory to help mechanize the labor of solving a system of linear equations. Curiously enough this particular application contributes to the theory of matrices by supplying a neat method for finding the inverse of a matrix. In this chapter we also examine the application of matrices to Markov processes.

Other applications will be made to special topics in Chapters 9, 10, 11, and 12.

1. systems of linear equations

An equation of the form

$$(1) \qquad\qquad a_1 x_1 + a_2 x_2 = b,$$

where a_1, a_2, and b are constants and a_1 and a_2 are not both zero, is called a *linear equation* in two unknowns, x_1 and x_2. For example,

$$(2) \qquad\qquad 2x_1 + 3x_2 = 6$$

is a linear equation in x_1 and x_2.

Now equation (2) is true for some values of the unknowns (for example, when $x_1 = 0$ and $x_2 = 2$) and false for some other values of the unknowns (for instance, when $x_1 = 1$ and $x_2 = 1$). The set of all vectors $X = \begin{bmatrix} x_1 \\ x_2 \end{bmatrix}$ for which (2) is true is called the *solution set* of (2).

We now consider the system of two linear equations in two unknowns:

$$
\begin{aligned}
a_{11}x_1 + a_{12}x_2 &= b_1 \\
a_{21}x_1 + a_{22}x_2 &= b_2.
\end{aligned}
$$

(3)

Let S_1 be the solution set of the first equation and S_2 the solution set of the second equation. We define the solution set of (3) to be the set of all vectors $X = \begin{bmatrix} x_1 \\ x_2 \end{bmatrix}$ that satisfy both equations of (3). It is obvious that the solution set of (3) is $S_1 \cap S_2$.

To generalize what we have said up to now, we let

$$
\begin{aligned}
a_{11}x_1 + a_{12}x_2 + \cdots + a_{1n}x_n &= b_1 \\
a_{21}x_1 + a_{22}x_2 + \cdots + a_{2n}x_n &= b_2 \\
&\ \vdots \\
a_{m1}x_1 + a_{m2}x_2 + \cdots + a_{mn}x_n &= b_m
\end{aligned}
$$

(4)

be a system of m linear equations in n unknowns. Here we understand that the a's and b's in (4) represent fixed numbers and we are to find a vector

$$
X = \begin{bmatrix} x_1 \\ x_2 \\ \vdots \\ x_n \end{bmatrix}
$$

that satisfies each of the equations in (4). If such a vector X exists, we call X a *solution* of (4) and say that (4) has a solution. The *solution set* of (4) is the collection of all such vectors.

Clearly the left-hand side of (4) can be represented by the $m \times n$ matrix of its coefficients and this is the matrix A given by

(5)
$$
A = \begin{bmatrix} a_{11} & a_{12} & \cdots & a_{1n} \\ a_{21} & a_{22} & \cdots & a_{2n} \\ \vdots & \vdots & & \vdots \\ a_{m1} & a_{m2} & \cdots & a_{mn} \end{bmatrix}.
$$

If we want to take into account the right-hand side of (4), we represent (4) by an $m \times (n + 1)$ *augmented matrix* (sometimes called the *detached coefficient tableau*):

$$A\star = \begin{bmatrix} a_{11} & a_{12} & \cdots & a_{1n} & b_1 \\ a_{21} & a_{22} & \cdots & a_{2n} & b_2 \\ \vdots & \vdots & & \vdots & \vdots \\ a_{m1} & a_{m2} & \cdots & a_{mn} & b_m \end{bmatrix}.$$

Suppose we represent $x_1, x_2, \ldots, x_n$ by a column vector

$$(6) \qquad\qquad X = \begin{bmatrix} x_1 \\ x_2 \\ \vdots \\ x_n \end{bmatrix}$$

and $b_1, b_2, \ldots, b_m$ by a column vector

$$(7) \qquad\qquad B = \begin{bmatrix} b_1 \\ b_2 \\ \vdots \\ b_m \end{bmatrix}.$$

Then by matrix multiplication, it is easily seen that the system of linear equations (4) can be written in the compact form

$$(8) \qquad\qquad AX = B.$$

Thus solving the system (4) is equivalent to finding all column vectors X such that the matrix equation (8) is satisfied.

To obtain some idea of the possible nature of the solution set of the system (4), let us consider first the simplest case of one equation in one unknown. In this situation the set (4) becomes the single equation

$$(9) \qquad\qquad a_{11}x_1 = b_1.$$

It is easy to see that if $a_{11} \neq 0$, then equation (9) has a unique solution given by $x_1 = b_1/a_{11}$. However, if $a_{11} = 0$ and $b_1 \neq 0$, then there is no number x_1 that satisfies (9). In this case we say that the system (9) *has no solution* or *a solution does not exist*. Now suppose $a_{11} = 0$ and $b_1 = 0$; then any number x_1 satisfies (9). In this situation, there are infinitely many solutions of (9), namely all the real numbers.

This very simple example illustrates that a system of equations may have: (a) a unique solution, (b) no solution, or (c) infinitely many solutions. So, for a given system of equations, we are faced with a twofold problem: first, to determine whether the system has a solution; and second, if the system has a solution, then we are interested in finding all the solutions.

We shall now study the *echelon method* of solving an arbitrary system of linear equations. We begin with

> **Definition 1.** Two systems of equations are called equivalent if they have identical solution sets.

The echelon method for determining the solution set of (4) is based upon producing an equivalent system in which the solution set is obvious. We use the following theorems to find equivalent systems.

> **Theorem 1.** If the position of any two equations of a given system is interchanged to form a new system, then the new system is equivalent to the original system.

Proof. By definition, the solution set of a system is the intersection of the solution sets of the equations of the system. Since the intersection of sets is commutative and associative, the theorem follows. ∎

> **Theorem 2.** If an equation of the system (4) is replaced by a nonzero multiple of itself plus a multiple of another equation of the system, then the new system is equivalent to the original system.

Proof. If in the system (4) we multiply the first equation by $c \neq 0$ and add to it d times the second equation, we obtain the new system

$$c(a_{11}x_1 + a_{12}x_2 + \cdots + a_{1n}x_n)$$
$$+ d(a_{21}x_1 + a_{22}x_2 + \cdots + a_{2n}x_n) = cb_1 + db_2$$

(10)
$$a_{21}x_1 + a_{22}x_2 + \cdots + a_{2n}x_n = b_2$$
$$\vdots \qquad \vdots \qquad \qquad \vdots \qquad \vdots$$
$$a_{m1}x_1 + a_{m2}x_2 + \cdots + a_{mn}x_n = b_m.$$

Clearly any vector X that satisfies the system (4) also satisfies the system (10). Conversely we can obtain the system (4) from the system (10) by the same type of operations with the equations of (10). Hence any solution of (10) is a solution of (4). A similar type of argument can be used for any pair of equations. ∎

We illustrate the method by considering some examples.

EXAMPLE 1. Solve the system of equations

$$x_1 - \ x_2 - \ x_3 = 1$$
(11) $\qquad 2x_1 - 3x_2 + \ x_3 = 10$
$$x_1 + \ x_2 - 2x_3 = 0.$$

Solution. We add to the second equation of this system -2 times the first equation, and obtain the new system:

$$x_1 - x_2 - \ x_3 = 1$$
(12) $\qquad\qquad - x_2 + 3x_3 = 8$
$$x_1 + x_2 - 2x_3 = 0,$$

which (by Theorem 2) is equivalent to (11). We now add -1 times the first equation of (12) to the third equation of (12) to get the new equivalent system:

$$x_1 - x_2 - \ x_3 = 1$$
(13) $\qquad\qquad - x_2 + 3x_3 = 8$
$$2x_2 - \ x_3 = -1.$$

We notice that x_1 has been eliminated from the second and third equations. We can now use the second equation of (13) to eliminate x_2 from each of the other two equations. To do this, we add -1 times the second equation to the first equation and then add 2 times the second equation to the third equation. We obtain

$$x_1 + 0 \ - 4x_3 = -7$$
(14) $\qquad\qquad - x_2 + 3x_3 = 8$
$$5x_3 = 15.$$

The last step is to divide both sides of the third equation of the system (14) by 5, obtaining the equation $x_3 = 3$, and to use this equation to eliminate x_3 from the first two equations of (14). This gives the system

$$x_1 + 0 + 0 = 5$$
(15) $\qquad\qquad x_2 + 0 = 1$
$$x_3 = 3,$$

which is equivalent to the system (11).

We thus have the solution $x_1 = 5$, $x_2 = 1$, and $x_3 = 3$. The student is advised to substitute these values of x_1, x_2, and x_3 in the equations of the system (11) to see that they are all satisfied.

We observe that we can avoid the labor of writing the symbols x_1, x_2, and x_3 if we can keep track of the coefficients that belong to each unknown. To do this we start with the augmented matrix for the system (11):

(11★)
$$A_1^\star = \begin{bmatrix} 1 & -1 & -1 & | & 1 \\ 2 & -3 & 1 & | & 10 \\ 1 & 1 & -2 & | & 0 \end{bmatrix}.$$

Note that the numbers in the first column are the coefficients of x_1, that those in the second column are the coefficients of x_2, and that those in the third column are the coefficients of x_3. The constants on the right-hand side of (11) are all found in the fourth column. The vertical line reminds us of the equal signs in the equations.

The operations used to reduce the system (11) to the equivalent system (15) can now be performed on the rows of the matrix (11★). For example if we add -2 times the first row to the second row, we get

(12★)
$$A_2^\star = \begin{bmatrix} 1 & -1 & -1 & | & 1 \\ 0 & -1 & 3 & | & 8 \\ 1 & 1 & -2 & | & 0 \end{bmatrix}.$$

Similarly, the matrices corresponding to (13), (14), and (15) are

(13★)
$$A_3^\star = \begin{bmatrix} 1 & -1 & -1 & | & 1 \\ 0 & -1 & 3 & | & 8 \\ 0 & 2 & -1 & | & -1 \end{bmatrix},$$

(14★)
$$A_4^\star = \begin{bmatrix} 1 & 0 & -4 & | & -7 \\ 0 & -1 & 3 & | & 8 \\ 0 & 0 & 5 & | & 15 \end{bmatrix},$$

and

(15★)
$$A_5^\star = \begin{bmatrix} 1 & 0 & 0 & | & 5 \\ 0 & 1 & 0 & | & 1 \\ 0 & 0 & 1 & | & 3 \end{bmatrix}.$$

In the augmented matrix (15★) we can easily restore the symbols x_1, x_2, and x_3 and deduce that $x_1 = 5$, $x_2 = 1$, and $x_3 = 3$.

Thus to solve the system of linear equations

$$AX = B,$$

we start with the augmented matrix

$$A\star = [A \mid B].$$

We then perform a suitable sequence of operations of the following type:

      (i) Interchange any two rows.
     (ii) Multiply any row by a nonzero number.
    (iii) Add to any row a multiple of another row.

If the sequence of operations is selected properly, the matrix $A\star$ can be converted to

$$A\star\star = [C \mid B\star],$$

where C has the following properties:

    (I) The first k rows are nonzero; the other rows are zero.
   (II) The first nonzero element in each nonzero row is 1, and it appears in a column to the right of the first nonzero element of any preceding row.
  (III) The first nonzero element in each nonzero row is the only nonzero element in its column.

If a matrix C has these three properties, then C is said to be in *reduced echelon form*. A typical matrix $A\star\star$ may have the form

$$\begin{bmatrix} 1 & 0 & 0 & c_{14} & c_{15} & 0 & 0 & b_1^\star \\ 0 & 1 & 0 & c_{24} & c_{25} & 0 & 0 & b_2^\star \\ 0 & 0 & 1 & c_{34} & c_{35} & 0 & 0 & b_3^\star \\ 0 & 0 & 0 & 0 & 0 & 1 & 0 & b_4^\star \\ 0 & 0 & 0 & 0 & 0 & 0 & 1 & b_5^\star \\ 0 & 0 & 0 & 0 & 0 & 0 & 0 & b_6^\star \end{bmatrix}.$$

EXAMPLE 2. Discuss the solution of the system of equations

(16)
$$\begin{aligned} x_1 + 5x_2 + 3x_3 &= 7 \\ 2x_1 + 11x_2 - 4x_3 &= 6. \end{aligned}$$

Solution. The augmented matrix for the system (16) is

(17)
$$A_1^\star = \begin{bmatrix} 1 & 5 & 3 & 7 \\ 2 & 11 & -4 & 6 \end{bmatrix}.$$

Adding -2 times the first row to the second row, we get

(18)
$$A_2^\star = \begin{bmatrix} 1 & 5 & 3 & 7 \\ 0 & 1 & -10 & -8 \end{bmatrix}.$$

Now adding -5 times the second row of $A_2^\star$ to the first row, we obtain

(19)
$$A_3^\star = \begin{bmatrix} 1 & 0 & 53 & 47 \\ 0 & 1 & -10 & -8 \end{bmatrix}.$$

Since $A_3^\star$ has the properties **(I)**, **(II)**, and **(III)**, no further transformations are required. The matrix $A_3^\star$ is the augmented matrix of

(20)
$$\begin{aligned} x_1 + 0 \ + 53x_3 &= 47 \\ x_2 - 10x_3 &= -8. \end{aligned}$$

We can assign an arbitrary value to x_3 and then compute x_1 and x_2 from (20) to get a solution of the system (16). For example, if we let $x_3 = 0$, we obtain the solution $x_1 = 47$, $x_2 = -8$, and $x_3 = 0$; and if we put $x_3 = 1$, we get the solution $x_1 = -6$, $x_2 = 2$, and $x_3 = 1$. We can express this dependence by rewriting (20) in the form

(21)
$$\begin{aligned} x_1 &= -53x_3 + 47 \\ x_2 &= 10x_3 - 8. \end{aligned}$$

Thus we see that the given system has infinitely many solutions, one for each value of x_3. This phenomenon is easily understood if we look at the geometric interpretation of the system of equations. The equation $x_1 + 5x_2 + 3x_3 = 7$ is the equation of a plane in three-dimensional space, and each vector X that is a solution of this equation represents a point on the plane. Similarly $2x_1 + 11x_2 - 4x_3 = 6$ is the equation of another plane. Hence X satisfies both equations if and only if it represents a point of the intersection of the two planes. Two planes either intersect in a line or are parallel. In the first case there are infinitely many solutions, and this is the case for the system (16). However, if the planes are parallel and distinct, there is no solution; but if the two planes coincide, there is again an infinity of solutions.

EXAMPLE 3.   Discuss the solution of the system of equations

(22)
$$\begin{aligned} x_1 + x_2 + \ x_3 &= 6 \\ 3x_1 - x_2 + 11x_3 &= 6 \\ 2x_1 + x_2 + \ 4x_3 &= 8. \end{aligned}$$

Solution.   The augmented matrix of this system is given by

(23)
$$A_1^\star = \begin{bmatrix} 1 & 1 & 1 & 6 \\ 3 & -1 & 11 & 6 \\ 2 & 1 & 4 & 8 \end{bmatrix}.$$

We perform the following row operations in the order indicated:

(1) Add -3 times row one to row two.
(2) Add -2 times row one to row three.
(3) Multiply row two by $-1/4$.
(4) Add -1 times row two to row one.
(5) Add row two to row three.

With a little labor, the energetic reader will find that these operations transform the matrix $A_1^\star$ into

$$(24) \qquad A_2^\star = \begin{bmatrix} 1 & 0 & 3 & 3 \\ 0 & 1 & -2 & 3 \\ 0 & 0 & 0 & -1 \end{bmatrix},$$

which is the augmented matrix of the system

$$(25) \qquad \begin{aligned} x_1 + 0\ + 3x_3 &= 3 \\ x_2 - 2x_3 &= 3 \\ 0 &= -1. \end{aligned}$$

The system (25) clearly has no solution, since the last equation is false for every choice of x_1, x_2, and x_3. Since the system (25) is equivalent to the system (22), we are forced to conclude that the system (22) has no solution.

The above examples illustrate that for a given system of equations, three possibilities can arise:

(a) The system has a unique solution (see Example 1).
(b) The system has more than one solution (see Example 2).
(c) The system has no solution (see Example 3).

EXAMPLE 4. Solve each of the following two systems of simultaneous equations:

$$(26) \qquad \begin{aligned} x_1 + 3x_2 - x_3 &= 5 \\ 2x_1 + 4x_2 + 6x_3 &= 4 \\ 2x_1 + 3x_2 + x_3 &= 3. \end{aligned}$$

$$(27) \qquad \begin{aligned} x_1 + 3x_2 - x_3 &= 2 \\ 2x_1 + 4x_2 + 6x_3 &= 4 \\ 2x_1 + 3x_2 + x_3 &= 6. \end{aligned}$$

Solution. We observe that these two systems differ only in their right-hand sides. Hence in the process of finding the solutions, the calculations on the left-hand side will be the same for both systems. Therefore we can solve both systems at the same time by considering the following matrix:

(28)
$$C = \begin{bmatrix} 1 & 3 & -1 & 5 & 2 \\ 2 & 4 & 6 & 4 & 4 \\ 2 & 3 & 1 & 3 & 6 \end{bmatrix}.$$

We perform (in the order indicated below) the following operations on this matrix.

(1) Add -2 times row one to row two and row three.
(2) Multiply row two by $-1/2$.
(3) Add -3 times row two to row one and add 3 times row two to row three.
(4) Multiply row three by $-1/9$.
(5) Add -11 times row three to row one, and 4 times row three to row two.

The energetic reader will find that these steps transform the matrix C into

$$D = \begin{bmatrix} 1 & 0 & 0 & -\dfrac{14}{9} & \dfrac{40}{9} \\ 0 & 1 & 0 & \dfrac{19}{9} & -\dfrac{8}{9} \\ 0 & 0 & 1 & -\dfrac{2}{9} & -\dfrac{2}{9} \end{bmatrix}.$$

For the system (26), the matrix D gives the solution

$$x_1 = -\frac{14}{9}, \qquad x_2 = \frac{19}{9}, \qquad x_3 = -\frac{2}{9}.$$

For the system (27), the matrix D gives the solution

$$x_1 = \frac{40}{9}, \qquad x_2 = -\frac{8}{9}, \qquad x_3 = -\frac{2}{9}.$$

Exercise 1

In problems 1 through 8, find all solutions of the given system of equations.

1. $x_1 + 2x_2 = 7$
   $3x_1 + 5x_2 = 11.$

2. $5x_1 + 7x_2 = 11$
   $13x_1 + 17x_2 = 19.$

3. $x_1 + 2x_2 + 3x_3 = 8$
$2x_1 + 5x_2 + 9x_3 = 16$
$3x_1 - 4x_2 - 5x_3 = 32.$

4. $x_1 + 2x_2 + 3x_3 = 4$
$3x_1 + 5x_2 + 7x_3 = 9$
$5x_1 + 8x_2 + 11x_3 = 14.$

5. $2x_1 + x_2 + 3x_3 = 6$
$3x_1 + 2x_2 + x_3 = 1$
$7x_1 + 3x_2 + 14x_3 = 27.$

6. $x_1 - x_2 + x_3 - x_4 = 0$
$2x_1 - x_2 + 3x_3 - 2x_4 = -1$
$3x_1 - 2x_2 - x_3 + 2x_4 = 4.$

7. $x_1 - 2x_2 + 3x_3 = 4$
$2x_1 + x_2 - 3x_3 = 5$
$-x_1 + 2x_2 + 2x_3 = 6$
$3x_1 - 3x_2 + 2x_3 = 7.$

8. $x_1 + x_2 + x_3 - x_4 = 2$
$x_1 + x_2 - x_3 + x_4 = 4$
$x_1 - x_2 + x_3 + x_4 = 6$
$-x_1 + x_2 + x_3 + x_4 = 8.$

9. Find x_1, x_2, and x_3 such that

$$[-10, 3, 17] = x_1[1, 2, 3] + x_2[4, 2, -3] + x_3[-5, 1, 2].$$

10. Find x_1, x_2, x_3, and x_4 such that

$$[5, 0, 3, -2] = x_1[1, 1, 1, 0] + x_2[0, 1, 1, 1] + x_3[1, 0, 1, 1] + x_4[1, 1, 0, 1].$$

11. For what value of the constant k does the following system have a unique solution:

$$5x_1 - x_2 + 2x_3 = 2$$
$$3x_1 + x_2 - 3x_3 = 7$$
$$x_1 + 5x_2 + x_3 = 5$$
$$kx_1 + x_2 - x_3 = 9.$$

12. If $X = [x_1, x_2, x_3]$ and A is the matrix

$$A = \begin{bmatrix} 0 & \dfrac{1}{3} & \dfrac{2}{3} \\ \dfrac{1}{2} & \dfrac{1}{2} & 0 \\ \dfrac{1}{3} & \dfrac{1}{3} & \dfrac{1}{3} \end{bmatrix},$$

find all solutions of the equation $XA = X$.

13. Give an example of a system of two equations in three unknowns that has no solution. What does this system represent geometrically?

14. Now I am 30 years older than my son. In another 5 years I will be 4 times as old as my son. Find our present ages.

15. The sum of the digits in a three-digit number is 18. The sum of the hundreds digit and the units digit is equal to the tens digit. If the hundreds digit and the units digit are interchanged, the number is decreased by 99. Find the number.

16. A man has invested a total of $20,000 in three different stocks that yield 4, 5, and 6 percent, respectively. The total income per year from the three stocks is $1060. The income from the 6 percent stock is twice his income from the 5 percent stock. Find the amount invested in each stock.

2. the inverse of a matrix

Definition 2. Nonsingular Matrix. A square matrix A is said to be nonsingular if it has an inverse; i.e., if there is a square matrix B such that

$$(29) \qquad\qquad AB = I,$$

where I is the identity matrix. When such a matrix B exists, we denote it by A^{-1}.

We remark that if A and B are square matrices of the same size and $AB = I$, then $BA = I$. Although the proof for this assertion is outside the scope of this book, we shall occasionally use this result in the following way. Suppose that we have two square matrices A and B and we wish to prove that B is an inverse of A. We can accomplish our end either by proving that $AB = I$ or by proving that $BA = I$, since either equation implies the other. Further, it is clear that if B is an inverse of A, then A is an inverse of B.

Theorem 3. If a matrix A is nonsingular, then the inverse is unique.

Proof. Suppose that A has two inverses, namely B and C. Then

$$AB = BA = I \qquad \text{and} \qquad AC = CA = I.$$

We see that

$$(30) \qquad B = BI = B(AC) = (BA)C = IC = C. \quad \blacksquare$$

Theorem 4. If A and B are nonsingular $n \times n$ matrices, then AB is nonsingular and the inverse of AB is given by

$$(AB)^{-1} = B^{-1}A^{-1}.$$

Proof. $(AB)(B^{-1}A^{-1}) = A(BB^{-1})A^{-1} = AIA^{-1} = AA^{-1} = I.$ Therefore $B^{-1}A^{-1}$ satisfies the requirement of the defining equation (29) for the unique inverse of AB.   ∎

Theorem 5. If A is nonsingular, then A^T is nonsingular and the inverse of A^T is given by

$$(A^T)^{-1} = (A^{-1})^T.$$

Proof. From the previous chapter, we know that

$$(AB)^T = B^T A^T.$$

Thus

$$A^T(A^{-1})^T = (A^{-1}A)^T = I^T = I.$$

Therefore, $(A^{-1})^T$ satisfies the requirement of the defining equation (29) for the unique inverse of A^T.   ∎

It is a fundamental fact of the real number system that every nonzero real number has a unique reciprocal (multiplicative inverse). However, as the following example illustrates, there are some nonzero matrices that do not have an inverse.

EXAMPLE 1. Prove that the matrix $A = \begin{bmatrix} 0 & 2 \\ 0 & 0 \end{bmatrix}$ does not have an inverse.

Solution. Suppose that $B = \begin{bmatrix} a & b \\ c & d \end{bmatrix}$ is an inverse of A. Then we must have

$$(31) \qquad \begin{bmatrix} 0 & 2 \\ 0 & 0 \end{bmatrix} \begin{bmatrix} a & b \\ c & d \end{bmatrix} = \begin{bmatrix} 1 & 0 \\ 0 & 1 \end{bmatrix}.$$

Multiplying the matrices on the left-hand side of (31), we obtain

$$(32) \qquad \begin{bmatrix} 2c & 2d \\ 0 & 0 \end{bmatrix} = \begin{bmatrix} 1 & 0 \\ 0 & 1 \end{bmatrix}.$$

By the definition of the equality of two matrices, we must have

$$0 = 1.$$

Since this is a contradiction, we infer that the nonzero matrix A does not have an inverse.

EXAMPLE 2.   Let

$$A = \begin{bmatrix} 23 & 30 \\ 13 & 17 \end{bmatrix} \quad \text{and} \quad B = \begin{bmatrix} 17 & -30 \\ -13 & 23 \end{bmatrix}.$$

Show that A and B are mutual inverses.

Solution.   Direct computation gives

$$AB = \begin{bmatrix} 23 & 30 \\ 13 & 17 \end{bmatrix} \begin{bmatrix} 17 & -30 \\ -13 & 23 \end{bmatrix} = \begin{bmatrix} 391 - 390 & 0 \\ 0 & 391 - 390 \end{bmatrix} = \begin{bmatrix} 1 & 0 \\ 0 & 1 \end{bmatrix}.$$

Consequently we can write either $A = B^{-1}$ or $B = A^{-1}$.

Given two matrices A and B, we can check (by direct computation) whether A and B are mutual inverses. We now present a method of calculating the inverse (if it exists) of a matrix.

EXAMPLE 3.   Find the inverse of

$$A = \begin{bmatrix} 2 & 3 \\ 1 & 2 \end{bmatrix}.$$

Solution.   We are looking for numbers a, b, c, and d such that

$$\begin{bmatrix} 2 & 3 \\ 1 & 2 \end{bmatrix} \begin{bmatrix} a & b \\ c & d \end{bmatrix} = \begin{bmatrix} 1 & 0 \\ 0 & 1 \end{bmatrix}$$

or

(33)
$$\begin{bmatrix} 2a + 3c & 2b + 3d \\ a + 2c & b + 2d \end{bmatrix} = \begin{bmatrix} 1 & 0 \\ 0 & 1 \end{bmatrix}.$$

For equation (33) to hold, we must have

(34)
$$2a + 3c = 1$$
$$a + 2c = 0$$

and

(35)
$$2b + 3d = 0$$
$$b + 2d = 1.$$

To solve the systems (34) and (35), we first observe that the coefficient matrix on the left-hand side of both these systems is the same and that these systems differ only in their right-hand side. So we may solve the two systems

at the same time by the method described in Example 4 of Section 1. We
thus consider the matrix

(36)
$$A_1^\star = \begin{bmatrix} 2 & 3 & 1 & 0 \\ 1 & 2 & 0 & 1 \end{bmatrix}.$$

We perform the following row operations on $A_1^\star$ in the order described below:

(1) Add -2 times the second row to the first row.
(2) Interchange the first and the second row.
(3) Add 2 times the second row to the first row.
(4) Multiply the second row by -1.

These operations carry $A_1^\star$ into

(37)
$$A_2^\star = \begin{bmatrix} 1 & 0 & 2 & -3 \\ 0 & 1 & -1 & 2 \end{bmatrix}.$$

From (37) we read off $a = 2$, $b = -3$, $c = -1$, and $d = 2$. Thus

$$A^{-1} = \begin{bmatrix} 2 & -3 \\ -1 & 2 \end{bmatrix}.$$

The student should check that $AA^{-1} = I$ by computing the product.

This procedure can be used to find the inverse of any square matrix A
(if the inverse exists). We summarize in

> **Theorem 6.** Let A be an arbitrary square matrix and let I be the
> identity matrix of the same size. If there is a sequence of row operations
> that transforms A into I, then this same sequence of row operations
> will transform $[A \mid I]$ into $[I \mid B]$, where B is the inverse of A. If it is
> impossible to transform A into I by row operations, then A does not
> have an inverse. This occurs if at any step in the process we obtain
> a matrix $[C \mid D]$ in which C has a row of zeros.

EXAMPLE 4. Find the inverse (if there is one) of the matrix.

(38)
$$A = \begin{bmatrix} 1 & 2 & 3 \\ 2 & 5 & 7 \\ 2 & 4 & 6 \end{bmatrix}.$$

Solution. We start with the augmented matrix

(39)
$$[A \mid I] = \begin{bmatrix} 1 & 2 & 3 & 1 & 0 & 0 \\ 2 & 5 & 7 & 0 & 1 & 0 \\ 2 & 4 & 6 & 0 & 0 & 1 \end{bmatrix}.$$

Adding -2 times the first row to the second row and the third row, we get

(40)
$$[A_1 \mid I] = \begin{bmatrix} 1 & 2 & 3 & 1 & 0 & 0 \\ 0 & 1 & 1 & -2 & 1 & 0 \\ 0 & 0 & 0 & -2 & 0 & 1 \end{bmatrix}.$$

Since the last row of A_1 in equation (40) has only zeros, it is clear that A_1 cannot be transformed into I by row operations. By Theorem 6, we conclude that A does not have an inverse.

EXAMPLE 5. Find the inverse of

(41)
$$A = \begin{bmatrix} 1 & 1 & 0 \\ 0 & 3 & 1 \\ 2 & 3 & 3 \end{bmatrix}.$$

Solution. As before, we start with the matrix

(42)
$$\begin{bmatrix} 1 & 1 & 0 & 1 & 0 & 0 \\ 0 & 3 & 1 & 0 & 1 & 0 \\ 2 & 3 & 3 & 0 & 0 & 1 \end{bmatrix}.$$

We perform the following sequence of row operations:

(1) Add -2 times the first row to the third row.
(2) Multiply the second row by $1/3$.
(3) Add -1 times the second row to the third row.
(4) Multiply the third row by $3/8$.
(5) Add $-1/3$ times the third row to the second row.
(6) Add -1 times the second row to the first row.

With these operations, (42) is transformed into

$$[I \mid B] = \begin{bmatrix} 1 & 0 & 0 & \frac{6}{8} & -\frac{3}{8} & \frac{1}{8} \\ 0 & 1 & 0 & \frac{2}{8} & \frac{3}{8} & -\frac{1}{8} \\ 0 & 0 & 1 & -\frac{6}{8} & -\frac{1}{8} & \frac{3}{8} \end{bmatrix}.$$

Hence the matrix

(43)
$$B = \frac{1}{8} \begin{bmatrix} 6 & -3 & 1 \\ 2 & 3 & -1 \\ -6 & -1 & 3 \end{bmatrix}$$

is the inverse of A. The doubtful reader should compute AB and BA using equations (41) and (43).

Exercise 2

In problems 1 through 8, find the inverse of the given matrix.

1. $\begin{bmatrix} 1 & 9 \\ 3 & 28 \end{bmatrix}$.

2. $\begin{bmatrix} 3 & 5 \\ 7 & 13 \end{bmatrix}$.

3. $\begin{bmatrix} 1 & 3 & -1 \\ 2 & 2 & -1 \\ 2 & 1 & -1 \end{bmatrix}$.

4. $\begin{bmatrix} 2 & 4 & 5 \\ 1 & 1 & 2 \\ 3 & 6 & 8 \end{bmatrix}$.

5. $\begin{bmatrix} 1 & 1 & 5 \\ 1 & 10 & -2 \\ 1 & 6 & 1 \end{bmatrix}$.

6. $\begin{bmatrix} 2 & 0 & -1 \\ -1 & 2 & -3 \\ 1 & -1 & 3 \end{bmatrix}$.

7. $\begin{bmatrix} 1 & 8 & 1 & 4 \\ 1 & 8 & 2 & 5 \\ 1 & 3 & 2 & 3 \\ 1 & 5 & 1 & 3 \end{bmatrix}$.

8. $\begin{bmatrix} 1 & 1 & -2 & 0 \\ 4 & 6 & -16 & -4 \\ 1 & 1 & -4 & 0 \\ -7 & -9 & 26 & 6 \end{bmatrix}$.

9. Suppose that $AC = AD$. Does it follow that $C = D$?

10. Prove that if A is a nonsingular matrix and $AC = AD$, then $C = D$.

11. Prove that if $AJ = A$ and A is nonsingular, then $J = I$.

★12. A matrix $A = [a_{ij}]$ is said to be a *diagonal matrix* if $i \neq j$ implies that $a_{ij} = 0$. Prove that a diagonal matrix is nonsingular if and only if $a_{ii} \neq 0$ for $i = 1, 2, \ldots, n$. Prove that if a diagonal matrix is non-singular, then the inverse matrix is a diagonal matrix.

13. We recall that for square matrices, the relation $AB = I$ implies $BA = I$ and conversely. By direct computation, check that $BA = I$ for (a) the matrices of Example 2 and (b) the matrices of Example 5.

3.   Markov chains

In this section we shall discuss the applications of matrix theory to Markov chains. We recall (from Chapter 6) that if there are four different states E_1, E_2, E_3, and E_4, then the Markov chain can be represented by the matrix

$$
\begin{array}{cccc}
& E_1 & E_2 & E_3 & E_4
\end{array}
$$

(44)
$$
P = \begin{array}{c} E_1 \\ E_2 \\ E_3 \\ E_4 \end{array}
\begin{bmatrix}
p_{11} & p_{12} & p_{13} & p_{14} \\
p_{21} & p_{22} & p_{23} & p_{24} \\
p_{31} & p_{32} & p_{33} & p_{34} \\
p_{41} & p_{42} & p_{43} & p_{44}
\end{bmatrix},
$$

where p_{ij} is the probability of changing the system from state E_i to state E_j. We also recall that the sum of the entries in each row is equal to 1.

Definition 3. A row vector $A = [a_1, a_2, \ldots, a_n]$ is called a probability vector if its components are nonnegative and their sum is 1.

Definition 4. A square matrix $P = [p_{ij}]$ is called a transition matrix if each of its rows is a probability vector.

We note that the matrix P given in (44) is a transition matrix.

Suppose at some arbitrary time the probability that the system is in state E_i is a_i. We can represent these probabilities by the probability vector $A = [a_1, a_2, \ldots, a_n]$, which is called the *probability distribution of the system* at that time. In particular, we let

$$
A^{(0)} = [a_1^{(0)}, a_2^{(0)}, \ldots, a_n^{(0)}]
$$

denote the *initial probability distribution* (the distribution when the process begins) and we let

$$
A^{(k)} = [a_1^{(k)}, a_2^{(k)}, \ldots, a_n^{(k)}]
$$

denote the probability distribution after the first k steps.

Theorem 7. If P is the transition matrix of a Markov chain process, then $A^{(k)} = A^{(0)} P^k$.

Proof. For simplicity, we shall prove our theorem for a two-state Markov chain. The procedure for an n-state Markov chain is similar.

Let E_1 and E_2 be the two states for the Markov chain and let the transition matrix for this Markov chain be

$$
\begin{array}{cc}
& E_1 \quad E_2
\end{array}
$$

(45)
$$
P = \begin{array}{c} E_1 \\ E_2 \end{array}
\begin{bmatrix}
p_{11} & p_{12} \\
p_{21} & p_{22}
\end{bmatrix}.
$$

A tree depicting the transition that occurs as a result of the kth step is shown in Fig. 1. Suppose that $A^{(k-1)} = [a_1^{(k-1)}, a_2^{(k-1)}]$ and $A^{(k)} = [a_1^{(k)}, a_2^{(k)}]$ are

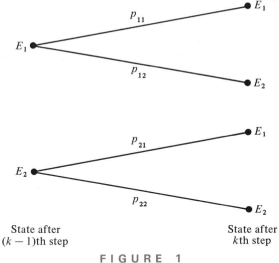

State after   State after
$(k-1)$th step kth step

FIGURE 1

the probability distributions after the $(k-1)$th step and the kth step, respectively. It is clear from the tree diagram that the probability for the process to be in state E_1 after the kth step is

(46) $$a_1^{(k)} = a_1^{(k-1)}p_{11} + a_2^{(k-1)}p_{21}.$$

Similarly, for E_2

(47) $$a_2^{(k)} = a_1^{(k-1)}p_{12} + a_2^{(k-1)}p_{22}.$$

From the definition of matrix multiplication, we see that equations (46) and (47) can be written in the form

(48) $$A^{(k)} = A^{(k-1)}P.$$

Setting $k = 1, 2, \ldots$ in (48), we obtain

(49)
$$A^{(1)} = A^{(0)}P$$
$$A^{(2)} = A^{(1)}P$$
$$\vdots \qquad \vdots$$
$$A^{(k)} = A^{(k-1)}P.$$

If we replace $A^{(1)}$ from the second equation of (49) by its value from the first equation, we see that

$$A^{(2)} = (A^{(0)}P)P = A^{(0)}P^2.$$

Similarly

$$A^{(3)} = A^{(2)}P = (A^{(0)}P^2)P = A^{(0)}(P^2P),$$

or

$$A^{(3)} = A^{(0)}P^3.$$

Continuing this process, we find that for any positive integer k

(50) $A^{(k)} = A^{(0)}P^k.$ ∎

EXAMPLE 1.   A man owns three cars, a Chevrolet, a Mercury, and a Dodge. Every day he drives one of his cars to his office. His driving habits are as follows: He never drives the same car to his office two days in a row. If he drives the Chevrolet one day, then the next day he is as likely to drive the Mercury as the Dodge. If he drives the Mercury one day, then the next day he drives the Chevrolet with probability 0.4. If he drives the Dodge one day, then the next day he drives the Chevrolet with probability 0.7. Given that he drives the Dodge on Monday, find the probability that he will drive the Mercury to his office on Wednesday of the same week.

Solution.   The states of the Markov chain are

{Chevrolet (C), Mercury (M), Dodge (D)}.

Since the man drives the Dodge to his office on Monday, we can denote the initial probability distribution by

$$A^{(0)} = [0, 0, 1].$$

The transition matrix for the Markov chain is

$$
\begin{array}{c}
\begin{array}{ccc} C & M & D \end{array} \\
P = \begin{array}{c} C \\ M \\ D \end{array}
\begin{bmatrix}
0 & 0.5 & 0.5 \\
0.4 & 0 & 0.6 \\
0.7 & 0.3 & 0
\end{bmatrix}.
\end{array}
$$

For this matrix P, we have

$$
P^2 = \begin{bmatrix}
0.55 & 0.15 & 0.30 \\
0.42 & 0.38 & 0.20 \\
0.12 & 0.35 & 0.53
\end{bmatrix}.
$$

By Theorem 7, the probability distribution $A^{(2)}$ after two steps is

$$A^{(2)} = A^{(0)}P^2 = [0, 0, 1]\begin{bmatrix} 0.55 & 0.15 & 0.30 \\ 0.42 & 0.38 & 0.20 \\ 0.12 & 0.35 & 0.53 \end{bmatrix}$$

$$= [0.12, 0.35, 0.53].$$

Thus, the probability that the man drives the Mercury to his office on Wednesday of the same week is 0.35.

We recall from Chapter 6 that if the system starts in state E_i, then $p_{ij}^{(k)}$ denotes the probability that the system will be in state E_j after k steps. These numbers can be arranged in a matrix, called the *k-step transition matrix,* which we denote by

(51)
$$P^{(k)} = \begin{bmatrix} p_{11}^{(k)} & p_{12}^{(k)} & \cdots & p_{1n}^{(k)} \\ p_{21}^{(k)} & p_{22}^{(k)} & \cdots & p_{2n}^{(k)} \\ \vdots & \vdots & & \vdots \\ p_{n1}^{(k)} & p_{n2}^{(k)} & \cdots & p_{nn}^{(k)} \end{bmatrix}.$$

Theorem 8. Let P be the transition matrix of a Markov chain process. Then the k-step transition matrix is equal to the kth power of P. In symbols, $P^{(k)} = P^k$.

Proof. We prove this theorem for a three-state process. The procedure is similar for an *n*-state process. Let

(52)
$$P = \begin{bmatrix} p_{11} & p_{12} & p_{13} \\ p_{21} & p_{22} & p_{23} \\ p_{31} & p_{32} & p_{33} \end{bmatrix}$$

be the transition matrix for the Markov chain process.

Let us choose $A^{(0)} = [1, 0, 0]$, which is equivalent to letting the process start in state E_1. By equation (50) we see that $A^{(k)}$ is the first row of the matrix P^k. Thus the entries in the first row of P^k give us the probabilities that the process will be in states E_1, E_2, and E_3 after k steps if it started in E_1. Similarly if we choose $A^{(0)} = [0, 1, 0]$, the second row of P^k gives us the probabilities that the process will be in states E_1, E_2, and E_3 after k steps if it started in E_2.

In the same way, the third row gives us these probabilities assuming that the process started in E_3. By comparing the notation of $P^{(k)}$, we see that the matrix $P^{(k)}$ is precisely the kth power of P. ∎

Definition 5. A nonzero vector $A = [a_1, a_2, \ldots, a_n]$ is called a fixed vector (also called a fixed point) of the square matrix B if $AB = A$.

Similarly the column vector C is called a fixed vector (also called a fixed point) of B if $BC = C$. However, we shall have no need for fixed column vectors in this text.

Since the zero vector $\mathbf{0}$ always satisfies $\mathbf{0}B = \mathbf{0}$, we do not include the zero vector $\mathbf{0}$ in the definition of a fixed point of a matrix.

In connection with Definition 5, we now prove a rather useful theorem.

Theorem 9. If A is a fixed point of a matrix B, then for any real number $a \neq 0$, the vector aA is also a fixed point of B.

Proof. Since A is a fixed point of B, we have by definition

(53)
$$AB = A.$$

Let $a \neq 0$ be a real number. Then

(54)
$$(aA)B = a(AB) = aA.$$

Equation (54) tells us that aA is a fixed point of B. ∎

EXAMPLE 2. Verify that $[8, 9]$ is a fixed point of the matrix

$$P = \begin{bmatrix} \dfrac{1}{4} & \dfrac{3}{4} \\ \dfrac{2}{3} & \dfrac{1}{3} \end{bmatrix},$$

and find a probability vector which is also a fixed point of P.

Solution. By direct computation, we see that

$$[8, 9] \begin{bmatrix} \dfrac{1}{4} & \dfrac{3}{4} \\ \dfrac{2}{3} & \dfrac{1}{3} \end{bmatrix} = [2 + 6, 6 + 3] = [8, 9].$$

Thus $[8, 9]$ is a fixed point of P. We recall that a vector A is a probability vector if its components are nonnegative and the sum of its components is 1. Consequently, the vector

$$\frac{1}{17}[8, 9] = \left[\frac{8}{17}, \frac{9}{17} \right]$$

is a probability vector. Since $[8, 9]$ is a fixed point of P, then by Theorem 9, the probability vector $[8/17, 9/17]$ is also a fixed point of P.

Definition 6. A transition matrix P is said to be regular if all the entries of P^m are positive for some positive integer m.

EXAMPLE 3. Show that the transition matrix

$$P = \begin{bmatrix} 0 & 1 \\ \frac{1}{4} & \frac{3}{4} \end{bmatrix}$$

is regular.

Solution. By direct computation, we see that

$$P^2 = \begin{bmatrix} 0 & 1 \\ \frac{1}{4} & \frac{3}{4} \end{bmatrix} \begin{bmatrix} 0 & 1 \\ \frac{1}{4} & \frac{3}{4} \end{bmatrix} = \begin{bmatrix} \frac{1}{4} & \frac{3}{4} \\ \frac{3}{16} & \frac{13}{16} \end{bmatrix}.$$

Since all the entries of P^2 are positive, the matrix P is regular.

EXAMPLE 4. Show that the matrix

$$P = \begin{bmatrix} 1 & 0 \\ 0 & 1 \end{bmatrix}$$

is not regular.

Solution. Here

$$P^2 = \begin{bmatrix} 1 & 0 \\ 0 & 1 \end{bmatrix}, \qquad P^3 = \begin{bmatrix} 1 & 0 \\ 0 & 1 \end{bmatrix}, \qquad \text{etc.}$$

In fact every power $P^m = P$ and hence P^m will have a zero in the first row. Therefore P is not a regular matrix.

The following important theorem gives the relationship between a regular transition matrix and its fixed point. The proof of this theorem is beyond the scope of this book and is therefore omitted.

Theorem 10. Let P be a regular transition matrix. Then

(1) P has a unique fixed probability vector A, and the components of A are all positive.

(2) $\lim_{k \to \infty} P^k = T$, where each row of T is the fixed vector A.

(3) If X is any probability vector, then $\lim_{k \to \infty} XP^k = A$.

This theorem shows that if the transition matrix P is regular, then in the long run (large n), the probability that any state E_j occurs is approximately equal to the component a_j of the unique fixed probability vector A of P. Thus we see that as the number of steps of the process increases, the effect of the initial probability distribution is diminished.

EXAMPLE 5. Illustrate Theorem 10 for the matrix

$$P = \begin{bmatrix} 0.4 & 0.6 \\ 0.2 & 0.8 \end{bmatrix}$$

Solution. We wish to find the probability vector $A = [x, y]$ such that $AP = A$. Since A is to be a probability vector, we must have $x + y = 1$, so that $y = 1 - x$. We thus wish to find x so that

(55) $$[x, 1 - x]\begin{bmatrix} 0.4 & 0.6 \\ 0.2 & 0.8 \end{bmatrix} = [x, 1 - x].$$

Multiplying the left-hand side of the matrix equation (55), we get

(56) $$[0.2x + 0.2, -0.2x + 0.8] = [x, 1 - x].$$

From (56) we obtain the pair of linear equations

(57) $$\begin{aligned} 0.2x + 0.2 &= x \\ -0.2x + 0.8 &= 1 - x. \end{aligned}$$

This gives us $x = 0.25$. Thus $A = [0.25, 1 - 0.25] = [0.25, 0.75]$ is the unique fixed probability vector of P.

It is instructive to compute some of the powers of P in order to see that the sequence $P, P^2, P^3, \ldots$ does indeed approach the matrix

$$T = \begin{bmatrix} 0.25 & 0.75 \\ 0.25 & 0.75 \end{bmatrix}.$$

A laborious computation gives

$$P^2 = \begin{bmatrix} 0.28 & 0.72 \\ 0.24 & 0.76 \end{bmatrix}, \qquad P^3 = \begin{bmatrix} 0.256 & 0.744 \\ 0.248 & 0.752 \end{bmatrix},$$

$$P^4 = \begin{bmatrix} 0.2512 & 0.7488 \\ 0.2496 & 0.7504 \end{bmatrix}, \qquad P^5 = \begin{bmatrix} 0.25024 & 0.74976 \\ 0.24992 & 0.75008 \end{bmatrix}.$$

EXAMPLE 6. The Asma Corporation (*A*), Bronkial Brothers (*B*), and the
Coufmore Company (*C*) simultaneously decide to introduce a new cigarette
at a time when each company has one-third of the market. During the year,
the following occurs:

(1) *A* retains 40 percent of its customers and loses 30 percent to *B* and
30 percent to *C*.

(2) *B* retains 30 percent of its customers and loses 60 percent to *A* and
10 percent to *C*.

(3) *C* retains 30 percent of its customers and loses 60 percent to *A* and
10 percent to *B*.

Assuming that this trend continues, what share of the market will each
company have at the end of two years? What share of the market will each
company have in the long run?

Solution. The transition matrix for this process is

$$
\begin{array}{c} \\ A \\ P = B \\ C \end{array}
\begin{array}{ccc} A & B & C \\ \begin{bmatrix} 0.4 & 0.3 & 0.3 \\ 0.6 & 0.3 & 0.1 \\ 0.6 & 0.1 & 0.3 \end{bmatrix} \end{array}.
$$

The initial distribution is

$$
X = \left[\frac{1}{3}, \frac{1}{3}, \frac{1}{3} \right].
$$

The share of the market that each company will have at the end of two
years is given by

$$
XP^2 = \left[\frac{1}{3}, \frac{1}{3}, \frac{1}{3} \right] \begin{bmatrix} 0.4 & 0.3 & 0.3 \\ 0.6 & 0.3 & 0.1 \\ 0.6 & 0.1 & 0.3 \end{bmatrix}^2
$$

$$
= \left[\frac{1}{3}, \frac{1}{3}, \frac{1}{3} \right] \begin{bmatrix} 0.52 & 0.24 & 0.24 \\ 0.48 & 0.28 & 0.24 \\ 0.48 & 0.24 & 0.28 \end{bmatrix}
$$

$$
= \left[\frac{37}{75}, \frac{19}{75}, \frac{19}{75} \right].
$$

Thus at the end of two years, the Asma Corporation will have 37/75
(approximately 49.33 percent) of the market, while Bronkial Brothers and
the Coufmore Company will each have 19/75 (approximately 25.33 percent)
of the market.

To find each company's share of the market in the long run, we apply Theorem 10. According to Theorem 10, we must find the probability vector $A = [x_1, x_2, x_3]$ such that $AP = A$. This condition,

(58)
$$[x_1, x_2, x_3] \begin{bmatrix} 0.4 & 0.3 & 0.3 \\ 0.6 & 0.3 & 0.1 \\ 0.6 & 0.1 & 0.3 \end{bmatrix} = [x_1, x_2, x_3],$$

together with the fact that A is a probability vector, leads to the system of equations:

$$\begin{aligned}
x_1 + x_2 + x_3 &= 1 \\
0.4x_1 + 0.6x_2 + 0.6x_3 &= x_1 \\
0.3x_1 + 0.3x_2 + 0.1x_3 &= x_2 \\
0.3x_1 + 0.1x_2 + 0.3x_3 &= x_3.
\end{aligned}$$

The unique solution for this system of equations is $x_1 = 1/2$, $x_2 = 1/4$, and $x_3 = 1/4$. Therefore, the vector $A = [1/2, 1/4, 1/4]$ is the unique fixed probability vector of the matrix P. Hence in the long run the Asma Corporation will capture approximately 50 percent of the market, while Bronkial Brothers and the Coufmore Company will each capture approximately 25 percent of the market.

Exercise 3

1. Which of the following are probability vectors?

(a) $\left[\dfrac{1}{2}, 0, \dfrac{1}{2}\right]$.

(b) $[1, 2, 3, 4]$.

(c) $\left[\dfrac{1}{2}, \dfrac{1}{3}\right]$.

(d) $\left[\dfrac{1}{3}, \dfrac{1}{2}, \dfrac{5}{6}, -\dfrac{2}{3}\right]$.

(e) $\left[1, \dfrac{1}{2}, \dfrac{1}{3}, \dfrac{1}{4}\right]$.

(f) $\left[2, \dfrac{1}{2}, \dfrac{3}{4}\right]$.

2. In problem 1, convert (if possible) each nonprobability vector into a probability vector by multiplying it by a suitable real number.

3. Which of the following are regular transition matrices?

(a) $\begin{bmatrix} 1 & 0 \\ \dfrac{1}{2} & \dfrac{1}{2} \end{bmatrix}$.

(b) $\begin{bmatrix} 0 & 1 \\ \dfrac{1}{2} & \dfrac{1}{2} \end{bmatrix}$.

(c) $\begin{bmatrix} 1 & 2 \\ 3 & 1 \end{bmatrix}.$

(d) $\begin{bmatrix} \dfrac{1}{4} & \dfrac{3}{4} \\ \dfrac{1}{2} & \dfrac{1}{3} \end{bmatrix}.$

(e) $\begin{bmatrix} 0 & 1 & 0 \\ 0 & 0 & 1 \\ \dfrac{1}{3} & \dfrac{2}{3} & 0 \end{bmatrix}.$

(f) $\begin{bmatrix} 0 & 0 & 1 \\ \dfrac{1}{4} & \dfrac{3}{4} & 0 \\ 0 & \dfrac{1}{2} & \dfrac{1}{2} \end{bmatrix}.$

(g) $\begin{bmatrix} 0 & \dfrac{1}{2} & \dfrac{1}{2} \\ \dfrac{1}{4} & \dfrac{1}{2} & \dfrac{1}{4} \\ \dfrac{1}{3} & \dfrac{1}{3} & \dfrac{1}{3} \end{bmatrix}.$

(h) $\begin{bmatrix} 0 & \dfrac{1}{2} & \dfrac{1}{2} \\ \dfrac{1}{3} & \dfrac{1}{2} & \dfrac{1}{3} \\ \dfrac{1}{3} & \dfrac{3}{4} & 0 \end{bmatrix}.$

4. Find the fixed probability vector for the regular transition matrices of problem 3.

5. Let $P = [p_{ij}]$ be a 3×3 transition matrix and let $A = [a_1, a_2, a_3]$ be a probability vector. Show that AP is also a probability vector.

6. State and prove a generalization of problem 5.

★7. If A and B are transition matrices, prove that AB is also a transition matrix.

8. Let $P = \begin{bmatrix} 1-a & a \\ b & 1-b \end{bmatrix}$ be a regular transition matrix. Find the fixed probability vector of P.

9. Mr. Smart has the following dating habits. If he dates on one weekend, he is 60 percent sure not to date the next weekend. On the other hand, the probability that he does not date on two successive weekends is 0.2. In the long run, how often does he date?

10. Suppose that of the sons of Republicans, 80 percent vote Republican and the rest vote Democratic; of the sons of Democrats, 60 percent vote Democratic, 20 percent vote Republican, and 20 percent vote Socialist; and of the sons of Socialists, 50 percent vote Socialist, 40 percent vote Democratic, and 10 percent vote Republican.

   (a) What is the probability that the grandson of a Democrat will vote Socialist?

   (b) What is the long range fraction of the population expected in each party?

11. A salesman has cities A, B, C, and D in his territory. He never stays in the same city more than a week. If he is in city A, he is equally likely to go to either of the other three cities the next week. If he is

in B, then the next week he is likely to be in cities A, C, and D with probability 1/2, 1/4, and 1/4, respectively. If he is in C, then the next week he will not go to B but is equally likely to be in A or D. If he is in city D, then the next week he will not be in A but will be in B or C with probability 2/3 and 1/3, respectively.

    (a) Set up this process as a Markov chain.

    (b) If he is in A this week, what is the probability that he will be in C the week after next.

    (c) In the long run, how often does he stay in each city?

12. Assume that the probability that a blonde mother will have a blonde, a brunette, or a red-headed daughter is 0.6, 0.2, and 0.2, respectively; the probability that a brunette mother will have a blonde, a brunette, or a red-headed daughter is 0.1, 0.7, and 0.2, respectively; and the probability that a red-headed mother will have a blonde, a brunette, or a red-headed daughter is 0.4, 0.2, and 0.4, respectively. What is the probability that a blonde woman will be the great-grandmother of a brunette. Assume that the population of women now is 50 percent brunettes, 30 percent blondes, and the rest redheads; what will be the distribution of women with respect to natural hair color

    (a) After three generations?

    (b) After a large number of generations?

4. absorbing Markov chains

The reader will recall from Chapter 6 that some Markov chain experiments contain states which it is impossible to leave. The random walk problem and the gambler's ruin discussed in Chapter 6 are experiments of this nature. We formally define such states in

> **Definition 7.** A state E_i of a Markov chain is called absorbing if the process remains in this state once it enters there.

From this definition it is easy to see that a state E_i is absorbing if and only if the ith row of the transition matrix P has a 1 on the main diagonal and zeros elsewhere.

> **Definition 8.** A Markov chain is said to be absorbing if it has at least one absorbing state and it is possible to go from every nonabsorbing state to an absorbing state.

EXAMPLE 1. Mr. Slipstik has $2 and is playing pool with the wealthy Prof. Bankball at the club. Mr. Slipstik bets $1 per game. His probability of winning a game is 2/5, and he decides to play until either he has a total of \$4 or loses all his money. Write the transition matrix for this absorbing Markov chain.

Solution. The states of this Markov chain are \$0, \$1, \$2, \$3, and \$4. It is obvious that the absorbing states are \$0 and \$4, while the other states are nonabsorbing. The transition matrix is given by

$$
P =
\begin{array}{c}
 \\
0 \\
1 \\
2 \\
3 \\
4
\end{array}
\begin{array}{ccccc}
0 & 1 & 2 & 3 & 4
\end{array}
\begin{bmatrix}
1 & 0 & 0 & 0 & 0 \\
\frac{3}{5} & 0 & \frac{2}{5} & 0 & 0 \\
0 & \frac{3}{5} & 0 & \frac{2}{5} & 0 \\
0 & 0 & \frac{3}{5} & 0 & \frac{2}{5} \\
0 & 0 & 0 & 0 & 1
\end{bmatrix}.
$$

Let us now consider a general five-state absorbing Markov chain, such that states E_0 and E_4 are absorbing and states E_1, E_2, and E_3 are nonabsorbing. Suppose that the transition matrix for this process is

(59)
$$
\begin{array}{c}
 \\
E_0 \\
E_1 \\
E_2 \\
E_3 \\
E_4
\end{array}
\begin{array}{ccccc}
E_0 & E_1 & E_2 & E_3 & E_4
\end{array}
\begin{bmatrix}
1 & 0 & 0 & 0 & 0 \\
p_{10} & p_{11} & p_{12} & p_{13} & p_{14} \\
p_{20} & p_{21} & p_{22} & p_{23} & p_{24} \\
p_{30} & p_{31} & p_{32} & p_{33} & p_{34} \\
0 & 0 & 0 & 0 & 1
\end{bmatrix}.
$$

By interchanging necessary rows and columns of this matrix, we can rewrite this transition matrix so that absorbing states E_0 and E_4 appear first. Thus, the above transition matrix can be written as

(60)
$$
P =
\begin{array}{c}
 \\
E_0 \\
E_4 \\
E_1 \\
E_2 \\
E_3
\end{array}
\begin{array}{ccccc}
E_0 & E_4 & E_1 & E_2 & E_3
\end{array}
\begin{bmatrix}
1 & 0 & 0 & 0 & 0 \\
0 & 1 & 0 & 0 & 0 \\
p_{10} & p_{14} & p_{11} & p_{12} & p_{13} \\
p_{20} & p_{24} & p_{21} & p_{22} & p_{23} \\
p_{30} & p_{34} & p_{31} & p_{32} & p_{33}
\end{bmatrix}
$$

$$
= \begin{bmatrix} I & 0 \\ \hline R & S \end{bmatrix},
$$

where I is a 2×2 identity matrix, $\mathbf{0}$ is a 2×3 zero matrix, R is a 3×2 matrix, and S is a 3×3 matrix.

If we have an $n \times n$ transition matrix with k absorbing states, we can, by a suitable interchange of rows and columns, write it in the form

(61)
$$P = \left[\begin{array}{c|c} I & \mathbf{0} \\ \hline R & S \end{array} \right],$$

where I is a $k \times k$ identity matrix, $\mathbf{0}$ is the $k \times (n - k)$ zero matrix, R is an $(n - k) \times k$ matrix, and S is a square matrix of order $n - k$.

The form (61) is usually called the *canonical (standard) form*.

For an absorbing Markov chain we are interested in the following three important questions:

  I. What is the expected (mean or average) number of times the process is in nonabsorbing state E_j if it starts in nonabsorbing state E_i?
  II. Starting from a given nonabsorbing state, what is the expected number of steps before absorption?
  III. What is the probability that the process will be absorbed in a particular absorbing state E_j if it starts in a nonabsorbing state E_i?

We begin by answering question **I** in

> **Theorem 11.** Let the canonical form for a given absorbing Markov chain with k absorbing states be given by (61). Then the entries of $N = [I - S]^{-1}$ give the expected number of times the process will be in each nonabsorbing state for each possible nonabsorbing starting state, where I is an identity matrix of order $n - k$.

Proof. We shall prove our theorem for the five-state absorbing Markov chain with two absorbing states. The proof for the general case is similar.

Let $\{E_0, E_1, E_2, E_3, E_4\}$ be the given five-state absorbing Markov chain having E_0 and E_4 as its absorbing states and E_1, E_2, and E_3 as its nonabsorbing states. Let the canonical form for the transition matrix of this chain be given by (60).

Let us denote by n_{ij}, the expected number of times the process is in nonabsorbing state E_j if it starts in nonabsorbing state E_i. Suppose the process is in state E_1 at the beginning. Then the number

$$x = p_{11}n_{11} + p_{12}n_{21} + p_{13}n_{31}$$

gives us the expected number of times the process comes back to state E_1

after it leaves state E_1. But since the process is already in state E_1, the
number

$$1 + x = 1 + p_{11}n_{11} + p_{12}n_{21} + p_{13}n_{31}$$

gives us the total expected number of times the process is in state E_1. Hence

$$n_{11} = 1 + p_{11}n_{11} + p_{12}n_{21} + p_{13}n_{31}.$$

Similarly

$$n_{12} = p_{11}n_{12} + p_{12}n_{22} + p_{13}n_{32}.$$

In general for $i = 1, 2, 3$ and $j = 1, 2, 3$, we have

(62) $$n_{ij} = e_{ij} + p_{i1}n_{1j} + p_{i2}n_{2j} + p_{i3}n_{3j},$$

where $e_{ij} = 1$ if $i = j$ and $e_{ij} = 0$ if $i \neq j$.

Equation set (62) may be written in matrix form:

(63) $$N = I + SN,$$

where $N = [n_{ij}]$, I is the 3×3 identity matrix, and S is given in (60).
From (63) we have $N - SN = I$, or

(64) $$[I - S]N = I,$$

and if we assume that the inverse of $[I - S]$ exists, then

(65) $$N = [I - S]^{-1}. \quad \blacksquare$$

The fact that $[I - S]$ is nonsingular can be proved, but the proof lies beyond the scope of this book. The matrix $N = [I - S]^{-1}$ is called the *fundamental matrix* for an absorbing chain.

We observe that the sum of the elements in a particular row of N gives the answer to our question **II**. For example,

$$n_{21} + n_{22} + n_{23}$$

gives the expected number of steps before absorption, if the process begins in state E_2. In general, if the process begins in state E_i, then the sum of the elements in the ith row of $[I - S]^{-1}$,

$$n_{i1} + n_{i2} + \cdots + n_{i(n-k)},$$

is the expected number of steps before absorption.

EXAMPLE 2. Compute the fundamental matrix of the absorbing Markov chain of Example 1 and interpret your result.

Solution. The transition matrix of Example 1 in canonical form is

$$
\begin{array}{c}
 \\
0 \\
4 \\
1 \\
2 \\
3
\end{array}
\begin{array}{ccccc}
0 & 4 & 1 & 2 & 3
\end{array}
\left[
\begin{array}{ccccc}
1 & 0 & 0 & 0 & 0 \\
0 & 1 & 0 & 0 & 0 \\
\frac{3}{5} & 0 & 0 & \frac{2}{5} & 0 \\
0 & 0 & \frac{3}{5} & 0 & \frac{2}{5} \\
0 & \frac{2}{5} & 0 & \frac{3}{5} & 0
\end{array}
\right].
$$

From this we have

$$
S = \begin{bmatrix}
0 & \frac{2}{5} & 0 \\
\frac{3}{5} & 0 & \frac{2}{5} \\
0 & \frac{3}{5} & 0
\end{bmatrix}
$$

and

$$
I - S = \begin{bmatrix}
1 & -\frac{2}{5} & 0 \\
-\frac{3}{5} & 1 & -\frac{2}{5} \\
0 & -\frac{3}{5} & 1
\end{bmatrix}.
$$

Computing the inverse of this matrix, we get

$$
N = [I - S]^{-1} = \begin{bmatrix}
\frac{19}{13} & \frac{10}{13} & \frac{4}{13} \\
\frac{15}{13} & \frac{25}{13} & \frac{10}{13} \\
\frac{9}{13} & \frac{15}{13} & \frac{19}{13}
\end{bmatrix}.
$$

The entries of N give the expected number of times Mr. Slipstik would have the corresponding amount before absorption if he starts with \$1, \$2, and \$3, respectively. For example, the entry 15/13 in the second row and first column is the mean number of times he would have \$1 if he starts with $2. Further, if he starts with $2, the sum of the entries in the second row of N, namely

$$\frac{15}{13} + \frac{25}{13} + \frac{10}{13} = \frac{50}{13},$$

gives us the expected number of games (between three and four) before absorption.

We now answer our question **III** in

> **Theorem 12.** Let the canonical form for a given absorbing Markov chain with k absorbing states be given by (61). Then the entry c_{ij} of the matrix $C = NR$ gives the probability that the process will be absorbed in absorbing state E_j if it starts in a nonabsorbing state E_i, where $N = [I - S]^{-1}$.

Proof. We shall again prove our theorem for the five-state absorbing Markov chain described earlier. The proof for the general case is similar.

Let E_i be a nonabsorbing state and E_j be an absorbing state. Suppose the process is in state E_1 and we wish to compute the probability c_{14} for the process to be absorbed in state E_4. The process can either go directly

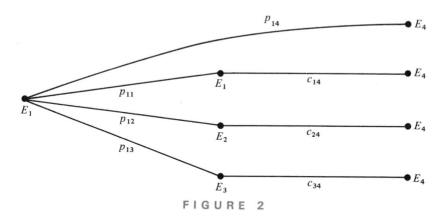

FIGURE 2

to E_4 with probability p_{14} or it can go to a nonabsorbing state E_k and from there eventually reach E_4. From the tree diagram in Fig. 2 it is clear that

$$c_{14} = p_{14} + p_{11}c_{14} + p_{12}c_{24} + p_{13}c_{34}.$$

In general, for $i = 1, 2, 3$ and $j = 0, 4$, we have

(66) $$c_{ij} = p_{ij} + p_{i1}c_{1j} + p_{i2}c_{2j} + p_{i3}c_{3j}.$$

Writing these equations in matrix form, we obtain

$$\begin{bmatrix} c_{10} & c_{14} \\ c_{20} & c_{24} \\ c_{30} & c_{34} \end{bmatrix} = \begin{bmatrix} p_{10} & p_{14} \\ p_{20} & p_{24} \\ p_{30} & p_{34} \end{bmatrix} + \begin{bmatrix} p_{11} & p_{12} & p_{13} \\ p_{21} & p_{22} & p_{23} \\ p_{31} & p_{32} & p_{33} \end{bmatrix} \begin{bmatrix} c_{10} & c_{14} \\ c_{20} & c_{24} \\ c_{30} & c_{34} \end{bmatrix}$$

or

(67) $C = R + SC.$

From (67) we have $C - SC = R$ or

(68) $[I - S]C = R,$

and hence

(69) $C = [I - S]^{-1}R = NR.$ ∎

EXAMPLE 3.   In Example 1, find **(a)** the probability that Mr. Slipstik will lose all his money and **(b)** the probability that he will have a total of \$4.

Solution.   In Example 1, we have

$$R = \begin{bmatrix} \dfrac{3}{5} & 0 \\ 0 & 0 \\ 0 & \dfrac{2}{5} \end{bmatrix},$$

and, as computed in Example 2,

$$N = \begin{bmatrix} \dfrac{19}{13} & \dfrac{10}{13} & \dfrac{4}{13} \\ \dfrac{15}{13} & \dfrac{25}{13} & \dfrac{10}{13} \\ \dfrac{9}{13} & \dfrac{15}{13} & \dfrac{19}{13} \end{bmatrix}.$$

Now

$$C = NR = \begin{matrix} & \begin{matrix} 0 & \quad 4 \end{matrix} \\ \begin{matrix} 1 \\ 2 \\ 3 \end{matrix} & \begin{bmatrix} \dfrac{57}{65} & \dfrac{8}{65} \\ \dfrac{45}{65} & \dfrac{20}{65} \\ \dfrac{27}{65} & \dfrac{38}{65} \end{bmatrix} \end{matrix}.$$

Thus starting with \$2, the probability that Mr. Slipstik will lose all his money is given by the entry in the second row and first column of the matrix C; namely the probability that he will lose all his money is $45/65 = 9/13$. From the entry in the second row and second column of C, we note that the probability that he will have a total of \$4 is $20/65 = 4/13$.

Exercise 4

1. By inspection, state which of the following are matrices for absorbing Markov chains. For each absorbing chain, give the number of absorbing states.

(a) $\begin{bmatrix} 0 & 1 \\ 1 & 1 \\ \frac{1}{2} & \frac{1}{2} \end{bmatrix}$.

(b) $\begin{bmatrix} 1 & 0 & 0 \\ \frac{1}{3} & \frac{1}{3} & \frac{1}{2} \\ 0 & 0 & 1 \end{bmatrix}$.

(c) $\begin{bmatrix} \frac{1}{4} & 0 & \frac{3}{4} & 0 \\ 0 & 1 & 0 & 0 \\ 0 & 0 & 1 & 0 \\ \frac{1}{2} & 0 & 0 & \frac{1}{2} \end{bmatrix}$.

(d) $\begin{bmatrix} \frac{2}{3} & \frac{1}{3} & 0 & 0 \\ 0 & 0 & 1 & 0 \\ \frac{1}{4} & \frac{1}{4} & \frac{1}{4} & \frac{1}{4} \\ 0 & 0 & 0 & 1 \end{bmatrix}$.

(e) $\begin{bmatrix} 0 & \frac{1}{2} & 0 & \frac{1}{4} & \frac{1}{4} \\ 0 & 1 & 0 & 0 & 0 \\ 0 & 0 & 1 & 0 & 0 \\ \frac{2}{3} & 0 & \frac{1}{3} & 0 & 0 \\ 0 & 0 & \frac{1}{2} & \frac{1}{2} & 0 \end{bmatrix}$.

(f) $\begin{bmatrix} 1 & 0 & 0 & 0 & 0 \\ \frac{1}{5} & \frac{1}{5} & \frac{1}{5} & \frac{1}{5} & \frac{1}{5} \\ 0 & 0 & 1 & 0 & 0 \\ 0 & \frac{1}{2} & 0 & \frac{1}{2} & 0 \\ 0 & 0 & 0 & 0 & 1 \end{bmatrix}$.

2. Write the absorbing matrices of problem 1 in canonical form.
3. For the absorbing matrices of problem 1, compute N and C.
4. For the random walk problem of Chapter 6, Section 2, Example 3, let $N = 5, p = 1/3$, and $q = 2/3$. Write a transition matrix for this problem. Compute N and C.
5. In Example 1 of this section suppose Mr. Slipstik is evenly matched with the Professor; i.e., his probability of winning is $1/2$. Find the probability that he will lose all his money. Find the expected number of games.
6. Repeat problem 5 above if Mr. Slipstik wins with probability $1/3$ and loses with probability $2/3$.

7. In problem 6, suppose that Mr. Slipstik has $6 and bets $2 on each game. He continues to play until either he has a total of $12 or is left with $2 for cab fare. Find the probability that he will have a total of $12.

8. The Right Co. and the Costless Co. both manufacture and sell very good gidgets in Slikville. Each year the market in Slikville will absorb exactly 5000 gidgets. The Right Co. currently gets 40 percent of the sales, while the Costless Co. handles the rest of the business. The Right Co. has found an untested advertising gimmick. With this advertising gimmick, the Right Co. assumes that each year it has probability 3/5 of increasing its sale of gidgets by 1000 and that it has probability 2/5 of decreasing its sale of gidgets by 1000.

   (a) In how many (expected) years will the Costless Co. either go broke or control all the gidget business in Slikville?

   (b) Answer part (a) for the Right Co.

   (c) Find the probability that each company will lose all its customers and go broke.

9. Prove Theorem 11 for an n-state absorbing Markov chain having k absorbing states, where $1 < k < n$.

10. Prove Theorem 12 for an n-state absorbing Markov chain having k absorbing states, where $1 < k < n$.

★ 11. Three tanks I, J, and K are joined in battle, with J and K united against I. The tanks fire together, and I always fires at J unless J has already been destroyed, in which case he fires at K. Tank I has probability $1/2$ of destroying the tank it fires at, while for J and K the probabilities of success are $1/4$ and $1/5$, respectively. They continue to fire until either I is destroyed or J and K are both destroyed. Using as states the surviving tanks, set up a Markov chain.

   (a) How many states are in this chain?

   (b) How many are absorbing?

   (c) Find the expected number of rounds fired.

   (d) Find the probability that I survives.

★ 12. Suppose that the tanks in problem 11 fight a three-way battle (rather than two against one) and that on each round each tank aims at the strongest surviving opponent. Answer questions (a), (b), (c), and (d) given in problem 11.

★ 5. genetics

We have already touched on some of the basic concepts of genetics in Chapter 5, Exercise 7, Problem 13 (page 157). We now review these ideas and go a little deeper.

Certain characteristics in individuals are governed by a pair of carriers called genes that are always present in both the parents and the offspring. In the simplest case each gene of a particular pair can assume two forms (called alleles) which are conveniently denoted by the letters A and a. For each such characteristic the pair of genes can occur in three different forms, AA, Aa, and aa, and the individual belongs to one of three genotypes in accordance with the particular pair present. In some cases the characteristic is such that the genotype is easy to determine. For example, in peas the genotypes AA, Aa, and aa have red, pink, and white blossoms, respectively. In other cases, such as right-handedness or eye color, the characteristic is visible in only two forms. In such cases the genotypes AA and Aa always exhibit the same observable form of the characteristic, while the genotype aa always exhibits the alternate form. For this reason A is called the dominant gene and a is called the recessive gene. An individual is called dominant (D) (with respect to the characteristic) if he has the AA genes, recessive (R) if he has the aa genes, and hybrid (H) (or heterozygous) if he has the Aa mixture.

The basic assumption of the Mendelian[1] Theory is that each offspring inherits two genes, one from each parent, and that the gene is selected at random from each parent. Thus if one parent is AA, then at least one gene of the offspring must be A. If one parent is Aa, then it transmits either the gene A with probability 1/2 or the gene a with probability 1/2.

EXAMPLE 1. In a controlled experiment we mate an individual of unknown genotype with a hybrid. We select an offspring and mate the offspring with a hybrid. Find the transition matrix P for the Markov process in which at each stage the offspring is mated with a hybrid. Does P^k have a limit as $k \longrightarrow \infty$?

Solution. The state of the system is the genotype of the offspring. Let p_k denote the probability that the kth offspring is dominant. Similarly let q_k be the probability that the kth offspring is hybrid, and let r_k be the probability that the kth offspring is recessive. To find the transition matrix for this process we consider each possibility for the genotype of the unknown parent.

If the unknown parent is dominant (AA) and is mated with a hybrid (Aa), then the probabilities for the genotypes AA, Aa, and aa give the row vector [1/2, 1/2, 0]. If the unknown parent is hybrid (Aa) and is mated with another hybrid (Aa), then the corresponding probability vector is [1/4, 1/2, 1/4]. Finally, if the unknown parent is recessive (aa), then the corresponding vector is [0, 1/2, 1/2].

---

[1] This theory of genetics had its origin in the experiments of Gregor Mendel (1822–1884).

Suppose now that $[p_0, q_0, r_0]$ is a probability vector for the genotype of the unknown parent and that $[p_1, q_1, r_1]$ is the probability vector for the genotype of the first offspring. Using the three vectors obtained above, we find that

$$(70) \qquad [p_1, q_1, r_1] = [p_0, q_0, r_0] \begin{bmatrix} \frac{1}{2} & \frac{1}{2} & 0 \\ \frac{1}{4} & \frac{1}{2} & \frac{1}{4} \\ 0 & \frac{1}{2} & \frac{1}{2} \end{bmatrix}.$$

Consequently the transition matrix for this process is the 3×3 matrix on the right-hand side of (70).

By direct computation we find that

$$P^2 = \begin{bmatrix} \frac{1}{2} & \frac{1}{2} & 0 \\ \frac{1}{4} & \frac{1}{2} & \frac{1}{4} \\ 0 & \frac{1}{2} & \frac{1}{2} \end{bmatrix}^2 = \begin{bmatrix} \frac{3}{8} & \frac{1}{2} & \frac{1}{8} \\ \frac{1}{4} & \frac{1}{2} & \frac{1}{4} \\ \frac{1}{8} & \frac{1}{2} & \frac{3}{8} \end{bmatrix}.$$

Since all entries of P^2 are positive, we can apply Theorem 10 and deduce that P^k does approach a limit as $k \longrightarrow \infty$. To find this limit, we solve the vector equation

$$(71) \qquad [x_1, x_2, x_3] = [x_1, x_2, x_3]P.$$

This leads to the three equations

$$x_1 = \frac{1}{2}x_1 + \frac{1}{4}x_2$$

$$x_2 = \frac{1}{2}x_1 + \frac{1}{2}x_2 + \frac{1}{2}x_3$$

$$x_3 = \qquad \frac{1}{4}x_2 + \frac{1}{2}x_3.$$

If we include the constraint $x_1 + x_2 + x_3 = 1$, then the system of four equations has the unique solution $x_1 = 1/4$, $x_2 = 1/2$, and $x_3 = 1/4$. Consequently by Theorem 10

$$(72) \qquad \lim_{k \to \infty} P^k = \begin{bmatrix} \frac{1}{4} & \frac{1}{2} & \frac{1}{4} \\ \frac{1}{4} & \frac{1}{2} & \frac{1}{4} \\ \frac{1}{4} & \frac{1}{2} & \frac{1}{4} \end{bmatrix}.$$

Equation (72) has this interpretation: No matter what genotype we begin with, after repeated mating of the offspring with a hybrid, the probabilities for the genotypes of the kth offspring are $p_k \approx 1/4$, $q_k \approx 1/2$, and $r_k \approx 1/4$, when k is sufficiently large.

The alert reader will notice that in Example 1, the genotype of one parent was carefully controlled. What is the situation if we have a large population, and the mating is random? At first glance, the problem appears to be so complicated that it is impossible to handle, and in the early development of the theory, this problem was a major stumbling block. Opponents of the Mendelian theory argued that were the Mendelian theory true, then in a large population the recessive genotype must disappear in due time. Consequently, the continued presence of recessive individuals clearly showed that the theory must be altered or rejected.

It was G. H. Hardy[1] who first resolved the difficulty with a very surprising result.

Theorem 13. In a large population in which (a) the proportion of the genotypes is the same for both the males and the females, (b) mating is completely random, and (c) half of the offspring are males, the distribution of the genotypes is theoretically stable after one generation.

The word *theoretically* is inserted to remind us that the theorem treats an ideal situation, whereas in a real population the true distribution of the genotypes may vary slightly from that predicted by the theory.

Proof. Let the row vector $[p_k, q_k, r_k]$ have the meaning already explained in Example 1, and let $[p_0, q_0, r_0]$ refer to the original population. Since three different genotypes, D, H, and R, are possible for each parent, there are nine different combinations. Let the symbol XY denote the mating in which the male has genotype X and the female has genotype Y. The nine combinations of parents can be reduced to six by observing that the probabilities for the genotypes of the offspring are the same for XY and YX.

The probabilities for the various combinations of parents and their offsprings are listed in Table 1 (which is really a tree without the branches).

[1] G. H. Hardy, "Mendelian Proportions in a Mixed Population," *Science,* N.S. vol. 28 (1908), pp. 49–50.

TABLE 1

Parent genotype	Probability of combination	Probability for genotype of offspring from parents		
		D	H	R
DD	p_0^2	1	0	0
DH or HD	$2p_0q_0$	$\dfrac{1}{2}$	$\dfrac{1}{2}$	0
DR or RD	$2p_0r_0$	0	1	0
HH	q_0^2	$\dfrac{1}{4}$	$\dfrac{1}{2}$	$\dfrac{1}{4}$
HR or RH	$2q_0r_0$	0	$\dfrac{1}{2}$	$\dfrac{1}{2}$
RR	r_0^2	0	0	1

From the table it is easy to determine the probabilities for the genotypes of the offspring. Indeed, we find that

$$p_1 = p_0^2 + p_0q_0 + \frac{q_0^2}{4} = \left(p_0 + \frac{q_0}{2}\right)^2,$$

(73) $$q_1 = p_0q_0 + 2p_0r_0 + \frac{1}{2}q_0^2 + q_0r_0 = 2\left(p_0 + \frac{q_0}{2}\right)\left(r_0 + \frac{q_0}{2}\right),$$

$$r_1 = \frac{q_0^2}{4} + q_0r_0 + r_0^2 = \left(r_0 + \frac{q_0}{2}\right)^2.$$

Since half of the offspring are males and half of the offspring are females, the row vector $[p_1, q_1, r_1]$ applies to both parents in the first generation and gives the probabilities to be used in the second column of Table 1 to determine $[p_2, q_2, r_2]$. Before computing this vector, we recall that by the definition of $[p_0, q_0, r_0]$

(74) $$p_0 + q_0 + r_0 = 1.$$

To compute p_2, we merely raise the subscripts by 1 for each term in the first equation in (73). We find that

$$p_2 = \left(p_1 + \frac{q_1}{2}\right)^2 = \left[\left(p_0 + \frac{q_0}{2}\right)^2 + \frac{2(p_0 + q_0/2)(r_0 + q_0/2)}{2}\right]^2$$

(75) $$= \left[\left(p_0 + \frac{q_0}{2}\right)\left(p_0 + \frac{q_0}{2} + r_0 + \frac{q_0}{2}\right)\right]^2$$

$$= \left(p_0 + \frac{q_0}{2}\right)^2 (p_0 + q_0 + r_0)^2 = \left(p_0 + \frac{q_0}{2}\right)^2 \cdot 1 = p_1.$$

Hence $p_2 = p_1$. A similar computation will give

(76) $$q_2 = q_1 \quad \text{and} \quad r_2 = r_1.$$

It follows from (75) and (76) that for any $k \geq 2$

(77) $$[p_k, q_k, r_k] = [p_1, q_1, r_1].$$

Hence "theoretically" the distribution of the genotypes is constant after one generation. ∎

As Hardy states, "there is not the slightest foundation for the idea that a dominant character should show a tendency to spread over a whole population, or that a recessive (character) should tend to die out .... I have of course considered only the very simplest hypotheses possible. Hypotheses other than that of purely random mating will give different results, and of course, if ... the character is not independent of that of sex, or has an influence on fertility the whole question may be greatly complicated."

To these remarks we add the obvious: If the AA parents on the average tend to have more children than the Aa parents or the aa parents, then certainly (77) no longer holds. The reader who wishes to pursue this topic further may consult G. Dahlberg, *Mathematical Methods for Population Genetics*, Wiley-Interscience, New York, 1948.

Exercise 5

1. Derive equation (70).
2. Draw the tree that corresponds to Table 1, and use this tree to derive equation set (73).
3. Under the conditions of Theorem 13 [equation set (73)] find the vectors $[p_1, q_1, r_1]$ and $[p_2, q_2, r_2]$ given that $[p_0, q_0, r_0]$ is
   (a) $[1/5, 0, 4/5]$.
   (b) $[1/10, 0, 9/10]$.
   (c) $[0, 1, 0]$.
4. Repeat problem (3) if $[p_0, q_0, r_0]$ is
   (a) $[1/3, 1/3, 1/3]$.
   (b) $[1/2, 0, 1/2]$.
   (c) $[1/4, 1/2, 1/4]$.
5. Use equations (73) and (74) to prove that $p_1 + q_1 + r_1 = 1$.
6. Complete the proof of Theorem 13 by showing that $q_2 = q_1$ and $r_2 = r_1$.
7. We modify Example 1 by mating with a dominant (in place of a hybrid) at each stage. Find the transition matrix P. Compute P^2 and P^3.

8. Show that equation set (73) can be put in the form AP, where A is a six-dimensional row vector and P is a 6×3 matrix.

9. In some species such as dogs or cats, two offsprings from the same parents can mate. We start with two parents selected at random and mate two of the offspring of opposite sex. Find the transition matrix for this process using as states DD, RR, HH, DR, DH, and RH.

10. For the Markov chain described in problem 9, find (a) $N = [I - S]^{-1}$ and (b) $C = NR$, where the symbols have the meaning described in Theorems 11 and 12.

11. For the Markov chain described in problem 9, find the probability that the process is absorbed by (a) having two dominant offspring and (b) having two recessive offspring, given that we start with one dominant and one hybrid. What is the expected number of generations for this to occur?

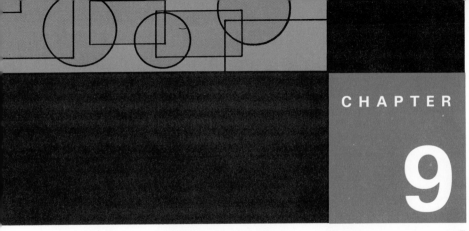

9

convex sets and linear programming

Many important problems in economics have the following form: Find the maximum (largest) value of a certain function of several variables when these variables are restricted by a set of inequalities. In general, such problems are very difficult, but fortunately in many cases the function and the inequalities are linear. In the latter case there is a relatively simple theory (called *linear programming*) that supplies the solution. The concept of a convex set is fundamental in this theory. Even without this practical application, convex sets are of great interest in their own right.

1. convex sets

We recall from Chapter 7 the following:

Theorem 1. Let $A = (a_1, a_2)$ and $B = (b_1, b_2)$ be two points in the plane (with a rectangular coordinate system). Then the line segment AB joining A and B is the set of points

$$C = (1 - t)A + tB = ((1 - t)a_1 + tb_1, \ (1 - t)a_2 + tb_2),$$

where $0 \leq t \leq 1$.

Definition 1. Convex Set. A subset S in the plane is said to be convex if for any two points $A,B \in S$, every point of the line segment AB is also in S.

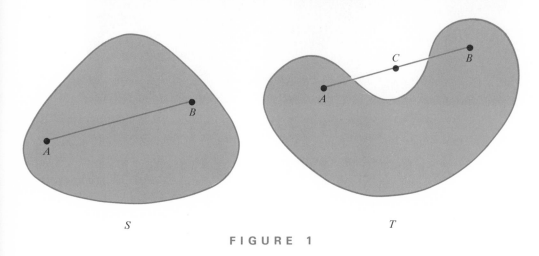

FIGURE 1

In Fig. 1, the set S is convex, while the set T is not convex, because C, which is on the line segment AB, is not in T.

When we discuss n-dimensional space R^n, where $n \geq 4$, we replace Theorem 1 by

Definition 2. Let $A = (a_1,a_2, \ldots ,a_n)$ and $B = (b_1,b_2, \ldots ,b_n)$ be two points in R^n. Then the line segment AB is the set of all points

$$C = (1 - t)A + tB$$
$$= ((1 - t)a_1 + tb_1, \ (1 - t)a_2 + tb_2, \ \ldots , \ (1 - t)a_n + tb_n)$$

where $0 \leq t \leq 1$.

Definition 1 remains the same in R^n except that we now take S to be a subset of R^n.

Theorem 2. If S and T are convex subsets of R^n, then $S \cap T$ is a convex subset of R^n.

Proof. Let A and B be any two points of $S \cap T$. Then $A ,B \in S$ and $A,B \in T$. Since S and T are both convex, the line segment AB is contained in S and is also contained in T. Hence AB is contained in $S \cap T$. Therefore $S \cap T$ is convex. ∎

We now define a linear function over R^n (see Appendix 1 for the definition of a function).

> **Definition 3.** A function f defined over R^n is said to be linear if it is of the form
>
> (1) $$f(X) = c_1x_1 + c_2x_2 + \cdots + c_nx_n,$$
>
> where $X = (x_1, x_2, \ldots, x_n) \in R^n$ and $c_1, c_2, \ldots, c_n$ are constants.

We observe that the right-hand side of (1) is a real number and is referred to as the *value* of the function f at the point $X = (x_1, x_2, \ldots, x_n) \in R^n$. We now state and prove a useful result:

> **Theorem 3.** The values of a linear function defined on a line segment AB lie between the values of the function at the end points.

To state this result with symbols, it is necessary to know at which end point the function f has the larger value. We can avoid this difficulty if we assign the letter A to that end point for which f is the larger. To be precise, we assume that

$$f(B) \leqq f(A).$$

With this agreement Theorem 3 asserts that if C is a point on the line segment AB, then

(2) $$f(B) \leqq f(C) \leqq f(A).$$

Proof. Let $A = (a_1, a_2, \ldots, a_n)$ and $B = (b_1, b_2, \ldots, b_n)$, and let f, a linear function defined on AB, be given by (1). Then

$$f(A) = c_1a_1 + c_2a_2 + \cdots + c_na_n,$$

and

$$f(B) = c_1b_1 + c_2b_2 + \cdots + c_nb_n.$$

We set $f(A) = a$ and $f(B) = b$, and by our agreement we have $b \leqq a$. Then at any point $C = (1 - t)A + tB$ on the line segment AB, the value of the function is

$$f(C) = f((1 - t)A + tB)$$
$$= c_1((1 - t)a_1 + tb_1) + \cdots + c_n((1 - t)a_n + tb_n)$$
$$= (1 - t)(c_1 a_1 + \cdots + c_n a_n) + t(c_1 b_1 + \cdots + c_n b_n)$$
$$= (1 - t)f(A) + tf(B) = (1 - t)a + tb.$$

This value can also be written in two different convenient forms. First,

(3)
$$f(C) = (1 - t)a + tb = a - (a - b)t,$$

and second,

(4)
$$f(C) = (1 - t)a + tb = b + (1 - t)(a - b).$$

Since $0 \leq t \leq 1$ and $a \geq b$, the right-hand side of Equation (3) gives $f(C) \leq a$ and the right-hand side of Equation (4) gives $f(C) \geq b$. Since $f(A) = a$ and $f(B) = b$, we have

(2)
$$f(B) \leq f(C) \leq f(A). \quad \blacksquare$$

It is well known (see Appendix 1) that every equation of the form

(5)
$$ax + by = c,$$

where a and b are not both zero, represents a straight line in R^2.

The linear equation (5) partitions the plane R^2 into three subsets A, B, and C:

(I) A: the set of those points $P(x,y)$ whose coordinates satisfy (5);
(II) B: the set of those points $P(x,y)$ whose coordinates satisfy the inequality $ax + by < c$;
(III) C: the set of those points $P(x,y)$ whose coordinates satisfy the inequality $ax + by > c$.

The set A, as mentioned before, is a straight line, and the sets B and C are called *open half planes*. The sets $A \cup B$ (the points whose coordinates satisfy $ax + by \leq c$) and $A \cup C$ (the points whose coordinates satisfy $ax + by \geq c$) are called *closed half planes* or simply *half planes*.

EXAMPLE 1.   Sketch the graph of the inequality

(6)
$$4x + 3y \geq 12.$$

Solution.   To sketch the graph of (6), we first draw the line $A: 4x + 3y = 12$. Let us consider a point on this line say $P(3/2, 2)$. Now we keep the second coordinate fixed at 2 and increase the first coordinate say from 3/2 to 2 and obtain a new point $R(2, 2)$. By substituting the coordinates of R in (6) we see that the point $R(2, 2)$ is a solution of (6). This process is indicated by the line segment PR in Fig. 2. However, if we decrease the first coordinate say from 3/2 to 1, we notice that the new point $Q(1,2)$ is not a solution of (6). This process is indicated by the horizontal line segment PQ in Fig. 2. This situation is typical. In general, starting with any point on the line $4x + 3y = 12$, and decreasing its first coordinate we get the left-hand side of (6) less than 12, but by increasing the first coordinate we get the left-hand side of (6) more than 12, and hence the latter gives a point in the solution set of (6). Thus the solution set of (6) is given by the half-plane to the right of the line $4x + 3y = 12$. The graph of the inequality (6) is therefore given by the shaded area in Fig. 2.

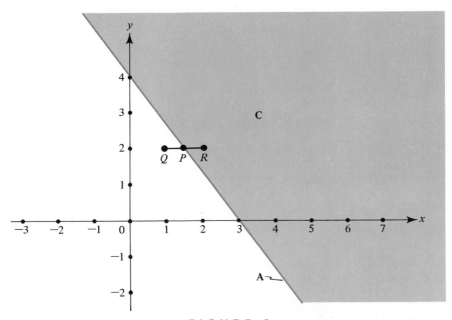

FIGURE 2

To summarize, if we wish to graph the solution of the inequality (6), we first draw A, the line $4x + 3y = 12$, and then choose an arbitrary point that does not belong to this line, say (0,0). We check whether (0,0) belongs to the solution set of (6). Since (0,0) does not satisfy (6), the points on the other side of A [i.e., the side of A which does not contain (0,0)] belong to the solution set of (6) as shown in Fig. 2. If (0,0) is on the line A, then the student must select some other point not on A as a test point.

Let us now consider a linear equation in n unknowns $x_1, x_2, \ldots, x_n$:

(7) $$a_1 x_1 + a_2 x_2 + \cdots + a_n x_n = b,$$

where not all a_i's are zero.

Definition 4. The set of all points in R^n that satisfy equation (7); i.e., the set of all solutions of the linear equation (7), is called a hyperplane.

We observe that a hyperplane in R^2 is a straight line and that a hyperplane in R^3 is a plane.

The solution set of the linear inequality

$$(8) \qquad a_1x_1 + a_2x_2 + \cdots + a_nx_n \leqq b,$$

consists of $A \cup B$, where A is the hyperplane (7) and B is the solution set of the linear inequality

$$(9) \qquad a_1x_1 + a_2x_2 + \cdots + a_nx_n < b.$$

Similarly the solution set of the inequality

$$(10) \qquad a_1x_1 + a_2x_2 + \cdots + a_nx_n \geqq b,$$

consists of $A \cup C$, where A is the set defined above and C is the solution set of the inequality

$$(11) \qquad a_1x_1 + a_2x_2 + \cdots + a_nx_n > b.$$

The sets B and C are called *open half-spaces* and the sets $A \cup B$ and $A \cup C$ are referred to as *closed half-spaces*, or simply *half-spaces*. The set A is called the *bounding hyperplane* of the closed half-space $A \cup B$. It is also the bounding hyperplane of $A \cup C$.

An important property of a half-space is stated in

Theorem 4. A half-space is a convex set.

Proof. Let S be the half-space defined by the inequality

$$(12) \qquad a_1x_1 + a_2x_2 + \cdots + a_nx_n \leqq b.$$

To prove that S is convex, we must show that if $P,Q \in S$, then the line segment $PQ \subset S$.

For $X = (x_1, x_2, \ldots, x_n) \in R^n$, let us write

$$(13) \qquad f(X) = a_1x_1 + a_2x_2 + \cdots + a_nx_n.$$

Now S consists of all points $A \in R^n$ such that $f(A) \leqq b$. Hence if P and

Q are points in S, then $f(P) \leq b$ and $f(Q) \leq b$. By Theorem 3, we see that
if $R \in PQ$, then either

$$f(P) \leq f(R) \leq f(Q)$$

or

$$f(Q) \leq f(R) \leq f(P).$$

In either case $f(R) \leq b$ and therefore each point R of the line segment PQ
is in S. ∎

Definition 5. The intersection S of a finite number of half-spaces in
R^n is called a polyhedral convex set. A point $P \in S$ is called an extreme
point or a corner point of S if P is the intersection point of n of the
bounding hyperplanes which determine S.

Definition 6. Let $X = [x_1, x_2, \ldots, x_n]$ and $Y = [y_1, y_2, \ldots, y_n]$ be two
n-dimensional row vectors. We say that $X \leq Y$ (read "X is less than
or equal to Y") if $x_i \leq y_i$ for $i = 1, 2, \ldots, n$. When X and Y are
two n-dimensional column vectors, the inequality $X \leq Y$ is defined
analogously.

Let

$$
\begin{aligned}
a_{11}x_1 + a_{12}x_2 + \cdots + a_{1n}x_n &\leq b_1 \\
a_{21}x_1 + a_{22}x_2 + \cdots + a_{2n}x_n &\leq b_2 \\
&\;\;\vdots \\
a_{m1}x_1 + a_{m2}x_2 + \cdots + a_{mn}x_n &\leq b_m
\end{aligned}
$$

(14)

be a system of m linear inequalities in n unknowns. By the solution of (14)
we mean the set of all points $X = (x_1, x_2, \ldots, x_n)$ whose coordinates satisfy
simultaneously all the inequalities in (14).

The system (14) of linear inequalities can be put in the compact form

(15) $AX \leq B$,

where $A = [a_{ij}]$ is the $m \times n$ matrix formed from the coefficients, and X, B
are the column vectors

$$
X = \begin{bmatrix} x_1 \\ x_2 \\ \vdots \\ x_n \end{bmatrix}, \qquad
B = \begin{bmatrix} b_1 \\ b_2 \\ \vdots \\ b_m \end{bmatrix}.
$$

Theorem 5. The solution set of (14) is a polyhedral convex set.

Proof. The solution set for each inequality in (14) is a half-space and by Theorem 4 this space is a convex set. The solution set for the system (14) is the intersection of these m half-spaces, and by Theorem 2 this intersection is again a convex set. Hence by Definition 5, the solution set of (14) is a polyhedral convex set. ∎

EXAMPLE 2. Graph the solution set of the system of inequalities

(16)
$$4x + 3y \leq 12$$
$$x - 2y \leq 2$$
$$-3x + y \leq 3.$$

Find the corner points of the polyhedral convex set defined by these inequalities.

Solution. Notice that when $n = 2$, we replace the point (x_1, x_2) by (x, y). For each inequality in (16) we first draw the line bounding the half-plane by finding two points on the line. We then choose an arbitrary point which does not belong to the line and test it in the given inequality. We then shade

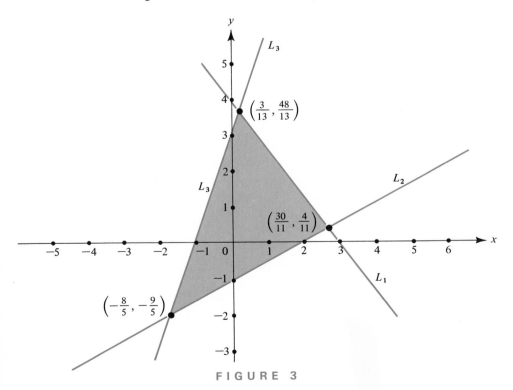

FIGURE 3

(perhaps with three different colors) the half-planes corresponding to each inequality. The convex set which has all three shades represents the solution of (16). For example, let L_1 be the line $4x + 3y = 12$. If $x = 0$, then $y = 4$; if $y = 0$, then $x = 3$. Consequently L_1 is the line passing through the points (0,4) and (3,0). We test some point, say (0,0). Since (0,0) does satisfy the inequality $4x + 3y \leq 12$, we shade that half-plane that is bounded by L_1 and contains the origin. We follow the same procedure for the lines L_2: $x - 2y = 2$ and L_3: $-3x + y = 3$. The solution of the system (16) is the shaded region given in Fig. 3. By solving simultaneously the equations of the bounding lines two at a time, we find that the corner points are: (30/11, 4/11), (−8/5, −9/5), and (3/13, 48/13). Observe that the shaded region shown in Fig. 3 is indeed a polyhedral convex set, as predicted by Theorem 5.

Exercise 1

In problems 1 through 8, sketch the graph of the solution set of $AX \leq B$, where A and B are given below. Also find the corner points.

1. $A = \begin{bmatrix} 3 & 5 \\ 2 & 2 \end{bmatrix}$, $B = \begin{bmatrix} 15 \\ 8 \end{bmatrix}$.

2. $A = \begin{bmatrix} -2 & 1 \\ 3 & 2 \end{bmatrix}$, $B = \begin{bmatrix} 2 \\ 4 \end{bmatrix}$.

3. $A = \begin{bmatrix} -1 & 0 \\ 0 & -1 \\ 1 & 1 \end{bmatrix}$, $B = \begin{bmatrix} 0 \\ 0 \\ 1 \end{bmatrix}$.

4. $A = \begin{bmatrix} 1 & 1 \\ -1 & 1 \\ 2 & 1 \end{bmatrix}$, $B = \begin{bmatrix} 1 \\ -1 \\ 1 \end{bmatrix}$.

5. $A = \begin{bmatrix} 1 & 1 \\ -1 & 1 \\ -2 & -1 \end{bmatrix}$, $B = \begin{bmatrix} 1 \\ -1 \\ -1 \end{bmatrix}$.

6. $A = \begin{bmatrix} 1 & 1 \\ 1 & -1 \\ -1 & 1 \\ -1 & -1 \end{bmatrix}$, $B = \begin{bmatrix} 1 \\ 2 \\ 3 \\ 4 \end{bmatrix}$.

7. $A = \begin{bmatrix} 1 & 1 \\ 2 & -1 \\ -2 & 1 \\ 2 & 2 \end{bmatrix}$, $B = \begin{bmatrix} -1 \\ -2 \\ -3 \\ -4 \end{bmatrix}$.

8. $A = \begin{bmatrix} -1 & 0 \\ 5 & 3 \\ -4 & 2 \\ -\frac{1}{3} & \frac{1}{3} \\ 1 & -1 \end{bmatrix}$, $B = \begin{bmatrix} 1 \\ 15 \\ 8 \\ 1 \\ 2 \end{bmatrix}$.

9. A grocer has shelf space for at most 100 cans of soft drinks of type A and type B. He wishes to have at least one can of each type. Graph the possible stock set and find the extreme points.

10. In problem 9, if a can of type A sells for 10 cents and a can of type

B sells for 15 cents, solve problem 9 if the owner wishes to receive at least $12 for selling all of his stock.

11. Solve problem 10 if the owner wishes to receive at least (a) $10, (b) $14, and (c) $15 for selling all of his stock.

2. linear programming

The problem of linear programming is as follows: Find the maximum value (or the minimum value) of a linear function

$$(17) \qquad f(X) = c_1x_1 + c_2x_2 + \cdots + c_nx_n,$$

called the *objective function,* subject to the *constraints* (restrictions)

$$(18) \qquad \begin{aligned} a_{11}x_1 + a_{12}x_2 + \cdots + a_{1n}x_n &\leq b_1 \\ a_{21}x_1 + a_{22}x_2 + \cdots + a_{2n}x_n &\leq b_2 \\ \vdots \qquad\qquad \vdots \qquad\qquad \vdots \qquad\quad \vdots \\ a_{m1}x_1 + a_{m2}x_2 + \cdots + a_{mn}x_n &\leq b_m, \\ x_i \geq 0; \quad i = 1, 2, \ldots, n. \end{aligned}$$

In other words, we wish to maximize (or minimize) the given linear function (17) defined over a convex polyhedral set S given by (18).

Each point of S is called a *feasible solution* of the problem, and a point in S at which f takes on its maximum (or minimum) value is called an *optimum solution.*

We illustrate this problem in

EXAMPLE 1. Find x_1 and x_2 which will make the value of $f(X) = 2x_1 + 3x_2$ as large as possible, where x_1 and x_2 are restricted by the following constraints:

$$3x_1 + 5x_2 \leq 15$$
$$x_1 \geq 0$$
$$x_2 \geq 0.$$

Solution. We first sketch the graph of the permissible values of x_1 and x_2; i.e., the polyhedral convex set defined by the constraints of this problem. Using the method explained in Section 1, we find that this set is the shaded region in Fig. 4. Without knowing much theory, all we have to do is to find the values of the given function at each of the points in the shaded region of Fig. 4 and see at which point the function takes on the maximum value. Unfortunately, there are infinitely many points in this set. So we need some method other than trial-and-error substitution to solve the problem.

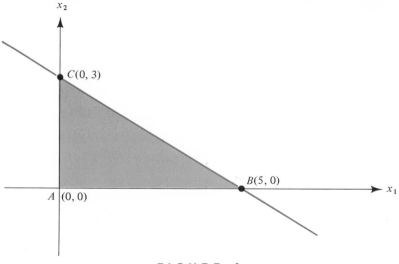

FIGURE 4

We therefore postpone the solution of this problem until after we prove the
following important theorem.

Theorem 6. A linear function

(17) $$f(X) = c_1x_1 + c_2x_2 + \cdots + c_nx_n$$

defined over a bounded polyhedral convex set S takes on its maximum
value at a corner point of S and its minimum value at some other corner
point of S.

Proof. We shall prove this theorem for R^2, but the theorem is true for R^n.

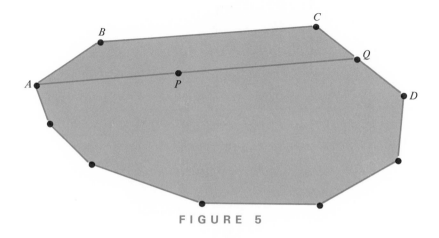

FIGURE 5

As illustrated in Fig. 5, let the corner points of S be $A, B, C, \ldots, G, \ldots$, and suppose that among the values of the function at these corner points $f(A)$ is the largest and $f(G)$ is the smallest. Let P be an arbitrary point of S. We shall prove that

$$(19) \qquad f(G) \leqq f(P) \leqq f(A).$$

To this end, we join A and P by a straight line segment and extend this segment to meet an edge (say CD in Fig. 5) of S at Q. We can select the lettering of the vertices so that $f(C) \leqq f(D)$. By Theorem 3, we have

$$(20) \qquad f(C) \leqq f(Q) \leqq f(D).$$

Since $f(G)$ and $f(A)$ are the smallest and largest corner point values, respectively, we get

$$(21) \qquad f(G) \leqq f(C) \leqq f(Q) \leqq f(D) \leqq f(A).$$

From (21) we conclude that

$$(22) \qquad f(G) \leqq f(Q) \leqq f(A).$$

Again applying Theorem 3, this time to the line segment AQ, we get

$$(23) \qquad f(Q) \leqq f(P) \leqq f(A).$$

Combining (22) and (23), we find that for any $P \in S$,

$$(24) \qquad f(G) \leqq f(P) \leqq f(A). \quad \blacksquare$$

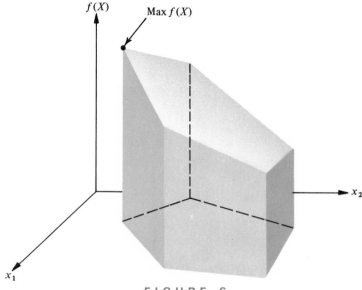

FIGURE 6

The reader should note that if f is not a linear function, then the maximum value need not occur at a corner point. The underlying reason for this distinction will be clear if we look at the graph in the case $n = 2$. In this case the graph of $f(X) = c_1 x_1 + c_2 x_2$ is a plane and, as illustrated in Fig. 6, the maximum must occur at a corner point. On the other hand, if f is not linear, then (as indicated in Fig. 7) the maximum may occur at an interior point.

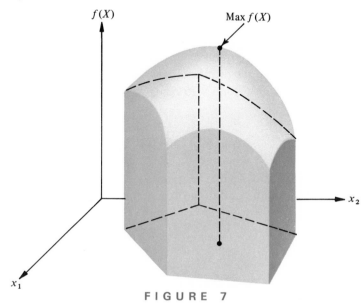

FIGURE 7

Theorem 6 gives a simple method of finding the maximum value and the minimum value of a linear function defined over a polyhedral convex set S:

(I) Find all the corner points of S.

(II) Evaluate the function at each of these points.

(III) The smallest of these values is the minimum and the largest of these values is the maximum for the function f.

We are now in a good position for the solution of Example 1. From Fig. 4 we see that the corner points of the convex set are $A(0,0)$, $B(5,0)$ and $C(0,3)$. We find the value of f at each of these points:

$$f(A) = f(0,0) = 2 \cdot 0 + 3 \cdot 0 = 0$$
$$f(B) = f(5,0) = 2 \cdot 5 + 3 \cdot 0 = 10$$
$$f(C) = f(0,3) = 2 \cdot 0 + 3 \cdot 3 = 9.$$

Theorem 6 now tells us that at $B(5,0)$ the function takes on the largest value and at $A(0,0)$ the function takes on the smallest value.

EXAMPLE 2. A restaurant owner has at most 100 pounds of ground beef which he can use for making hamburgers and slices of meat loaf. Each hamburger contains 1/4 pound of ground beef and each slice of meat loaf contains 1/5 pound of ground beef. He makes a profit of 10 cents on each hamburger, while he makes a profit of 15 cents on each slice of meat loaf. Labor cost for producing a hamburger and a slice of meat loaf are 5 and 12 cents, respectively. If he is willing to pay at most $30 for the labor costs of producing these two items, find the number of hamburgers and the number of slices of meat loaf that will maximize his profits.

Solution. Let x_1 denote the number of hamburgers and x_2 the number of slices of meat loaf. Since his profit on each hamburger is 10 cents and his profit on each slice of meat loaf is 15 cents, we are asked to maximize the profit function

(25) $$P(X) = 0.1x_1 + 0.15x_2.$$

Obviously $x_1 \geq 0$ and $x_2 \geq 0$. The quantity of ground beef available and the labor costs give us the additional constraints, namely

$$\frac{1}{4}x_1 + \frac{1}{5}x_2 \leq 100$$

and

$$0.05x_1 + 0.12x_2 \leq 30.$$

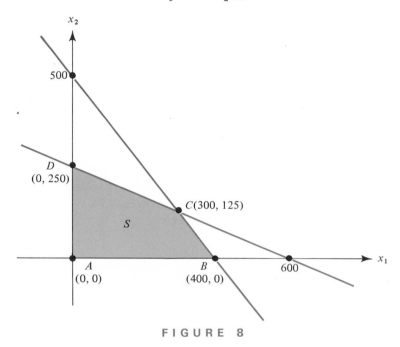

FIGURE 8

So the problem reduces to the following: Maximize the linear function P given by (25) with the set of constraints:

$$\frac{1}{4}x_1 + \frac{1}{5}x_2 \leqq 100$$

(26)
$$0.05x_1 + 0.12x_2 \leqq 30$$
$$x_1 \geqq 0$$
$$x_2 \geqq 0.$$

The constraints (26) define a polyhedral convex set S shown shaded in Fig. 8. The corner points of this convex set are $A(0, 0)$, $B(400, 0)$, $C(300, 125)$, and $D(0, 250)$. The value of P at each of these points is

$$P(A) = P(0, 0) = (0.1)0 + (0.15)0 = 0,$$
$$P(B) = P(400, 0) = (0.1)400 + (0.15)0 = 40,$$
$$P(C) = P(300, 125) = (0.1)300 + (0.15)125 = 30 + 18.75 = 48.75,$$
$$P(D) = P(0, 250) = (0.1)0 + (0.15)250 = 37.5.$$

Thus to make a maximum profit of $48.75, he should make 300 hamburgers and 125 pieces of meat loaf.

Exercise 2

1. Find the maximum value and the minimum value of the function $P(X) = \$15x_1 + \$25x_2$ subject to the following sets of constraints:

(a)
$$x_1 \geqq 0$$
$$x_2 \geqq 0$$
$$3x_1 + 4x_2 \leqq 15.$$

(b) $x_1 \geqq 0$
$$x_2 \geqq 0$$
$$x_1 \leqq 5$$
$$x_2 \leqq 3.$$

(c)
$$x_1 \geqq 0$$
$$x_2 \geqq 0$$
$$x_1 + x_2 \leqq 25$$
$$2x_1 + 2x_2 \geqq 10.$$

(d)
$$x_1 \geqq 0$$
$$x_2 \geqq 0$$
$$x_1 + 2x_2 \leqq 50$$
$$-x_1 + 3x_2 \leqq 25.$$

(e)
$$x_1 \geqq 0$$
$$x_2 \geqq 0$$
$$x_1 + x_2 \leqq 50$$
$$2x_1 - x_2 \leqq 40$$
$$-3x_1 + x_2 \leqq 10.$$

(f)
$$x_1 \geqq 0$$
$$x_2 \geqq 0$$
$$5x_1 + 2x_2 \leqq 300$$
$$x_1 + x_2 \leqq 90$$
$$x_2 \leqq 70$$
$$5x_1 - x_2 \geqq -20.$$

2. Find the maximum value and the minimum value of the function $P(X) = \$25x_1 + \$15x_2$ subject to the constraints of problem 1.

3. Find the maximum value and the minimum value of the function $P(X) = \$15x_1 - \$25x_2$ subject to the constraints of problem 1.

4. Find the maximum value and the minimum value of the function $P(X) = \$25x_1 - \$15x_2$ subject to the constraints of problem 1.

5. Let $f(X) = c_1x_1 + c_2x_2 + \cdots + c_nx_n + k$, where $c_1, c_2, \ldots, c_n$ and k are constants, and $f(X)$ is defined over a polyhedral convex set S. Prove that $f(X)$ takes on a maximum value at a corner point of S, and takes on a minimum value at (some other) corner point of S.

6. A contractor has the following materials available: 100 units of concrete, 160 units of wood, and 400 units of glass. He builds houses of type A and type B. A house of type A requires 1 unit of concrete, 2 units of wood, and 2 units of glass, while a house of type B requires 1 unit of concrete, 1 unit of wood, and 5 units of glass. A house of type A sells for $10,000, and a house of type B sells for $11,000. How many houses of each type should the contractor make in order to obtain the maximum amount of money?

7. A businessman owns a motel consisting of 300 single rooms and an attached restaurant which can accommodate 100 people. He knows that 30 percent of the male guests and 50 percent of the female guests will eat in the restaurant. Suppose he makes a profit of $1.50 per day from every guest who eats in his restaurant and $1.00 per day from every guest who does not eat in his restaurant. Find the number of male and female guests he should have in order to obtain the maximum profit.

8. In problem 7, find the businessman's maximum profit if he makes a profit of $1.50 per day from every guest who eats in his restaurant while he suffers a loss of $1.00 per day from every guest who does not eat in his restaurant.

9. Major Motors, Inc. must produce at least 5000 luxury cars and 12,000 medium-priced cars. They also must produce at most 30,000 units of compact cars. The company owns two factories A and B at different locations. Factory A produces 20, 40, and 60 units of luxury, medium-priced, and compact cars, respectively, per day; factory B produces 10, 30, and 50 luxury, medium-priced, and compact cars, respectively, per day. If A costs $60,000 per day to operate and B costs $40,000 per day to operate, find the number of days each factory should be run to minimize the cost and meet the requirements.

10. Suppose a person needs a minimum of 60 units of carbohydrates, 40 units of proteins, and 35 units of fats per month. Food A contains 5, 3, and 5 units of carbohydrates, proteins, and fats, respectively, per pound; food B contains 2, 2, and 1 units of carbohydrates, proteins,

and fats, respectively, per pound. If A costs $1.50 per pound and B costs $0.70 per pound, how many pounds of each should he buy per month to minimize his cost and still meet his minimum requirements?

★ 3.   the simplex method

In the preceding section, we developed a method of solving linear programming problems. The method involved the calculation of all the corner points of S [determined by the set of inequalities (18)] and then the evaluation of the linear function (17) at each of the corner points of S. However, as the number of variables increases, the process of finding all the corner points of S may become very tedious and time-consuming. In this section, we develop an alternative method called the *simplex method* for solving linear programming problems. Underlying the method is the elimination procedure for solving systems of simultaneous linear equations discussed in Chapter 8. This method is rapidly gaining popularity, because this process is readily adapted to computers. We first review the elimination method in

EXAMPLE 1.   Solve the system of linear equations

$$
\begin{aligned}
3x_1 + 5x_2 + x_3 \qquad\qquad &= 14 \\
x_1 + 7x_2 \qquad + x_4 \quad\;\; &= 10 \\
-x_1 - 2x_2 \qquad\qquad + x_5 &= 0
\end{aligned}
$$

(27)

for x_1, x_2, and x_5 in terms of x_3 and x_4.

Solution.   As in Chapter 8, we can write this system of equations in matrix form:

(28)
$$
A = \begin{array}{c}
\begin{array}{ccccc} x_1 & x_2 & x_3 & x_4 & x_5 \end{array} \\
\left[\begin{array}{ccccc|c}
3 & 5 & 1 & 0 & 0 & 14 \\
1 & ⑦ & 0 & 1 & 0 & 10 \\
-1 & -2 & 0 & 0 & 1 & 0
\end{array}\right].
\end{array}
$$

Let us eliminate x_2 from each equation except the second. To this end, we divide each entry in the second row of A by 7, the coefficient of x_2 in the second equation. This coefficient is called a *pivot*. (We usually encircle the pivot. The row containing the pivot is called the *pivot row* and the column containing the pivot is called the *pivot column*.) After dividing by 7 we have a new second row in which the x_2-entry is 1. We now add -5 times the new second row to the first row and also add 2 times the new second row to the third row. We then have

(29)

$$B = \begin{bmatrix} x_1 & x_2 & x_3 & x_4 & x_5 & \\ \boxed{\dfrac{16}{7}} & 0 & 1 & \dfrac{-5}{7} & 0 & \dfrac{48}{7} \\ \dfrac{1}{7} & 1 & 0 & \dfrac{1}{7} & 0 & \dfrac{10}{7} \\ \dfrac{-5}{7} & 0 & 0 & \dfrac{2}{7} & 1 & \dfrac{20}{7} \end{bmatrix}.$$

Notice that the system of equations represented by (29) is equivalent to (27). This process of reducing the pivot column to a column having 1 in the pivot entry and zeros everywhere else will be called the *complete elimination process* with the specified pivot. Thus, in our example the complete elimination process with the pivot 7 transforms the matrix (28) into the matrix (29).

We now repeat the above process for the unknown x_1; i.e., we eliminate x_1 from each of the equations in (29) except the first. We select $16/7$ as our new pivot in B. We divide each entry of the first row of (29) by $16/7$ to obtain the new first row. Adding $-(1/7)$ times the new first row to the second row and adding $5/7$ times the new first row to the third row, we obtain the equivalent matrix:

(30)

$$C = \begin{bmatrix} x_1 & x_2 & x_3 & x_4 & x_5 & \\ 1 & 0 & \dfrac{7}{16} & -\dfrac{5}{16} & 0 & 3 \\ 0 & 1 & -\dfrac{1}{16} & \dfrac{3}{16} & 0 & 1 \\ 0 & 0 & \dfrac{5}{16} & \dfrac{1}{16} & 1 & 5 \end{bmatrix}.$$

Thus our original system (27) is equivalent to

(31)

$$x_1 + 0 + \frac{7}{16}x_3 \quad -\frac{5}{16}x_4 + 0 \ = 3$$

$$x_2 - \frac{1}{16}x_3 \quad +\frac{3}{16}x_4 + 0 \ = 1$$

$$\frac{5}{16}x_3 \quad +\frac{1}{16}x_4 + x_5 = 5.$$

The system (31) can be rewritten as

$$x_1 = 3 - \frac{7}{16}x_3 + \frac{5}{16}x_4$$

(32)

$$x_2 = 1 + \frac{1}{16}x_3 - \frac{3}{16}x_4$$

$$x_5 = 5 - \frac{5}{16}x_3 - \frac{1}{16}x_4.$$

The system (32) gives us the desired result. By assigning values to x_3 and x_4 independently, it is obvious that we have infinitely many solutions for the system (27).

It is apparent from this example that the method is quite general. Thus if we have m linear equations in $m + n$ unknowns, then we can solve these equations for m variables in terms of the remaining n variables.

We proceed now to apply these ideas to the solution of a linear programming problem. Let us begin with a specific problem which is closely related to Example 1.

EXAMPLE 2. Maximize the linear function

(33)
$$f(X) = x_1 + 2x_2$$

subject to the constraints

(34)
$$3x_1 + 5x_2 \leqq 14$$
$$x_1 + 7x_2 \leqq 10$$
$$x_1 \geqq 0$$
$$x_2 \geqq 0.$$

Solution. We recall that by the definition of "less than or equal to," the statement $3x_1 + 5x_2 \leqq 14$, means that there exists a real number $x_3 \geqq 0$ such that

$$3x_1 + 5x_2 + x_3 = 14.$$

Similarly there exists a real number $x_4 \geqq 0$ such that

$$x_1 + 7x_2 + x_4 = 10.$$

In this way we have introduced two more unknowns, x_3 and x_4. These unknowns (variables) are called *slack variables.*

Thus, the given linear programming problem is equivalent to the following:

Among all the possible solutions of the system of equations

(35)
$$
\begin{aligned}
3x_1 + 5x_2 + x_3 \quad\quad\quad &= 14 \\
x_1 + 7x_2 \quad\quad + x_4 &= 10 \\
-x_1 - 2x_2 \quad\quad\quad +f &= 0,
\end{aligned}
$$

determine a solution for which

(36) $\quad\quad\quad x_1 \geqq 0, \quad\quad x_2 \geqq 0, \quad\quad x_3 \geqq 0, \quad\quad x_4 \geqq 0,$

and for which f is a maximum.

The simplex method is a procedure for arriving at an optimal solution, when it exists. We begin by writing the system of equations (35) in a matrix form (also called a *simplex tableau*):

(37)
$$
A = \begin{array}{c@{\;}c}
\begin{array}{ccccc} x_1 & x_2 & x_3 & x_4 & f \end{array} & \\
\left[\begin{array}{ccccc|c}
3 & 5 & 1 & 0 & 0 & 14 \\
1 & 7 & 0 & 1 & 0 & 10 \\
\hline
-1 & -2 & 0 & 0 & 1 & 0
\end{array}\right] &
\end{array}.
$$

It is obvious that we already have one solution for the system, namely

(38) $\quad x_1 = 0, \quad\quad x_2 = 0, \quad\quad x_3 = 14, \quad\quad x_4 = 10, \quad\quad f = 0.$

However, this solution is not optimal, since from the last row of (37), we have

$$f = 0 + x_1 + 2x_2 - 0x_3 - 0x_4,$$

and, certainly, we can increase the value of f by increasing either x_1 or x_2. At the moment each of the unknowns x_1 and x_2 has the value 0. We can choose either one of these two variables and allow it to assume a positive value. It is certainly profitable to select x_2 for this purpose, because each unit increase in the value of x_2 increases the value of f by two units, while a corresponding unit increase in x_1 increases the value of f by only one unit. Therefore, let us increase x_2, while keeping $x_1 = 0$ temporarily.

We cannot, however, build x_2 up too big. For example, the first equation in (35) says that, with x_1 still at 0, x_2 cannot grow beyond 14/5 without making x_3 negative, thus violating one of the constraints in (36). Likewise, the second equation in (35) says that x_2 may not be increased at this stage to more than 10/7 without doing violence to x_4. Clearly, then, the allowable increase in x_2 may not be more than the smaller of the two ratios 10/7 and 14/5, namely 10/7.

When $x_1 = 0$ and $x_2 = 10/7$, we find that

$$x_3 = 14 - (5)\left(\frac{10}{7}\right) = \frac{48}{7}$$

$$x_4 = 10 - (7)\left(\frac{10}{7}\right) = 0$$

$$f = 0 + 2\left(\frac{10}{7}\right) = \frac{20}{7}.$$

We now have a new feasible solution:

(39) $x_1 = 0,$ $x_2 = \dfrac{10}{7},$ $x_3 = \dfrac{48}{7},$ $x_4 = 0,$ $f = \dfrac{20}{7}.$

In this new feasible solution, the value of f is substantially greater than it was before.

Now to avoid this tiresome line of reasoning at each stage we formulate a rule that gives us an improved feasible solution by systematically transforming the matrix (37) with the aid of the elimination procedure described in Example 1. Thus in the present situation we use the entry 7 as a pivot element and complete the elimination process with this pivot. Then (just as in Example 1) the matrix A in (37) is transformed into

(40)
$$
B =
\begin{bmatrix}
\dfrac{16}{7} & 0 & 1 & -\dfrac{5}{7} & 0 & \dfrac{48}{7} \\[2mm]
\dfrac{1}{7} & 1 & 0 & \dfrac{1}{7} & 0 & \dfrac{10}{7} \\[2mm]
\hline
-\dfrac{5}{7} & 0 & 0 & \dfrac{2}{7} & 1 & \dfrac{20}{7}
\end{bmatrix}
\begin{array}{l} x_1 \quad x_2 \quad x_3 \quad x_4 \quad f \end{array}.
$$

The key idea in this transformation is the selection of the pivot element. The pivot is selected by the following rule: (1) locate the "most negative" entry[1] in the last row and on the left side of the vertical bar; (2) divide each positive element in the column containing the most negative entry into the extreme right member of its row; and (3) select as a pivot the divisor that gives the smallest quotient.

Although the new feasible solution obtained in (39) gives a larger value for f than our first solution, the presence of negative numbers in the last row of (40) indicates that it may be possible to increase f still further. The most negative entry in the last row of (40) is $-(5/7)$ (in fact this is the only

---

[1] That negative entry for which the absolute value is greatest.

negative entry). We observe that all the remaining entries in the column containing $-(5/7)$ are positive. So, we calculate the ratios:

$$\frac{10/7}{1/7} = 10; \qquad \frac{48/7}{16/7} = 3.$$

The smallest ratio, 3, identifies $16/7$ as the next pivot. Rewriting (40) with the pivot element encircled, we have

(41)
$$B = \begin{array}{c} \begin{array}{ccccc} x_1 & x_2 & x_3 & x_4 & f \end{array} \\ \left[\begin{array}{ccccc|c} \boxed{\dfrac{16}{7}} & 0 & 1 & -\dfrac{5}{7} & 0 & \dfrac{48}{7} \\ \dfrac{1}{7} & 1 & 0 & \dfrac{1}{7} & 0 & \dfrac{10}{7} \\ \hline -\dfrac{5}{7} & 0 & 0 & \dfrac{2}{7} & 1 & \dfrac{20}{7} \end{array} \right] \end{array}.$$

Now if we complete the elimination process with the pivot $16/7$, the matrix B is transformed into

(42)
$$C = \begin{array}{c} \begin{array}{ccccc} x_1 & x_2 & x_3 & x_4 & f \end{array} \\ \left[\begin{array}{ccccc|c} 1 & 0 & \dfrac{7}{16} & -\dfrac{5}{16} & 0 & 3 \\ 0 & 1 & -\dfrac{1}{16} & \dfrac{3}{16} & 0 & 1 \\ \hline 0 & 0 & \dfrac{5}{16} & \dfrac{1}{16} & 1 & 5 \end{array} \right] \end{array}.$$

This matrix gives us an improved feasible solution. Since there are no negative entries in the last row, we can conclude that no further improvement in the value of f is possible. Indeed, the last row of (42) tells us that

(43)
$$f = 5 - \frac{5}{16}x_3 - \frac{1}{16}x_4.$$

Since x_3 and x_4 are both nonnegative, it is clear that the maximum value of f is equal to 5, and this is obtained when $x_3 = x_4 = 0$. Consequently, our optimal solution is

(44) $\qquad x_1 = 3, \qquad x_2 = 1, \qquad x_3 = 0, \qquad x_4 = 0, \qquad f = 5.$

EXAMPLE 3. A half-hour television show consists of a comedian, a magician, and a singer, with at least three minutes for commercials. The comedian insists on being on the air at least twice as long as the singer. The producer wants the combined time taken by the comedian and the singer to be at least as much as the time taken by the magician. In Neurosisville, 40, 60, and 50 people will watch the show for each minute the comedian, the magician, and the singer, respectively, are on the air. Find the time that must be allotted to each performer in order to get the maximum number of viewers in Neurosisville.

Solution.   Let the time allotted to the comedian, the magician, and the singer be x_1, x_2, and x_3 minutes, respectively. The problem is as follows:

Maximize the number of viewers

(45)
$$V = 40x_1 + 60x_2 + 50x_3$$

subject to the constraints

(46)
$$\begin{aligned} x_1 + x_2 + x_3 &\leq 27 \\ 2x_3 &\leq x_1 \\ x_2 &\leq x_1 + x_3 \\ x_i &\geq 0; \qquad i = 1, 2, 3. \end{aligned}$$

Introducing the slack variables, x_4, x_5, x_6, we see that the given linear programming problem is equivalent to the following:

Among all the possible solutions of the system of equations

(47)
$$\begin{aligned} x_1 + \quad x_2 + \quad x_3 + x_4 &\qquad\qquad\qquad = 27 \\ -x_1 \qquad\quad + \ 2x_3 \qquad + x_5 &\qquad\qquad = 0 \\ -x_1 + \quad x_2 - \quad x_3 \qquad\qquad + x_6 &\qquad = 0 \\ -40x_1 - 60x_2 - 50x_3 \qquad\qquad\qquad\quad + V &= 0, \end{aligned}$$

determine a solution for which

(48)
$$x_i \geq 0; \qquad i = 1, 2, \ldots, 6,$$

and for which V is a maximum.

We write the system of equations (47) in the matrix form:

(49)
$$A = \begin{bmatrix} x_1 & x_2 & x_3 & x_4 & x_5 & x_6 & V & \\ 1 & 1 & 1 & 1 & 0 & 0 & 0 & 27 \\ -1 & 0 & 2 & 0 & 1 & 0 & 0 & 0 \\ -1 & ① & -1 & 0 & 0 & 1 & 0 & 0 \\ -40 & -60 & -50 & 0 & 0 & 0 & 1 & 0 \end{bmatrix}.$$

To locate the pivot element in (49), we observe that the most negative entry in the last row is -60, and this appears in the second column. Dividing each positive entry of the second column into the extreme right member of its row, we obtain the ratios $0/1 = 0$ and $27/1 = 27$. The smallest ratio, 0, identifies 1 as the pivot element. The pivot is encircled in (49). The complete elimination process with pivot 1 transforms (49) into

$$
(50) \qquad B =
\begin{array}{c}
\begin{array}{ccccccc} x_1 & x_2 & x_3 & x_4 & x_5 & x_6 & V \end{array} \\
\left[
\begin{array}{cccccc|c}
2 & 0 & 2 & 1 & 0 & -1 & 0 \;\; 27 \\
-1 & 0 & ② & 0 & 1 & 0 & 0 \;\; 0 \\
-1 & 1 & -1 & 0 & 0 & 1 & 0 \;\; 0 \\
\hline
-100 & 0 & -110 & 0 & 0 & 60 & 1 \;\; 0
\end{array}
\right]
\end{array}.
$$

Since there are negative entries in the last row, we can continue the process. The entry 2 in the second row and third column of (50) is our new pivot. Using this pivot, we perform another complete elimination process and obtain

$$
(51) \qquad C =
\begin{array}{c}
\begin{array}{ccccccc} x_1 & x_2 & x_3 & x_4 & x_5 & x_6 & V \end{array} \\
\left[
\begin{array}{cccccc|c}
③ & 0 & 0 & 1 & -1 & -1 & 0 \;\; 27 \\
-\dfrac{1}{2} & 0 & 1 & 0 & \dfrac{1}{2} & 0 & 0 \;\; 0 \\
-\dfrac{3}{2} & 1 & 0 & 0 & \dfrac{1}{2} & 1 & 0 \;\; 0 \\
\hline
-155 & 0 & 0 & 0 & 55 & 60 & 1 \;\; 0
\end{array}
\right]
\end{array}.
$$

Again, the presence of the negative entry in the last row of (51) suggests that one more step is necessary. The new pivot is easily seen to be the entry 3 encircled in the first column of (51). The final complete elimination process yields

$$
(52) \qquad D =
\begin{array}{c}
\begin{array}{ccccccc} x_1 & x_2 & x_3 & x_4 & x_5 & x_6 & V \end{array} \\
\left[
\begin{array}{cccccc|c}
1 & 0 & 0 & \dfrac{1}{3} & -\dfrac{1}{3} & -\dfrac{1}{3} & 0 \;\; 9 \\
0 & 0 & 1 & \dfrac{1}{6} & \dfrac{1}{3} & -\dfrac{1}{6} & 0 \;\; \dfrac{9}{2} \\
0 & 1 & 0 & \dfrac{1}{2} & 0 & \dfrac{1}{2} & 0 \;\; \dfrac{27}{2} \\
\hline
0 & 0 & 0 & \dfrac{155}{3} & \dfrac{10}{3} & \dfrac{25}{3} & 1 \;\; 1395
\end{array}
\right]
\end{array}.
$$

Since all the entries in the last row of (52) are nonnegative, D is the final matrix. The last row of (52) tells us that

$$(53) \qquad V = 1395 - \frac{155}{3}x_4 - \frac{10}{3}x_5 - \frac{25}{3}x_6.$$

Since x_4, x_5, and x_6 are all nonnegative, it is clear that the maximum value of V is 1395, and this maximum is obtained when $x_4 = x_5 = x_6 = 0$. Consequently, from the matrix D, the optimal solution to our problem is

$$(54) \quad x_1 = 9, \quad x_2 = \frac{27}{2}, \quad x_3 = \frac{9}{2}, \quad x_4 = x_5 = x_6 = 0, \quad V = 1395.$$

In the next example, we shall see how the simplex method is applied to minimize a linear function f. The problem of minimizing a linear function f is equivalent to maximizing the linear function $-f$.

EXAMPLE 4. In Example 3, suppose that the time allotted for the commercials is at most six minutes. The comedian, the magician, and the singer cost the producer \$50, \$30, and \$20 per minute, respectively, and the commercials are free. Find the time that should be allotted to each performer so that the cost of the program is a minimum.

Solution. Let x_1, x_2, and x_3 be the same as in Example 3. The problem at hand is as follows:

Minimize the cost

$$(55) \qquad C = 50x_1 + 30x_2 + 20x_3$$

subject to the constraints

$$(56) \qquad \begin{aligned} x_1 + x_2 + x_3 &\leq 27 \\ x_1 &\geq 2x_3 \\ x_1 + x_3 &\geq x_2 \\ x_1 + x_2 + x_3 &\geq 24 \\ x_i &\geq 0; \qquad i = 1, 2, 3. \end{aligned}$$

Multiplying equation (55) by -1, and introducing the slack variables, we obtain the equivalent problem:

Maximize

$$(57) \qquad f = -C = -50x_1 - 30x_2 - 20x_3$$

subject to the constraints

$$x_1 + x_2 + x_3 + x_4 = 27$$

(58)
$$x_1 \qquad\qquad = 2x_3 + x_5$$
$$x_1 \qquad + x_3 \qquad = x_2 + x_6$$
$$x_1 + x_2 + x_3 \qquad = 24 + x_7$$

and

(59)
$$x_i \geqq 0; \qquad i = 1, 2, \ldots, 7.$$

As before, we rewrite the system of equations (57) and (58) in the matrix form:

(60)
$$
\begin{array}{cccccccc}
x_1 & x_2 & x_3 & x_4 & x_5 & x_6 & x_7 & f \\
\end{array}
$$
$$
\left[
\begin{array}{cccccccc|c}
1 & 1 & 1 & 1 & 0 & 0 & 0 & 0 & 27 \\
-1 & 0 & 2 & 0 & 1 & 0 & 0 & 0 & 0 \\
-1 & 1 & -1 & 0 & 0 & 1 & 0 & 0 & 0 \\
1 & 1 & 1 & 0 & 0 & 0 & -1 & 0 & 24 \\
\hline
50 & 30 & 20 & 0 & 0 & 0 & 0 & 1 & 0 \\
\end{array}
\right].
$$

Looking at the last row of (60) we observe that all of the entries are nonnegative. Since our rule for selecting the pivot element assumes that there is a negative element in the last row of the matrix, one may be tempted to conclude that we already have a maximal solution, namely $x_1 = x_2 = x_3 = 0$ and $f = 0$. But substituting these values in the fourth row [the fourth equation in (58)], we find that $x_7 = -24$ and this violates the condition $x_7 \geqq 0$. Thus this solution to the system of equations is not a feasible solution for the problem. Consequently the rule for finding a pivot element must be modified whenever the rule leads to a solution that is not feasible. The formation of a set of rules that always leads to a feasible solution is too complex for us to consider here. The reader should observe that for the problem at hand the first and fourth inequality in (56) give both an upper bound and a lower bound for $x_1 + x_2 + x_3$. It is this lower bound that causes the difficulty in this example.

To continue with the solution for Example 4, we modify the matrix given by (60) as follows: Add -50 times the fourth row to the last row, and add -1 times the fourth row to the first row. We then have

(61)
$$
\begin{array}{cccccccc}
x_1 & x_2 & x_3 & x_4 & x_5 & x_6 & x_7 & f \\
\end{array}
$$
$$
\left[
\begin{array}{cccccccc|c}
0 & 0 & 0 & 1 & 0 & 0 & 1 & 0 & 3 \\
-1 & 0 & ② & 0 & 1 & 0 & 0 & 0 & 0 \\
-1 & 1 & -1 & 0 & 0 & 1 & 0 & 0 & 0 \\
1 & 1 & 1 & 0 & 0 & 0 & -1 & 0 & 24 \\
\hline
0 & -20 & -30 & 0 & 0 & 0 & 50 & 1 & -1200 \\
\end{array}
\right].
$$

We now use our rule for locating the pivot element and complete the elimination process. We then arrive at the following sequence of matrices:

(62)

$$
\begin{array}{ccccccccc}
 & x_1 & x_2 & x_3 & x_4 & x_5 & x_6 & x_7 & f \\
\left[\begin{array}{cccccccc|c}
0 & 0 & 0 & 1 & 0 & 0 & 1 & 0 & 3 \\
-\dfrac{1}{2} & 0 & 1 & 0 & \dfrac{1}{2} & 0 & 0 & 0 & 0 \\
-\dfrac{3}{2} & ① & 0 & 0 & \dfrac{1}{2} & 1 & 0 & 0 & 0 \\
\dfrac{3}{2} & 1 & 0 & 0 & -\dfrac{1}{2} & 0 & -1 & 0 & 24 \\
\hline
-15 & -20 & 0 & 0 & 15 & 0 & 50 & 1 & -1200
\end{array}\right]
\end{array}
$$

(63)

$$
\begin{array}{ccccccccc}
 & x_1 & x_2 & x_3 & x_4 & x_5 & x_6 & x_7 & f \\
\left[\begin{array}{cccccccc|c}
0 & 0 & 0 & 1 & 0 & 0 & 1 & 0 & 3 \\
-\dfrac{1}{2} & 0 & 1 & 0 & \dfrac{1}{2} & 0 & 0 & 0 & 0 \\
-\dfrac{3}{2} & 1 & 0 & 0 & \dfrac{1}{2} & 1 & 0 & 0 & 0 \\
③ & 0 & 0 & 0 & -1 & -1 & -1 & 0 & 24 \\
\hline
-45 & 0 & 0 & 0 & 25 & 20 & 50 & 1 & -1200
\end{array}\right]
\end{array}
$$

(64)

$$
\begin{array}{ccccccccc}
 & x_1 & x_2 & x_3 & x_4 & x_5 & x_6 & x_7 & f \\
\left[\begin{array}{cccccccc|c}
0 & 0 & 0 & 1 & 0 & 0 & 1 & 0 & 3 \\
0 & 0 & 1 & 0 & \dfrac{1}{3} & -\dfrac{1}{6} & -\dfrac{1}{6} & 0 & 4 \\
0 & 1 & 0 & 0 & 0 & \dfrac{1}{2} & -\dfrac{1}{2} & 0 & 12 \\
1 & 0 & 0 & 0 & -\dfrac{1}{3} & -\dfrac{1}{3} & -\dfrac{1}{3} & 0 & 8 \\
\hline
0 & 0 & 0 & 0 & 10 & 5 & 35 & 1 & -840
\end{array}\right]
\end{array}
$$

From the last row of (64) we see that

(65)
$$f = -840 - 10x_5 - 5x_6 - 35x_7.$$

Since x_5, x_6, and x_7 are all nonnegative, we see at once that the maximum value of f is -840 and that this occurs when $x_5 = x_6 = x_7 = 0$. Hence the optimal solution is

(66)
$$x_1 = 8, \qquad x_2 = 12, \qquad x_3 = 4, \qquad x_4 = 3,$$
$$x_5 = x_6 = x_7 = 0, \qquad f = -840.$$

Consequently the minimum cost is $840, with $x_1 = 8$, $x_2 = 12$, and $x_3 = 4$.

Exercise 3

1. Find the pivot element in each of the following matrices:

(a) $\begin{bmatrix} 2 & 4 & 0 & | & 2 \\ 3 & 1 & 0 & | & 1 \\ \hline -2 & -1 & 1 & | & 0 \end{bmatrix}$.

(b) $\begin{bmatrix} 1 & 1 & 0 & | & 4 \\ 3 & 2 & 0 & | & 2 \\ \hline -3 & -3 & 1 & | & 0 \end{bmatrix}$.

(c) $\begin{bmatrix} 3 & 1 & 2 & 0 & | & 5 \\ -4 & 7 & 2 & 0 & | & 6 \\ \hline 1 & -3 & -2 & 1 & | & 2 \end{bmatrix}$.

(d) $\begin{bmatrix} 4 & 3 & 4 & 0 & | & 2 \\ 2 & -1 & -2 & 0 & | & 4 \\ 3 & 2 & 4 & 0 & | & 5 \\ \hline -3 & -2 & -1 & 1 & | & 3 \end{bmatrix}$.

2. Transform each of the matrices of problem 1 into a new matrix, using the complete elimination process with the pivot element found in problem 1.

3. Fill in the details in Example 3 and Example 4.

4. Solve problems 6 through 10 of Exercise 2 by the simplex method.

5. A coffee company mixes x_1, x_2, and x_3 pounds of coffee of type I, II, and III, respectively, to obtain a mixture of 100 pounds. The coffee of type I, II, and III costs $1, 50 cents, and 25 cents per pound, respectively. Suppose that the mixture must contain at least 30 pounds of coffee of type I and $x_1 + x_2 \geq x_3$. Find x_1, x_2, and x_3 so that the company has maximum profit if the mixture sells for 75 cents per pound.

6. Major Motors, Inc. produces cars at factory I and factory II. Factory I produces at most 120 cars per week, and factory II produces at most 250 cars per week. The company wishes to transport these cars to two dealers A and B. Dealer A requires at least 200 cars per week and dealer B requires at least 150 cars per week. If the cost of transporting cars from factory I to dealer A and B is $120 and $100 per car, respectively, and the cost of transporting cars from factory II to dealers A and B is $130 and $150 per car, respectively, find the number of cars the company must transport from each of the two factories to each of the two dealers so that the cost of transportation is minimum.

7. The mathematics department of Uproar University has x_1, x_2, and x_3 full, associate, and assistant professors, respectively, and the administration insists that $x_1 + x_2 \leq x_3$. The full, associate, and assistant professors teach 6, 8, and 10 hours per week, respectively. The University requires the department to schedule at least 480 hours of teaching per

week. The weekly salary of full, associate, and assistant professors is $300, $250, and $200, respectively. The total weekly salary budget for the department is at most $15,000. If the research papers produced by full, associate, and assistant professors per year is 3, 2, and 1, respectively, find the number of professors in each rank the department must have so that the department produces the maximum number of research papers.

8. In problem 7, if $x_1 + x_3 \geq 2x_2$, find x_1, x_2, and x_3 so that the budget of the department is a minimum.

9. A manufacturer uses machines A, B, and C to produce x_1 items of product I, x_2 items of product II, and x_3 items of product III per week. Each item of product I requires two hours on machine A, five hours on machine B, and three hours on machine C. Each item of product II requires six hours on machine A, two hours on machine B, and six hours on machine C. Each item of product III required five hours on machine A, four hours on machine B, and four hours on machine C. The machines are operational eight hours a day, five days a week. If he makes a profit of $3, $4.50, and $4 on each item of products I, II, and III, respectively, find x_1, x_2, and x_3 so as to maximize his profit.

the theory of games

The theory of games forms a complicated and difficult branch of mathematics, and we can give only a brief introduction to this topic.

The mathematical analysis of games had its beginnings in 1921 in a brief note by Emile Borel.[1] However, the firm foundation along with much of the superstructure was the work of John von Neumann,[2] who proved the fundamental theorem of game theory (see Theorem 1, page 317).

Although the games treated in this chapter are rather elementary, the reader must realize that the ultimate objective is an accurate analysis of more complicated and more important games. For example, the competition for the automobile market can be regarded as a game in which there are four (giant) players and the stakes are extremely high. In other applications of game theory to economics, there may be only two players, as in the negotations for the sale of a commodity between a single buyer and a single merchant.

A farmer trying to decide when to plant his crop is playing a game against nature. So also are the oil companies when they try to decide where to drill for oil, and how far to drill, before they abandon the well.

The theory of games is still in its infancy and most of the practical

---

[1] "La théorie du jeu et les équations integrales à noyau symétrique gauche," *Comptes Rendus,* vol. 173 (1921), pp. 1304–1308.

[2] Von Neumann, among the greatest mathematicians of this century, undoubtedly saved his life by leaving Germany and taking refuge in this country. The flood of great scientists who likewise found safety in the United States just before the Second World War included such men as Albert Einstein, Enrico Fermi, Hermann Weyl, and Carl Ludwig Siegel. These distinguished refugees gave a tremendous lift to U.S. science.

problems are still unsolved. However, the few successes achieved so far justify the continued attempt to extend the theory.

1. matrix games

Many games can be put into a matrix form. The essential features for such games are as follows:

(I) There are two players, traditionally known as R and C (the reasons for these names will soon be clear).

(II) On each play of the game, R may make any one of m moves, $R_1, R_2, \ldots, R_m$. Similarly C may make any one of n moves, C_1, $C_2, \ldots, C_n$.

(III) After R and C have each made a move, the winner is determined according to the rules of the game, and a suitable payment is made. If R plays the move R_i and C plays the move C_j, then the amount that C pays R is denoted by a_{ij}, where if a_{ij} is negative, then R pays C the amount $|a_{ij}|$.

Under these conditions it is clear that the matrix $[a_{ij}]$ can be used to represent the game. Such a matrix representation gives all the information that we need to analyze the game.

EXAMPLE 1. Each of the two players R and C has three cards: a 2, a 3, and a 4. Each man selects a card, and simultaneously they lay the card selected face up on the table. If they play the same card, no payment is made. If they play different cards, then the man who played the lower (face value) card wins (in dollars) the sum of the face values of the two cards. Give the matrix representation of this game.

Solution. The move R_i (or C_i) is the play of the card $i + 1$. Then according to the rules of the game, the matrix representation for this game is

(1)
$$A = \begin{array}{c} \\ 2 \\ 3 \\ 4 \end{array} \begin{array}{ccc} 2 & 3 & 4 \\ \left[\begin{array}{ccc} 0 & 5 & 6 \\ -5 & 0 & 7 \\ -6 & -7 & 0 \end{array}\right]. \end{array}$$

We use this example to illustrate some terminology. For obvious reasons such a game is called a *two-person game*. It is also called a *zero-sum*[1] game,

---

[1] The non-zero-sum games are too complicated to be discussed here. The interested reader is referred to *Games and Decisions*, by R. D. Luce and H. Raiffa, Wiley, New York, 1957.

since the total payment on each play is zero; i.e., R wins the same amount that C loses, or C wins the same amount that R loses. The entry a_{ij} is called the *payoff*, and the matrix is called the *game matrix* (or the *payoff matrix*).

Conversely any $m \times n$ matrix $A = [a_{ij}]$ can be regarded as the game matrix for a suitable two-person zero-sum game in which the player R chooses any one of the m rows of A and simultaneously the player C chooses any one the n columns of A. The entry in the chosen row and column is the payoff.

We assume that the game may be played repeatedly, and the sole guiding motivation for R is to maximize his winnings, and that for C is to minimize his losses.

EXAMPLE 2. In the game described in Example 1, how should R play in order to maximize his winnings? How should C play to minimize his losses?

Solution. The game matrix of Example 1 is

$$(1) \qquad\qquad A = \begin{bmatrix} 0 & 5 & 6 \\ -5 & 0 & 7 \\ -6 & -7 & 0 \end{bmatrix}.$$

Of course the player R would like to win \$7 by playing the second row. But he knows that since C is involved in the game, C might play the first column. Thus R would end up losing \$5 instead of gaining \$7. He therefore examines each of his moves to check just how much he can win even if C makes the best available countermove. Hence R must consider the worst payoff for each of his moves. To aid the analysis, R encircles the smallest element in every row and thus obtains

$$A = \begin{bmatrix} \boxed{0} & 5 & 6 \\ \boxed{-5} & 0 & 7 \\ -6 & \boxed{-7} & 0 \end{bmatrix}.$$

Of these encircled numbers, he must obviously choose the largest, i.e., the entry zero in the first row. Thus R must always play the first row, because he has nothing to lose and perhaps may gain something if C is not careful.

Now C's interests are directly in conflict with those of R, but his reasonings are similar to R's. Thus C boxes (since R chose circles) the worst payoff for each of his moves. That is, he boxes the largest element in every column and thus obtains

$$A = \begin{bmatrix} \boxed{0} & \boxed{5} & 6 \\ -5 & 0 & \boxed{7} \\ -6 & -7 & 0 \end{bmatrix}.$$

Of these boxed numbers, C must obviously choose the smallest, i.e., the zero entry in the first column. Hence C must always play the first column.

The "best move" for the players is as follows: R always plays the first row and C always plays the first column. If both players use their best moves, then the payoff is zero; i.e., the game is *fair*.

EXAMPLE 3. Mr. X, a mutual mathematician friend of R and C, suggests the following version of matching pennies to them. If R and C both show heads, R wins \$4 from C, and if they both show tails, then R wins \$9 from C. However, if R shows heads and C shows tails, then C wins \$12 from R, and if R shows tails and C shows heads, then C wins \$3 from R. Just as Mr. X is assuring them that this game is fair (students will see in Section 4 that he was right), the dean calls him for a conference. Help R and C by **(1)** finding the matrix representation for this game, **(2)** suggesting how R should play in order to maximize his winnings, and **(3)** suggesting how C should play in order to minimize his losses.

Solution. It is easily seen that the matrix representation for this game is

$$A = \begin{array}{c} \\ H \\ T \end{array} \begin{array}{c} H \qquad T \\ \begin{bmatrix} 4 & -12 \\ -3 & 9 \end{bmatrix} \end{array}.$$

Now, if we follow the same argument used in Example 2, it would seem that the best thing for R to do is always to play the second row, and the best thing for C to do is always to play the first column. If they both follow this suggestion, then it is clear that C would win \$3 on every play. Such a game could hardly be called fair. So, if Mr. X is right (and we swear that he is right), then R should not play the second row every time.

Suppose, indeed, that R knows that C will play the first column. Then R can win \$4 by playing the first row. On the other hand, if C knows that R will play the first row, then C can shift to the second column and win \$12. Consequently, R must try to confuse C by playing the first row some times and the second row on other occasions. Similarly C should try to confuse R by not always playing the same column. The question that naturally arises is, on what percentage of the plays should R select the first row? Before we can answer this question, we need to introduce some more terminology and learn a little more theory. We regret that we cannot help R and C at this time, but we will help them both in Section 4.

By a *strategy* of R for a given matrix game A, we mean the decision by R to select rows of A in some manner. This may be a selection of some particular row in every play or a selection of rows in a random manner with a definite probability for each row. In either case we can denote the strategy of R by a row probability vector $P = [p_1, p_2, \ldots, p_m]$; i.e., R chooses the first row with probability p_1, the second row with probability p_2, etc. Similarly, we can denote a strategy of C by a column probability vector:

$$Q = \begin{bmatrix} q_1 \\ q_2 \\ \vdots \\ q_n \end{bmatrix}.$$

We recall from Chapter 8 that if P and Q are probability vectors, then

$$p_i \geqq 0 \text{ for } i = 1, 2, \ldots m; \qquad q_i \geqq 0 \text{ for } i = 1, 2, \ldots n,$$

and

(2) $p_1 + p_2 + \cdots + p_m = 1$

(3) $q_1 + q_2 + \cdots + q_n = 1.$

These conditions on P and Q will be used frequently in this chapter.

A strategy $P = [p_1, p_2, \ldots, p_m]$ for R is called a *pure strategy* if for some index i, $p_i = 1$ and all other p_j's are equal to zero. If P is not a pure strategy, then it is called a *mixed strategy*. Thus, a pure strategy for R means that R plays a particular row every time. Pure and mixed strategies for C are defined analogously. Because of the meaning usually given to the word *strategy,* there could be some confusion. But we would like to emphasize that when we say P is a strategy for R, we are not implying that this is his best strategy.

Let us consider a game defined by the 2×2 matrix

$$A = \begin{bmatrix} a_{11} & a_{12} \\ a_{21} & a_{22} \end{bmatrix}.$$

Let the strategy for R be $P = [p_1, p_2]$ and the strategy for C be $Q = \begin{bmatrix} q_1 \\ q_2 \end{bmatrix}$.

Now the probability that R wins the amount a_{11} (or loses the amount $|a_{11}|$ if a_{11} is negative) is $p_1 q_1$. Similarly the probabilities that R wins the amounts a_{12}, a_{21}, and a_{22} are $p_1 q_2$, $p_2 q_1$, and $p_2 q_2$, respectively. Let us denote by $E(P,Q)$, the expectation of R (the expected value of the amount R wins) when R uses the strategy P and C uses the strategy Q. Then

(4) $E(P,Q) = p_1 q_1 a_{11} + p_1 q_2 a_{12} + p_2 q_1 a_{21} + p_2 q_2 a_{22}.$

We can write the right-hand side of (4) as a product of matrices:

(5) $E(P,Q) = PAQ.$

Although we used a 2 × 2 matrix for simplicity, it is easily seen that if R and C use strategies P and Q, respectively, for an m × n matrix game A, then the expectation E(P,Q) of R is given by (5).

Exercise 1

In problems 1 through 3, represent each of the games by a two-person zero-sum matrix game.

1. R and C choose, independently, one of the numbers 0, 1, 2, and 3. If both choose the same number, then C pays to R the amount of the chosen number; otherwise R pays to C the amount of his own number.
2. R and C choose, independently, one of the numbers 2, 0, and −1. Let R's choice be denoted by x and that of C by y. Then C pays R the amount (x − y)(x + y).
3. R has an urn containing four balls numbered 1, 2, 3, and 4. C has a fair die. Let x be the number on the ball drawn at random by R, and let y be the number that C rolls with the die. If x + y is even, then R pays C $2; otherwise R wins $3 from C.
4. In problem 3, what are the strategies of R and C? Compute the expectation of R.

In problems 5 through 8, A defines a two-person matrix game. P and Q are the strategies of R and C, respectively. In each case compute the expectation of R.

5. $A = \begin{bmatrix} 3 & -2 \\ 1 & 2 \end{bmatrix}$; $P = \begin{bmatrix} \frac{1}{2}, \frac{1}{2} \end{bmatrix}$, $Q = \begin{bmatrix} \frac{1}{3} \\ \frac{2}{3} \end{bmatrix}$.

6. $A = \begin{bmatrix} 3 & 1 & -4 \\ -2 & 3 & 2 \end{bmatrix}$; $P = \begin{bmatrix} \frac{1}{2}, \frac{1}{2} \end{bmatrix}$, $Q = \begin{bmatrix} \frac{1}{5} \\ \frac{4}{5} \\ 0 \end{bmatrix}$.

7. $A = \begin{bmatrix} 3 & -2 & -1 \\ 2 & -3 & 1 \\ -2 & 1 & 3 \end{bmatrix}$; $P = \begin{bmatrix} \frac{1}{5}, \frac{2}{5}, \frac{2}{5} \end{bmatrix}$, $Q = \begin{bmatrix} \frac{1}{3} \\ \frac{4}{9} \\ \frac{2}{9} \end{bmatrix}$.

8. $A = \begin{bmatrix} 1 & -2 & -1 & -3 \\ -2 & 0 & 2 & 1 \\ 1 & -2 & 0 & 1 \end{bmatrix}$; $P = \left[\dfrac{1}{2}, \dfrac{1}{3}, \dfrac{1}{6}\right]$, $Q = \begin{bmatrix} \dfrac{1}{4} \\ 0 \\ \dfrac{1}{4} \\ \dfrac{1}{2} \end{bmatrix}$.

2.   the value of a game

In the preceding section, we saw that if P and Q are strategies of R and C, respectively, for a matrix game A, then the expected gain of R is $E(P,Q) = PAQ$. Now if we assume that R is a rational person (an indulgent father purposely trying to lose to his child will be considered as irrational), he will try to find a strategy $P\star$, called an *optimum strategy for R*, so that his expected gain is as large as possible, regardless of what C does. That is, R wishes to find a maximum number v_R and a (optimum) strategy $P\star$ such that

(6) $$P\star AQ \geqq v_R$$

for every strategy Q of C.

Similarly, the best thing for C to do is to find a strategy $Q\star$, called an *optimum strategy for C*, so that regardless of what R does, C's payments to R(or R's winnings) are as small as possible. In other words, C wishes to find a smallest number v_C and a (optimum) strategy $Q\star$ such that

(7) $$PAQ\star \leqq v_C$$

for every strategy P of R.

The fundamental theorem of game theory is

> **Theorem 1.** For every matrix game A, there exist optimum strategies $P\star$ and $Q\star$ for R and C, respectively, and numbers v_R and v_C such that (6) and (7) hold. Furthermore, $v_R = v_C$.

The proof is somewhat difficult, so we omit it.[1]

Writing $v_R = v_C = v$, the inequalities (6) and (7) can be rewritten as

(8) $$P\star AQ \geqq v$$

---

[1] The energetic reader can find a proof of this theorem in *Introduction to the Theory of Games* by J. C. C. McKinsey, McGraw-Hill, New York, 1952, pp. 32–37.

for every strategy Q of C and

(9) $$PAQ\star \leqq v$$

for every strategy P of C.

Definition 1. Let A be a matrix game. If the inequalities (8) and (9) hold, then $P\star$ and $Q\star$ are called optimum strategies for R and C, respectively, and v is called the value of the game. If $v = 0$, the game is said to be fair.

We now prove some interesting theorems regarding the value of the game. For an $m \times n$ matrix game A, we shall always use

$$P\star = [p_1^\star, p_2^\star, \ldots, p_m^\star] \quad \text{and} \quad Q\star = \begin{bmatrix} q_1^\star \\ q_2^\star \\ \vdots \\ q_n^\star \end{bmatrix}$$

to denote the optimum strategies for R and C, respectively,

Theorem 2. If A is a matrix game, then

$$E(P\star, Q\star) = P\star AQ\star = v.$$

The proof is left for the next exercise.

Theorem 3. Let A be an $m \times n$ matrix game with $P\star$ and $Q\star$ as optimum strategies for R and C, respectively. Let v be a number, and let

$$V = [v, v, \ldots, v] \quad \text{and} \quad V' = \begin{bmatrix} v \\ v \\ \vdots \\ v \end{bmatrix}$$

be an n-component row vector and an m-component column vector, respectively. Then the number v is the value of the game if and only if

(10) $$P\star A \geqq V$$

and

(11) $$AQ\star \leqq V'.$$

Proof. We first prove that if v is the value of the game, then (10) and (11) hold. By definition, we have

(12) $$P \star A Q \geqq v$$

for every strategy Q of C.
    Let us write

(13) $$P \star A = [x_1, x_2, \ldots, x_n].$$

Now if C uses the strategy

$$Q_1 = \begin{bmatrix} 1 \\ 0 \\ 0 \\ \vdots \\ 0 \end{bmatrix},$$

then from (12) we get

(14) $$P \star A Q_1 = [x_1, x_2, \ldots, x_n] \begin{bmatrix} 1 \\ 0 \\ 0 \\ \vdots \\ 0 \end{bmatrix} \geqq v.$$

Multiplying the matrices on the left-hand side of (14), we obtain $x_1 \geqq v$. Similarly, if we assume that C uses the strategy

$$Q_2 = \begin{bmatrix} 0 \\ 1 \\ 0 \\ 0 \\ \vdots \\ 0 \end{bmatrix},$$

then from $P \star A Q_2$ we obtain $x_2 \geqq v$.
    In general, using the pure strategy Q_i for $i = 1, 2, \ldots, n$, we find that

(15) $$x_i \geqq v.$$

From (13), (15), and the definition of V, it follows that

$$P \star A \geqq V.$$

The proof that $A Q \star \leqq V'$ is similar.

Conversely, let us assume that (10) and (11) hold. Multiplying (10) on the right by $Q\star$, we get

(16) $$P\star AQ\star \geqq VQ\star = v.$$

Similarly, multiplying (11) on the left by $P\star$, we obtain

(17) $$P\star AQ\star \leqq P\star V' = v.$$

From (16) and (17), it follows that

(18) $$P\star AQ\star = v.$$

Hence by Theorem 2, v is the value of the game. ∎

We cannot prove that an optimum strategy is unique because, in the next section, we shall see some examples in which R and C each have two different optimum strategies. However, for the value of a game we have

Theorem 4. Every matrix game has a unique value.

Proof. Let A be an $m \times n$ matrix game, and let u and v be two values of the game. We shall prove that $u = v$.

To this end, let us suppose that R and C have optimum strategies $P\star$ and $Q\star$, respectively, associated with the value v and have optimum strategies $P\star\star$ and $Q\star\star$, respectively, associated with the value u. Then, by Theorem 3, we have

(19) $$P\star A \geqq V,$$
(20) $$AQ\star \leqq V',$$
(21) $$P\star\star A \geqq U,$$
and
(22) $$AQ\star\star \leqq U',$$

where $V = [v, v, \ldots, v]$ and $U = [u, u, \ldots, u]$ are n-component row vectors, and

$$V' = \begin{bmatrix} v \\ v \\ \vdots \\ v \end{bmatrix} \quad \text{and} \quad U' = \begin{bmatrix} u \\ u \\ \vdots \\ u \end{bmatrix}$$

are m-component column vectors.

Multiplying (19) on the right by $Q\star\star$ and multiplying (22) on the left by $P\star$, we get

(23)
$$P\star AQ\star\star \geqq VQ\star\star = v$$

and

(24)
$$P\star AQ\star\star \leqq P\star U' = u.$$

From (23) and (24) we have $v \leqq P\star AQ\star\star \leqq u$, and hence

(25)
$$v \leqq u.$$

Similarly, multiplying (20) on the left by $P\star\star$ and (21) on the right by $Q\star$, we obtain

(26)
$$u \leqq v.$$

Thus (25) and (26) together imply that $u = v$. ∎

To solve a given matrix game A means to find optimum strategies for R and C and the value of the game. Before we give a method for solving an arbitrary $m \times n$ matrix game A in Section 5, we shall discuss special cases in Sections 3 and 4.

Exercise 2

1. Prove Theorem 2. *Hint:* Use Theorem 1.
2. For V, V', $P\star$, and $Q\star$ defined in Theorem 3, prove that **(a)** $VQ\star = v$ and **(b)** $P\star V' = v$.
3. In Theorem 3, prove that $AQ\star \leqq V'$.

In problems 4 through 8, verify that $P\star$ and $Q\star$ are optimum strategies for R and C, respectively, for the given matrix game A. *Hint:* First find a tentative value of the game by using the formula $v = P\star AQ\star$. Then form the appropriate V and V' and prove that v is the value of the game by showing that $P\star A \geqq V$ and $AQ\star \leqq V'$.

4. $A = \begin{bmatrix} -4 & 1 & -3 \\ 3 & 2 & 4 \\ 5 & -2 & -1 \end{bmatrix}$; $P\star = [0, 1, 0]$, $Q\star = \begin{bmatrix} 0 \\ 1 \\ 0 \end{bmatrix}$.

5. $A = \begin{bmatrix} 1 & -1 \\ -1 & 1 \end{bmatrix}$; $P\star = \left[\frac{1}{2}, \frac{1}{2}\right]$, $Q\star = \begin{bmatrix} \frac{1}{2} \\ \frac{1}{2} \end{bmatrix}$.

6. $A = \begin{bmatrix} 1 & 0 & 0 \\ 0 & 1 & 0 \\ 0 & 0 & 1 \end{bmatrix}$; $P\star = \left[\dfrac{1}{3}, \dfrac{1}{3}, \dfrac{1}{3}\right]$, $Q\star = \begin{bmatrix} \dfrac{1}{3} \\ \dfrac{1}{3} \\ \dfrac{1}{3} \end{bmatrix}$.

7. $A = \begin{bmatrix} 1 & 0 & 2 & 0 \\ -6 & -4 & 5 & -2 \\ 2 & 0 & 3 & 0 \\ 3 & -3 & 2 & -1 \end{bmatrix}$; $P\star = \left[\dfrac{3}{8}, 0, \dfrac{5}{8}, 0\right]$, $Q\star = \begin{bmatrix} 0 \\ \dfrac{1}{4} \\ 0 \\ \dfrac{3}{4} \end{bmatrix}$.

8. $A = \begin{bmatrix} 3 & -\dfrac{1}{2} & 2 \\ \dfrac{3}{2} & -1 & -\dfrac{1}{2} \\ -2 & \dfrac{1}{2} & \dfrac{3}{2} \end{bmatrix}$; $P\star = \left[\dfrac{5}{12}, 0, \dfrac{7}{12}\right]$, $Q\star = \begin{bmatrix} \dfrac{1}{6} \\ \dfrac{5}{6} \\ 0 \end{bmatrix}$.

9. Let A be a matrix game with value v and optimum strategies $P\star$ and $Q\star$ for R and C, respectively. Prove that for any $k > 0$, the game kA has value kv with the same optimum strategies for R and C as before.

10. If every entry in a matrix game A is increased by an amount k, show that the value of the game is also increased by k but that the optimum strategies remain the same.

11. If every entry in a matrix game is positive, show that the value of the game is also positive.

3. strictly determined games

Definition 2. A matrix game is said to be strictly determined if the matrix has an entry which is a minimum in its row and simultaneously a maximum in its column. Such an entry is called a saddle point.

EXAMPLE 1. Find a saddle point in the matrix

$$A = \begin{bmatrix} 6 & 3 & 4 & 2 \\ 3 & -1 & 3 & 1 \\ 5 & 6 & -3 & 0 \end{bmatrix}.$$

Solution. Let us rewrite the matrix A, encircling the minimum entry in each row:

$$A = \begin{bmatrix} 6 & 3 & 4 & ②\\ 3 & ⊖1 & 3 & 1\\ 5 & 6 & ⊖3 & 0 \end{bmatrix}.$$

We then inspect A to see if one of these encircled entries is maximum in its column. We observe that 2 is the maximum entry in the fourth column. Thus, 2 is a saddle point and the matrix game defined by A is strictly determined.

EXAMPLE 2. Find a saddle point in the matrix

$$A = \begin{bmatrix} 6 & 3 & 4 & 2\\ 3 & -1 & 3 & 1\\ 5 & 6 & -3 & 4 \end{bmatrix}.$$

Solution. Following the procedure of Example 1, we see that A does not have a saddle point. Thus, the matrix game defined by A is a nonstrictly determined game.

The above examples illustrate that while some matrix games are strictly determined, others are not. If a matrix game $A = [a_{ij}]$ is strictly determined and an entry, say a_{ks}, is a saddle point, then by rearranging and renumbering the rows and columns of A we can assume that a_{11} is a saddle point of A. The solution of a strictly determined game is contained in

Theorem 5. Let a_{11} be a saddle point of an $m \times n$ matrix game A. Then an optimum strategy for R is always to play the first row, and an optimum strategy for C is always to play the first column. Further, a_{11} is the value of the game.

Proof. Let $P\star = [p_1^\star, p_2^\star, \ldots, p_m^\star]$ be an optimal strategy for R and let v be the value of the game. Then by Theorem 3 each component of $P\star A$ is greater than or equal to v; i.e.,

(27)
$$\begin{aligned} p_1^\star a_{11} + p_2^\star a_{21} + \cdots + p_m^\star a_{m1} &\geq v\\ p_1^\star a_{12} + p_2^\star a_{22} + \cdots + p_m^\star a_{m2} &\geq v\\ &\vdots\\ p_1^\star a_{1n} + p_2^\star a_{2n} + \cdots + p_m^\star a_{mn} &\geq v. \end{aligned}$$

Since a_{11} is a saddle point, it is the largest element in the first column. Hence

(28) $a_{11} \geq a_{i1}, \qquad i = 2, 3, \ldots, m.$

Thus, from the first inequality of (27) and (28) we obtain

$$a_{11} = 1 \cdot a_{11} = (p_1^\star + p_2^\star + \cdots + p_m^\star)a_{11}$$

(29)
$$= p_1^\star a_{11} + p_2^\star a_{11} + \cdots + p_m^\star a_{11}$$

$$\geqq p_1^\star a_{11} + p_2^\star a_{21} + \cdots + p_m^\star a_{m1} \geqq v.$$

Therefore

(30)
$$a_{11} \geqq v.$$

Similarly, letting

$$Q^\star = \begin{bmatrix} q_1^\star \\ q_2^\star \\ \vdots \\ q_n^\star \end{bmatrix}$$

be an optimum strategy for C and using Theorem 3 and the fact that a_{11} is the smallest element in the first row of A, we get

(31)
$$a_{11} \leqq v.$$

From (30) and (31), we conclude that $a_{11} = v$. Thus, the value of the game is a_{11}.

From (29) we notice that if $p_1^\star = 1$ and $p_2^\star = p_3^\star = \cdots = p_m^\star = 0$, then we do get $v = a_{11}$. Thus an optimum strategy for R is $P\star = [1, 0, 0, \ldots, 0]$. Similarly an optimum strategy for C is

$$Q^\star = \begin{bmatrix} 1 \\ 0 \\ 0 \\ \vdots \\ 0 \end{bmatrix}. \quad \blacksquare$$

EXAMPLE 3. Solve the matrix game

$$A = \begin{bmatrix} 3 & 4 & 3 & 6 \\ 3 & 7 & 3 & 5 \\ 2 & -5 & 1 & 4 \end{bmatrix}.$$

Solution. We observe that the game defined by the matrix A is a strictly determined game. The value of the game is 3.

In this game, $[1, 0, 0]$ and $[0, 1, 0]$ are optimum strategies for R. Similarly

$$\begin{bmatrix} 1 \\ 0 \\ 0 \\ 0 \end{bmatrix} \quad \text{and} \quad \begin{bmatrix} 0 \\ 0 \\ 1 \\ 0 \end{bmatrix}$$

are optimum strategies for C. This example shows that an optimum strategy is not unique.

Exercise 3

In problems 1 through 6, find those matrix games that are strictly determined.

1. $A = \begin{bmatrix} 1 & -1 \\ -1 & 1 \end{bmatrix}$.

2. $A = \begin{bmatrix} 1 & -2 & 3 \\ 3 & -1 & 2 \end{bmatrix}$.

3. $A = \begin{bmatrix} 2 & 3 & 1 & 4 \\ -1 & 1 & -2 & 2 \\ -2 & 3 & -1 & 5 \end{bmatrix}$.

4. $A = \begin{bmatrix} -1 & 3 & -2 & 2 \\ 1 & -1 & 5 & -7 \\ 4 & 4 & 6 & 5 \end{bmatrix}$.

5. $A = \begin{bmatrix} 3 & -4 & 2 & 5 \\ 2 & 3 & -3 & -4 \\ 1 & 2 & 3 & -2 \end{bmatrix}$.

6. $A = \begin{bmatrix} 2 & 1 & 3 & 1 \\ 1 & -1 & 0 & -2 \\ 3 & 1 & 5 & 1 \\ 4 & -2 & 1 & 0 \end{bmatrix}$.

7. For each of the strictly determined games in problems 1 through 6, find the value of the game and optimum strategies for the players R and C.

8. In Example 3 verify that

$$\begin{bmatrix} \frac{1}{2}, \frac{1}{2}, 0 \end{bmatrix} \quad \text{and} \quad \begin{bmatrix} \frac{1}{3} \\ 0 \\ \frac{2}{3} \\ 0 \end{bmatrix}$$

are also optimum strategies for R and C, respectively.

9. In Example 3, show that

$$[a, 1 - a, 0] \quad \text{and} \quad \begin{bmatrix} b \\ 0 \\ 1 - b \\ 0 \end{bmatrix}$$

are also optimum strategies for R and C, respectively, where $0 \leq a \leq 1$ and $0 \leq b \leq 1$.

★**10.** Let A be a matrix game. Suppose that $P\star$ and $P\star\star$ are both optimum strategies for R and that $Q\star$ and $Q\star\star$ are both optimum strategies for C. Show that

$$P = aP\star + (1 - a)P\star\star \qquad \text{and} \qquad Q = bQ\star + (1 - b)Q\star\star$$

are also optimum strategies for R and C, respectively, where $0 \leq a \leq 1$ and $0 \leq b \leq 1$.

11. Suppose that in a 2×2 matrix game both entries of a particular row or a particular column are equal. Prove that the game is strictly determined.

★**12.** If a_{11} and a_{ij} are two saddle points of the matrix game $A = [a_{ij}]$, prove that

$$a_{11} = a_{1j} = a_{ij} = a_{i1}.$$

13. Suppose R and C play a game in which each of the two players shows one or two fingers simultaneously. They agree that R pays C an amount equal to the total number of fingers shown. Write a matrix for this game and determine the value of the game and the optimum strategies.

14. Show that the game described in problem 2 of Exercise 1 is strictly determined. Find its value. Is the game fair?

4. 2×2 matrix games

Let

(32)
$$A = \begin{bmatrix} a & b \\ c & d \end{bmatrix}$$

be a matrix game. If A is a strictly determined game, then its solution is described in Section 3. To decide whether or not a 2×2 matrix game is strictly determined, we can use

Theorem 6.   The matrix game

(32)
$$A = \begin{bmatrix} a & b \\ c & d \end{bmatrix}$$

is nonstrictly determined if and only if each of the entries on one of the diagonals is greater than each of the entries on the other diagonal. In symbols, the game is nonstrictly determined if and only if, either

	(i)	$a, d > b$	and	$a, d > c$

or

	(ii)	$b, c > a$	and	$b, c > d.$

Proof.   If either (i) or (ii) holds, then it is easy to see that none of the elements a, b, c, and d is simultaneously the smallest element in its row and the largest element in its column. Hence the game A has no saddle point and therefore the game is nonstrictly determined.

Conversely, suppose the matrix game A is nonstrictly determined. Then by problem 11 of Exercise 3, $a \neq b$. We consider two cases:

Case 1. Suppose $a > b$. In this case, we must have $d > b$, or else b would be a saddle point. Suppose further that $a \leq c$; then $c > d$, or else c would be a saddle point. Now $d > b$ and $c > d$ together imply that d is a saddle point, which is a contradiction. Hence $a > c$. Similarly $d > c$.

Case 2. Suppose $a < b$. The proof is similar to the proof in case 1.   ▌

To solve a nonstrictly determined 2×2 matrix game, we apply

Theorem 7.   Let the matrix game

(32)
$$A = \begin{bmatrix} a & b \\ c & d \end{bmatrix}$$

be nonstrictly determined. Set

(33) $p_1^\star = \dfrac{d - c}{a + d - b - c}, \qquad p_2^\star = \dfrac{a - b}{a + d - b - c},$

(34) $q_1^\star = \dfrac{d - b}{a + d - b - c}, \qquad q_2^\star = \dfrac{a - c}{a + d - b - c},$

and

(35)
$$v = \frac{ad - bc}{a + d - b - c}.$$

Then v is the value of the game A and

$$P\star = [p_1^\star, p_2^\star] \qquad \text{and} \qquad Q\star = \begin{bmatrix} q_1^\star \\ q_2^\star \end{bmatrix}$$

are optimum strategies for R and C, respectively.

Proof. Since the game A is nonstrictly determined, it follows from Theorem 6 that $a + d - b - c \neq 0$, and hence the denominator is not zero in (33), (34), and (35).

Now let $Q\star$ and v be defined as above. It is easy to verify that if $P = [p, 1 - p]$ is any strategy for R, then

$$(36) \qquad\qquad PAQ\star = v.$$

Similarly one can prove that if Q is any strategy for C, then

$$(37) \qquad\qquad P\star AQ = v.$$

Hence by Definition 1, the theorem follows. ∎

EXAMPLE 1. Solve the game of matching pennies, described in Example 3 of Section 1.

Solution. The matrix for the game of matching pennies is

$$A = \begin{bmatrix} 4 & -12 \\ -3 & 9 \end{bmatrix}.$$

By Theorem 6, we note that this is a nonstrictly determined game. Thus Theorem 7 applies. Here

$$p_1^\star = \frac{9 - (-3)}{4 + 9 - (-12) - (-3)} = \frac{12}{28} = \frac{3}{7}$$

$$p_2^\star = \frac{4 - (-12)}{4 + 9 - (-12) - (-3)} = \frac{16}{28} = \frac{4}{7}.$$

Consequently an optimum strategy for R is

$$\left[\frac{3}{7}, \frac{4}{7} \right].$$

Similarly an optimum strategy for C is

$$\begin{bmatrix} \dfrac{3}{4} \\ \dfrac{1}{4} \end{bmatrix}.$$

The value of the game is

$$v = \frac{(4)(9) - (-12)(-3)}{28} = 0.$$

Hence this is a fair game, as predicted by Mr. X.

Definition 3. Let $A = [a_{ij}]$ be an $m \times n$ matrix. If each element of the ith row of A is greater than or equal to the corresponding element of the kth row; i.e., if

(38) $a_{ij} \geqq a_{kj}, \qquad j = 1, 2, \ldots, n,$

then the kth row is called a recessive row and the ith row is said to dominate the kth row. If each element of the jth column is less than or equal to the corresponding element of the kth column; i.e., if

(39) $a_{ij} \leqq a_{ik}, \qquad i = 1, 2, \ldots, m,$

then the kth column is called a recessive column, and the jth column is said to dominate the kth column.

The reader should note carefully the reversal of the inequality sign in going from (38) to (39).

Now suppose that an $m \times n$ matrix game A has its ith row dominating the kth row. Then it is clear that player R would always rather play the ith row than the kth row, since he is guaranteed to win the same or a greater amount in every possible play of the game. Hence any recessive row may be dropped from A without affecting the solution of the game. For analogous reasons any recessive column may be omitted from the game.

EXAMPLE 2.   Solve the matrix game

$$A = \begin{bmatrix} 2 & 2 & 1 \\ -1 & 1 & 0 \\ 0 & 1 & 3 \end{bmatrix}.$$

Solution.   We note that the first row dominates the second row. Thus the second row is recessive and may be omitted from the game. Hence the game is reduced to

$$A\star = \begin{bmatrix} 2 & 2 & 1 \\ 0 & 1 & 3 \end{bmatrix}.$$

We now observe that the first column dominates the second column. Therefore the second column is recessive and may be dropped. Thus the game may be reduced to the 2×2 matrix game

$$A\star\star = \begin{bmatrix} 2 & 1 \\ 0 & 3 \end{bmatrix}.$$

The game $A\star\star$ is nonstrictly determined. Thus applying Theorem 7, we find that

$$\begin{bmatrix} \dfrac{3}{4}, \dfrac{1}{4} \end{bmatrix} \quad \text{and} \quad \begin{bmatrix} \dfrac{1}{2} \\ \dfrac{1}{2} \end{bmatrix}$$

are optimum strategies for R and C, respectively, for $A\star\star$ and that the value of $A\star\star$ is $v = 3/2$. The game is biased in favor of R.

It should be obvious to the reader that to obtain optimal strategies for A from those for $A\star\star$ it is sufficient to add a zero component to each of the vectors at a suitable place. Further, the value of A is the same as the value of $A\star\star$. Consequently the solution to the original game A is

$$v = \frac{3}{2}, \qquad P\star = \begin{bmatrix} \dfrac{3}{4}, 0, \dfrac{1}{4} \end{bmatrix}, \qquad \text{and} \qquad Q\star = \begin{bmatrix} \dfrac{1}{2} \\ 0 \\ \dfrac{1}{2} \end{bmatrix}.$$

Exercise 4

In problems 1 through 8, find the value of each game and optimum strategies for both players.

1. $\begin{bmatrix} 2 & 3 \\ 4 & 5 \end{bmatrix}.$

2. $\begin{bmatrix} 4 & 3 \\ 2 & 5 \end{bmatrix}.$

3. $\begin{bmatrix} -2 & 0 \\ 1 & -1 \end{bmatrix}.$

4. $\begin{bmatrix} 8 & -3 \\ 5 & 7 \end{bmatrix}.$

5. $\begin{bmatrix} 2 & 3 & -2 \\ 2 & 2 & 4 \end{bmatrix}.$

6. $\begin{bmatrix} 2 & 1 \\ -4 & 3 \\ -5 & 2 \end{bmatrix}.$

7. $\begin{bmatrix} 1 & -2 & -3 \\ 4 & 1 & 2 \\ 3 & 4 & -2 \end{bmatrix}$.

8. $\begin{bmatrix} 3 & -1 & -2 & 2 \\ 2 & -2 & -2 & -3 \\ 2 & 3 & 1 & -4 \\ 1 & 2 & 0 & -6 \end{bmatrix}$.

9. Solve the matrix game given in problem 3 of Exercise 1. Are the random strategies of R and C optimal?

10. Two players R and C show three or four fingers simultaneously If the sum of the fingers shown is even, then C pays the sum to R; if the sum is odd, then R loses the sum to C. Find the solution of this game. Is the game biased?

11. Rocky and Craig play the following card game: Rocky has two cards, a black 4 and a red 9. He puts one of these cards face down, and Craig is supposed to guess it. Every time Craig guesses the card correctly, he wins the amount equal to the number shown on the card; otherwise he loses the amount x to Rocky. What should x be for this game to be fair? What are optimum strategies for both players?

12. Each of the two players R and C has a nickel, dime, and quarter. They each show a coin simultaneously. If the total amount (in pennies) shown is even, then R wins C's coin; otherwise C wins R's coin. Find the value of this game and optimum strategies for both players.

5.　*m × n* matrix games

Let $A = [a_{ij}]$ be an $m \times n$ matrix game which is nonstrictly determined and does not contain recessive rows or columns. We may assume that all the entries of the matrix A are positive, for if some or all of the entries of A are negative, we can add a suitable number $k > 0$ to all of the entries of A so that all the new entries are positive. By problem 10 of Exercise 2, we know that the value of the game is increased by k, while the optimum strategies remain the same. We shall show that we can find optimum strategies and the value of the game by solving certain related problems in linear programming.

Now let v be the value of the game $A = [a_{ij}]$. Since we assumed that all the entries of A are positive, by problem 11 of Exercise 2, we know that v is positive. Let

$$Q = \begin{bmatrix} q_1 \\ q_2 \\ \vdots \\ q_n \end{bmatrix}$$

be a strategy for C. If Q is an optimum strategy and v is the value of the game, then by Theorem 3 we have

(40)
$$
\begin{aligned}
a_{11}q_1 + a_{12}q_2 + \cdots + a_{1n}q_n &\leqq v \\
a_{21}q_1 + a_{22}q_2 + \cdots + a_{2n}q_n &\leqq v \\
&\vdots \\
a_{m1}q_1 + a_{m2}q_2 + \cdots + a_{mn}q_n &\leqq v.
\end{aligned}
$$

Since v is positive, dividing both sides of each of the inequalities in (40) by v we obtain

(41)
$$
\begin{aligned}
a_{11}\frac{q_1}{v} + a_{12}\frac{q_2}{v} + \cdots + a_{1n}\frac{q_n}{v} &\leqq 1 \\
a_{21}\frac{q_1}{v} + a_{22}\frac{q_2}{v} + \cdots + a_{2n}\frac{q_n}{v} &\leqq 1 \\
&\vdots \\
a_{m1}\frac{q_1}{v} + a_{m2}\frac{q_2}{v} + \cdots + a_{mn}\frac{q_n}{v} &\leqq 1.
\end{aligned}
$$

If we now write $x_1 = q_1/v,\, x_2 = q_2/v, \ldots, x_n = q_n/v$, then the inequalities in (41) become

(42)
$$
\begin{aligned}
a_{11}x_1 + a_{12}x_2 + \cdots + a_{1n}x_n &\leqq 1 \\
a_{21}x_1 + a_{22}x_2 + \cdots + a_{2n}x_n &\leqq 1 \\
&\vdots \\
a_{m1}x_1 + a_{m2}x_2 + \cdots + a_{mn}x_n &\leqq 1.
\end{aligned}
$$

Since $q_1 + q_2 + \cdots + q_n = 1$, we also have

(43)
$$
x_1 + x_2 + \cdots + x_n = \frac{q_1 + q_2 + \cdots + q_n}{v} = \frac{1}{v}.
$$

We recall that a strategy Q for C is an optimum strategy if it makes v a minimum. That is, to find $Q\star$, it is sufficient to find $x_1, x_2, \ldots, x_n$ so that v is a minimum, subject to the constraints (42). But by (43), minimizing v is equivalent to maximizing the function

$$
f(X) = x_1 + x_2 + \cdots + x_n.
$$

Thus, the problem of finding an optimum strategy $Q\star$ for C is equivalent to the following linear programming problem: Maximize

$$
f(X) = x_1 + x_2 + \cdots + x_n
$$

$$a_{11}x_1 + a_{12}x_2 + \cdots + a_{1n}x_n \leqq 1$$
$$a_{21}x_1 + a_{22}x_2 + \cdots + a_{2n}x_n \leqq 1$$

(44)

$$a_{m1}x_1 + a_{m2}x_2 + \cdots + a_{mn}x_n \leqq 1$$
$$x_i \geqq 0; \qquad i = 1, 2, \ldots, n.$$

Theorem 8. Suppose that M is the maximum value of the function $f(X) = x_1 + x_2 + \cdots + x_n$ when the variables are subject to the constraints (44) and suppose that this maximum occurs at the point $(x_1^\star, x_2^\star, \ldots, x_n^\star)$. Then the value of the matrix game $[a_{ij}]$ is given by $v = 1/M$. Further if the components of $Q^\star$ are given by $q_i^\star = vx_i^\star$ for $i = 1, 2, \ldots, n$, then $Q^\star$ is an optimal strategy for C.

Following the same type of argument, the energetic student can prove that the problem of finding an optimum strategy $P^\star$ for R is equivalent to an appropriate minimum linear programming problem.

EXAMPLE 1. Solve the matrix game

$$A^\star = \begin{bmatrix} -2 & 0 \\ 2 & -4 \\ -4 & 1 \end{bmatrix}.$$

Solution. Since there are negative entries in this matrix, we add a suitable constant (say, 6) to each element of the given matrix to obtain the new matrix game

$$A = \begin{bmatrix} 4 & 6 \\ 8 & 2 \\ 2 & 7 \end{bmatrix}.$$

We solve the following linear programming problem: Maximize

$$f(X) = x_1 + x_2,$$

subject to the constraints

(45)

$$4x_1 + 6x_2 \leqq 1$$
$$8x_1 + 2x_2 \leqq 1$$
$$2x_1 + 7x_2 \leqq 1$$
$$x_1 \geqq 0$$
$$x_2 \geqq 0.$$

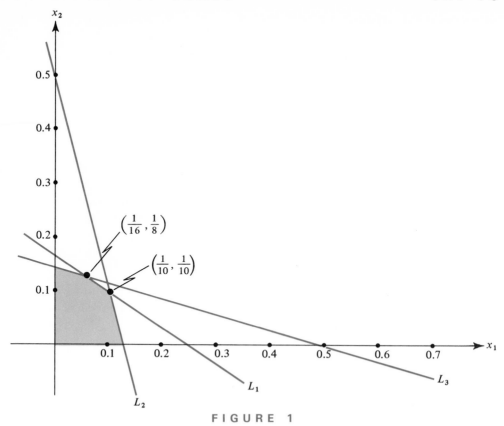

FIGURE 1

The polyhedral convex set defined by the constraint inequalities (45) is indicated by the shaded region in Fig. 1. The maximum of the linear function $f(X) = x_1 + x_2$ occurs at $x_1 = 1/10$ and $x_2 = 1/10$. The maximum value of f is $x_1 + x_2 = (1/10) + (1/10) = 1/5$. Thus the value of the game is $v = 1/(x_1 + x_2) = 5$. Now if $Q\star = \begin{bmatrix} q_1^\star \\ q_2^\star \end{bmatrix}$ is an optimal strategy for C, then $q_1^\star = x_1 v = (1/10)5 = 1/2$ and $q_2^\star = x_2 v = (1/10)5 = 1/2$. Hence an optimum strategy for C is

$$Q\star = \begin{bmatrix} \dfrac{1}{2} \\ \dfrac{1}{2} \end{bmatrix}.$$

To find an optimum strategy $P\star = [p_1^\star, p_2^\star, p_3^\star]$ for R we set

$$y_1 = \frac{p_1^\star}{v}, \qquad y_2 = \frac{p_2^\star}{v}, \qquad y_3 = \frac{p_3^\star}{v},$$

where v is the value of the game. Then we are lead to the following equivalent problem: Minimize

$$g(Y) = y_1 + y_2 + y_3$$

subject to the constraints

(46)
$$4y_1 + 8y_2 + 2y_3 \geq 1$$
$$6y_1 + 2y_2 + 7y_3 \geq 1$$

together with the inequalities

(47)
$$y_i \geq 0, \quad i = 1, 2, 3.$$

We solve this problem by the simplex method. Introducing the slack variables $y_4 \geq 0$ and $y_5 \geq 0$, we change the constraint inequalities to

(48)
$$4y_1 + 8y_2 + 2y_3 = 1 + y_4$$
$$6y_1 + 2y_2 + 7y_3 = 1 + y_5.$$

Solving the first equation in (48) for y_1 in terms of y_2, y_3, and y_4, and substituting this value of y_1 in the second equation of (48) and in the function g, we have the equivalent linear programming problem: Maximize

$$f = -g(Y) = -\frac{1}{4} + y_2 - \frac{1}{2}y_3 - \frac{1}{4}y_4$$

subject to the constraint equations:

(49)
$$y_1 + 2y_2 + \frac{1}{2}y_3 - \frac{1}{4}y_4 \qquad = \frac{1}{4}$$
$$10y_2 - 4y_3 - \frac{3}{2}y_4 + y_5 = \frac{1}{2},$$

together with the inequalities

(50)
$$y_i \geq 0, \quad i = 1, 2, \ldots, 5.$$

We write the system (49) and f in matrix form with the pivot element encircled (see Chapter 9):

(51)
$$B = \begin{array}{c} \begin{array}{cccccc} y_1 & y_2 & y_3 & y_4 & y_5 & f \end{array} \\ \left[\begin{array}{cccccc|c} 1 & 2 & \frac{1}{2} & -\frac{1}{4} & 0 & 0 & \frac{1}{4} \\ 0 & \boxed{10} & -4 & -\frac{3}{2} & 1 & 0 & \frac{1}{2} \\ \hline 0 & -1 & \frac{1}{2} & \frac{1}{4} & 0 & 1 & -\frac{1}{4} \end{array} \right] \end{array}.$$

With the pivot 10, the matrix B is transformed by the complete elimination process into

(52)
$$
D = \begin{array}{c} \begin{array}{cccccc} y_1 & y_2 & y_3 & y_4 & y_5 & f \end{array} \\ \left[\begin{array}{cccccc|c} 1 & 0 & \dfrac{13}{10} & \dfrac{1}{20} & -\dfrac{1}{5} & 0 & \dfrac{3}{20} \\[2mm] 0 & 1 & -\dfrac{2}{5} & -\dfrac{3}{20} & \dfrac{1}{10} & 0 & \dfrac{1}{20} \\[2mm] \hline 0 & 0 & \dfrac{1}{10} & \dfrac{1}{10} & \dfrac{1}{10} & 1 & -\dfrac{1}{5} \end{array} \right]. \end{array}
$$

Thus the maximum value of $-g$ is $-1/5$ and this value occurs at $y_1 = 3/20$, $y_2 = 1/20$, and $y_3 = 0$. Hence the minimum value of g is $1/5$, and consequently the value of the game is $v = 5$. Computing the components of $P\star$ we find:

$$
p_1^\star = y_1 v = \left(\frac{3}{20} \right)(5) = \frac{3}{4},
$$

$$
p_2^\star = y_2 v = \left(\frac{1}{20} \right)(5) = \frac{1}{4},
$$

and

$$
p_3^\star = y_3 v = (0)(5) = 0.
$$

Hence an optimum strategy for R is given by

$$
P\star = \left[\frac{3}{4}, \frac{1}{4}, 0 \right].
$$

To verify that the value of the matrix game defined by A is indeed 5, the student should show that $P\star A Q\star = 5$.

Since 6 was added to every entry of the original matrix game $A\star$, the value of the original matrix game is $5 - 6 = -1$. Thus the original matrix game is biased in favor of C.

Exercise 5

In problems 1 through 6, solve the given matrix game.

1. $\begin{bmatrix} 2 & 3 \\ 3 & 1 \\ 4 & 0 \end{bmatrix}$.

2. $\begin{bmatrix} 1 & 3 & 5 \\ 2 & 1 & -1 \end{bmatrix}$.

3. $\begin{bmatrix} 1 & -1 & 0 \\ 0 & 3 & 0 \\ 0 & 0 & 1 \end{bmatrix}$.

4. $\begin{bmatrix} -3 & 2 & -1 \\ 1 & 1 & -2 \\ -2 & -3 & 1 \end{bmatrix}$.

5. $\begin{bmatrix} 6 & 5 & -2 & 3 \\ 2 & 2 & 1 & -2 \\ 1 & 1 & 0 & -3 \\ 1 & -5 & 0 & 2 \end{bmatrix}$.

6. $\begin{bmatrix} -1 & 2 & -3 & 2 \\ 2 & 1 & -2 & -1 \\ -2 & 2 & -4 & 0 \\ 1 & -1 & 3 & -2 \end{bmatrix}$.

7. Solve the game described in problem 1 of Exercise 1.

8. In the well-known "stone-scissors-paper" game often played by children, two players *R* and *C* each select one of the three words *stone, scissors,* or *paper*. If the two players name the same item, the game is a tie; *stone* beats (breaks) *scissors*; *scissors* beats (cuts) *paper*; and *paper* beats (covers) *stone*. If each win is worth 1 cent, find the value of the game.

9. A real estate developer has bought a large tract of land in Mudland County. He is considering using some of the land for apartments, some for a shopping center, and some for houses. It is not certain whether the Mudland County Government will build a highway near his property or not. His financial advisor provides an estimate for the percentage profit to be made in each case and these percentages are given in the following table:

Builder	Government	
	Highway	*No highway*
Apartments	25%	5%
Shopping center	20%	15%
Houses	10%	20%

What percentage of the land should he use for each of the three categories, assuming that the government of Mudland County is an active opponent?

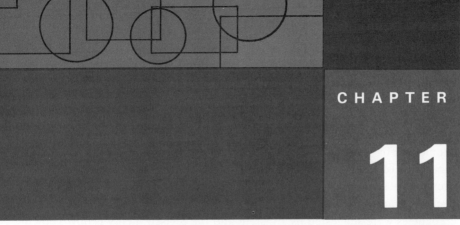

applications to the social sciences

In recent years there has been a tremendous effort to apply mathematics to the social sciences. Although there have been some minor successes, the results are not nearly so impressive as those obtained in physics, chemistry, and engineering. Undoubtedly animals and human beings, both singly and in social groups, are much more complicated than atoms, molecules, and structural materials.

1. theories and models

Briefly stated, a theory is an attempt to explain some phenomenon in nature. A theory may be very simple or extremely complex. A theory is considered valuable if it predicts the outcome of some experiment before it is performed. The number and importance of such successful predictions supplies a rough measure of the value of the theory. When the theory fails to predict correctly the outcome of an experiment, then the theory must be modified or perhaps abandoned.

As a very simple example consider the statement "Money is the root of all evil." Clearly this is a theory, because it attempts to explain a natural phenomenon. The relevant experiment is to remove money from some small community and observe whether evil also disappears. Now such an experiment has been tried more than once, but all records seem to indicate that evil persists, although an accurate evaluation of the results is not easy. The

339

theory is usually modified to read "The love of money is the root of all evil." This revised form of the theory may represent the facts more clearly, but now the relevant experiment is hard to perform, because we do not know how to measure the love of money or to remove the love of money from some small community so that we can observe the results.

One of the most successful theories carries the title "Newtonian mechanics." For hundreds of years, it predicted accurately the outcomes of thousands of widely diverse experiments. As the accuracy of certain measurements was increased, discrepancies between the theory and experimental results began to appear. The modifications necessary to explain these discrepancies were suggested by Einstein and the revised theory is now known as Einsteinian mechanics.

Occasionally a theory is so successful that the student tends to regard it as a true representation of the objects it attempts to describe. In such a case the student becomes somewhat disturbed when the relevant experiments give data that do not match the predictions of theory. In recent times the term "mathematical model" has been used in place of "mathematical theory." By using the term *model* the author is deliberately calling attention to the fact that his "model" is only a crude approximation to the true explanation. Thus the author sounds a subtle warning for the reader that he should examine the model (or theory) with the proper mixture of respect, belief, and suspicion. When a model has gained universal respect by an impressive list of successful and accurate predictions, then it has earned the right to the title of *theory*.[1]

2. a model for paired-associate learning

The processes of the mind are so complex that any mathematical analysis must start with an extremely simple situation. One such is paired-associate learning (PAL), which we now describe.

The experimenter creates a set of ordered pairs (s,r) of items, where s is selected from a set S, the *stimulus set,* and r is from a set R, the *response set*. The subject in the experiment is some cooperative person who is willing to try to learn the pairs (s,r) created by the experimenter. An element s (called a *stimulus element*) is shown to the subject, and the subject is asked to give the corresponding r (called a *response element*). Each such exposure to an element of S is called a *presentation*. Of course, on the first presentation of s the subject can only guess the associated r, but after he has responded he is then informed of the correct answer. Thus on any subsequent exposure

---

[1] The distinction between *model* and *theory* presented here is by no means universally accepted. Some authors use the words interchangeably, while others regard a model as a small portion of a theory. Still other authors are not clear about the distinction.

to s, he may either guess again, or perhaps give the correct associated r because he has learned it from the previous presentations.

The mechanism for conducting the experiment may vary widely. One simple method is to use a set of flash cards, one for each pair (s,r), with s on the front side and r on the reverse side. One complete showing of each card in the deck is a *trial* (or *cycle*) and after each trial the cards should be shuffled to ensure randomness.

Paired-associate learning occurs quite naturally when one is learning the vocabulary of a foreign language. Here S might be a set of French words and R the set of appropriate English equivalents. Or in another application S might be a set of mathematical concepts, and with each $s \in S$, we may associate a certain important theorem. However, such natural pairs should not be used in an experiment because of the possible intrusion of unsymmetric elements (some words or theorems may be harder to learn than others). Consequently the experimenter is well advised to select some simple and innocuous elements for S and R.

In a typical PAL experiment by Gordon Bower,[1] the set S consisted of 10 different pairs of consonants and the set R was the set of digits $\{1,2\}$. The correct response 1 was assigned to 5 randomly selected elements of S, while the remaining 5 were paired with 2. Each correct response was noted by recording a 0, and each incorrect response was noted by recording a 1. One cycle gave rise to a sequence of 10 integers $[y_1, y_2, \ldots, y_{10}]$, where each y_i is 0 or 1. Twenty-nine subjects volunteered for the experiment, and each subject continued to respond until he had completed two full cycles without an error. When this occurred, it was assumed that the subject was no longer guessing but had in fact actually learned the correct response for each element of S.

The sequences to be analyzed are not the sequences $[y_1, y_2, \ldots, y_{10}]$ described above. Rather, we fix our attention on a single item from S and ask how the subject responds to that item in a succession of cycles (trials). Thus each item and each subject generates a sequence $[x_1, x_2, \ldots, x_n, \ldots]$, where $x_n = 0$ if the response is correct on the nth presentation of that item, and $x_n = 1$ if it is not.

We note that theoretically it is possible for a sequence $[x_1, x_2, \ldots, x_n, \ldots]$ to be infinitely long in case the subject never learns the correct responses for the entire list. This causes a slight difficulty in the mathematical treatment, but with normal subjects an infinite sequence never appears.

The task of the model is to make some reasonable predictions about these 290 subject-item sequences (10 sequences for each of the 29 subjects). In presenting a model one usually states the assumptions in the form of axioms.

---

[1] "Application of a Model to Paired-Associate Learning," *Psychometrika*, vol. 26 (1961), pp. 255–280.

These axioms are then supplemented with explanations of the terms and reasons for selecting the axioms. Such a formal procedure makes it easy for the critic to select the axiom or axioms which he wishes to attack, alter, or replace. In presenting the axioms formulated by Bower, we take the liberty of dropping some of the terms commonly used in psychology, and we restate the axioms in a purely mathematical form.

Axiom 1. On each presentation of an item s from S only two states, C and C', are possible for the subject. If he is in state C for that item, then he knows the proper response (the r that is paired with s). If he is in state C', then he does not know the proper response and he guesses.

Axiom 2. At the beginning of the experiment the subject is in state C' for each item in S.

Axiom 3. If the subject is in state C for a particular item on the nth trial (or presentation), then he remains in state C for that item on the $(n + 1)$th trial. If he is in state C', then the probability of transition from C' to C is $c > 0$. This probability c is the same for each presentation, for each item, and for each subject.

Axiom 4. If the subject is in state C', then the probability that he guesses the correct response is $g = 1/N$, where N is the number of response alternatives (the number of elements in the set R).

The model described by these four axioms is called the *one-element* model, or the *all-or-none* model. The latter name stems from the all-or-nothing character of Axiom 1. In this model the subject has either learned the correct response or he has not, and there is no allowance for a gradual learning process or for forgetting an item after it has been learned.

Axiom 3 is somewhat unrealistic. Some subjects learn more rapidly than others. Further, this axiom ignores the effects of boredom and fatigue. However, if the number of subjects tested is very large, then these deviations may be unimportant.

Axiom 4 is a reasonable one. However, if we want to allow for a gradual learning process, we might let g be some increasing function of n, the number of the trial.

The transition matrix for this process is

(1)
$$P = \begin{array}{c} \\ C \\ C' \end{array} \begin{array}{cc} C & C' \\ \begin{bmatrix} 1 & 0 \\ c & 1 - c \end{bmatrix} \end{array}.$$

It is easy to prove that for each $n \geq 1$,

$$(2) \qquad P^n = \begin{bmatrix} 1 & 0 \\ 1 - (1 - c)^n & (1 - c)^n \end{bmatrix}.$$

By Axiom 2, the subject starts in state C' so the initial probability vector is $[0, 1]$. Using this vector with equation (2) we see that *after* the nth trial the probability that the subject is in state C' is $(1 - c)^n$. If we use C_n and C_n' to denote states C and C' *immediately preceding* the nth trial, we have

$$(3) \qquad P[C_n'] = (1 - c)^{n-1}$$

and

$$(4) \qquad P[C_n] = 1 - (1 - c)^{n-1}.$$

We recall that each subject and each stimulus element generates a sequence $[x_1, x_2, \ldots, x_n, \ldots]$, where $x_n = 1$ if the subject makes an error on the nth trial and $x_n = 0$ if the response is correct. An error can occur for a particular item on the nth trial if and only if the subject is in C_n' and guesses wrong. Hence

$$(5) \qquad P[x_n = 1] = (1 - c)^{n-1}(1 - g)$$

and

$$P[x_n = 0] = 1 - P[x_n = 1] = 1 - (1 - c)^{n-1}(1 - g).$$

We now derive a number of formulas concerning events that can occur during a PAL experiment. These formulas give **(a)** the expected value of the number of errors a subject makes before learning an item, **(b)** the expected value of the number of the trial on which the subject makes the last error, **(c)** the probability of a run of errors of length k that starts on the nth trial, and **(d)** the expected value of the number of runs of length k. By comparing the values computed from these formulas with the values obtained from an experiment, we can form some judgment about the validity of the axioms.

First we consider T_M, the total number of errors made by a subject on a particular stimulus item in M trials. Since $x_n = 1$ if and only if the subject makes an error on the nth trial, we have

$$(6) \qquad T_M = x_1 + x_2 + \cdots + x_n + \cdots + x_M.$$

To find the expected number of errors we shall need the formula

$$(7) \qquad 1 + a + a^2 + \cdots + a^{M-1} = \frac{1 - a^M}{1 - a}, \qquad a \neq 1.$$

The proof of (7) will be indicated in the next exercise. Once (7) is established, we can let M approach ∞ (infinitely many trials). If $|a| < 1$, then a^M approaches 0 as M approaches ∞, and we obtain the useful formula

(8) $$\sum_{n=0}^{\infty} a^n = 1 + a + a^2 + \cdots = \frac{1}{1-a}, \qquad |a| < 1.$$

Returning now to the problem at hand and equation (6), we see that the expected number of errors in the first M trials is

$$E(T_M) = \sum_{n=1}^{M} E(x_n) = \sum_{n=1}^{M} 0 \cdot P[x_n = 0] + \sum_{n=1}^{M} 1 \cdot P[x_n = 1].$$

Using (5) in this last sum, we find that

(9) $$E(T_M) = \sum_{n=1}^{M} (1-c)^{n-1}(1-g) = (1-g)\sum_{n=1}^{M}(1-c)^{n-1}$$

$$= (1-g)(1 + (1-c) + (1-c)^2 + \cdots + (1-c)^{M-1}).$$

We now apply formula (7) to the last sum in (9), with $a = 1 - c$, and we obtain

(10) $$E(T_M) = (1-g)\frac{1-(1-c)^M}{1-(1-c)} = \frac{1-g}{c}[1 - (1-c)^M].$$

If now we let M approach infinity in equation (10), then the term $(1-c)^M$ approaches zero. This gives

Theorem 1. If T is the number of errors that a subject makes before learning an item, then the expected value of T is given by

(11) $$E(T) = \frac{1-g}{c} = \frac{1-1/N}{c}.$$

If each experiment is stopped after M trials, then (11) is replaced by

(10) $$E(T_M) = \frac{1-g}{c}[1 - (1-c)^M].$$

Equation (10) or (11) gives us a method of estimating c. In any experiment N (the number of different response elements) is known and we can find

$E(T)$, the average number of errors made for each item by each subject. Then we can use (10) or (11) to compute c. Further, we can check the model by rerunning the experiment with different sets R and S or with a different collection of subjects. The computed values of c should be nearly the same under these conditions. However, if the number of elements in R or S is sharply increased, we may find a decrease in c.

Once c has been computed, we can check the model by looking at other data that involve c. Some of the possibilities are given in the following theorems.

Theorem 2. If T is the number of errors that a subject makes, then

$$(12) \qquad P[T = 0] = gb,$$

where

$$(13) \qquad b = \frac{c}{1 - g(1 - c)}.$$

Proof. Suppose that the subject learns the correct response after the nth trial. Since the subject makes no errors, it follows that he guessed correctly on the first $n - 1$ trials and did not learn on each of these trials. The probability of this event is $g^{n-1}(1 - c)^{n-1}$. On the nth trial the subject again guesses correctly and then learns the correct response. The probability of this event is gc. Hence the probability that the subject makes no errors and learns on the nth trial is $g^n(1 - c)^{n-1}c$. Using this formula for $n = 1, 2, \ldots$ we have

$$(14) \quad P[T = 0] = \sum_{n=1}^{\infty} g^n(1 - c)^{n-1}c$$

$$= gc[1 + g(1 - c) + g^2(1 - c)^2 + g^3(1 - c)^3 + \cdots].$$

Using formula (8) with $a = g(1 - c)$, we find

$$(15) \qquad P[T = 0] = g\frac{c}{1 - g(1 - c)}. \quad \blacksquare$$

The derivation of the more general formula for the probability of exactly k errors,

$$(16) \qquad P[T = k] = \frac{c(1 - g)^k(1 - c)^{k-1}}{[1 - g(1 - c)]^{k+1}}, \qquad k \geq 1,$$

is too complicated for this text. Using b, defined by equation (13), we can simplify (16) to

(17)
$$P[T = k] = \frac{(1 - b)^k b}{1 - c}, \qquad k \geqq 1.$$

Theorem 3. For b, defined by (13), we have

(18)
$$P[\text{no errors after the } k\text{th trial} \mid C_k'] = b.$$

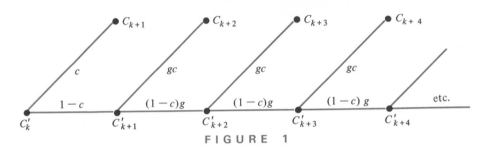

FIGURE 1

Proof. We examine that portion of the tree diagram that corresponds to correct responses only (see Fig. 1). At each branch point, the subject either learns with probability c (and thereafter makes no wrong response) or does not learn but guesses correctly with probability $g(1 - c)$. From the tree it is clear that

$$P[\text{no errors after the } k\text{th trial} \mid C_k']$$
$$= c + (1 - c)gc + (1 - c)^2 g^2 c + (1 - c)^3 g^3 c + \cdots$$
$$= c \sum_{n=0}^{\infty} [g(1 - c)]^n = c \frac{1}{1 - g(1 - c)} = b. \qquad \blacksquare$$

Theorem 4. Let L denote the number of the trial on which the last error occurs for a particular subject and a given stimulus. Then

(19)
$$P[L = 0] = gb,$$

and if $k \geqq 1$, then

(20)
$$P[L = k] = (1 - c)^{k-1}(1 - g)b.$$

Proof. Equation (19) has already been proved because it is merely a restatement of equation (12) of Theorem 2. Thus if the subject makes his last error on the zeroth trial ($L = 0$), then he makes no errors ($T = 0$).

Suppose now that $k \geq 1$. If the subject makes an error on the kth trial, he must be in state C'_k and clearly the probability for these two events is $(1 - g)(1 - c)^{k-1}$. The probability that he makes no further errors was determined to be b in Theorem 3. Hence

$$P[L = k] = (1 - c)^{k-1}(1 - g)b. \quad \blacksquare$$

Theorem 5. The expected value of L is given by

(21) $$E(L) = E(T)\frac{b}{c}.$$

Proof. By the definition of expected value

(22) $$E(L) = \sum_{k=0}^{\infty} kP[L = k] = \sum_{k=0}^{\infty} k(1 - c)^{k-1}(1 - g)b$$

$$= (1 - g)b \sum_{k=1}^{\infty} k(1 - c)^{k-1}.$$

Here we need the formula

(23) $$\sum_{k=1}^{\infty} ka^{k-1} = 1 + 2a + 3a^2 + 4a^3 + \cdots = \frac{1}{(1 - a)^2},$$

which holds if $|a| < 1$. Using (23) in (22) with $a = 1 - c$, we find

(24) $$E(L) = (1 - g)b\frac{1}{(1 - (1 - c))^2} = (1 - g)b\frac{1}{c^2} = \frac{1 - g}{c} \cdot \frac{b}{c}. \quad \blacksquare$$

We next investigate runs of errors. The simplest case is the occurrence of a pair of errors. The key formula is

(25) $$P[x_{n+1} = 1 | x_n = 1] = (1 - c)(1 - g).$$

For a proof we observe that if $x_n = 1$, then certainly the subject is in C'_n. To make the next error, it must be that he did not learn and he guessed wrong. The product of the probabilities of these two events is $(1 - c)(1 - g)$. We shall use β for this quantity. Suppose that the subject guesses correctly on the nth trial. The same argument that gave (25) also gives

(26) $$P[x_{n+1} = 1 | x_n = 0, \text{ and } C'_n] = (1 - c)(1 - g) = \beta.$$

Theorem 6. Let $P[r_{k,n}]$ denote the probability of a run of exactly k errors, beginning with an error on the nth trial. Then

(27) $$P[r_{k,1}] = \beta^{k-1}(1 - \beta)(1 - g),$$

and if $n \geq 2$

(28) $$P[r_{k,n}] = (1 - c)^{n-2}g\beta^k(1 - \beta).$$

Proof. Suppose $n = 1$. The probability of making an error on the first trial is $1 - g$. We then apply equation (25) with $n = 1$ to find that the probability of making an error on the second trial, given that an error was made on the first trial, is $(1 - c)(1 - g) = \beta$. Continuing this way, the probability of making an error on each of the first k trials is $(1 - g)\beta^{k-1}$. The subject can now give the correct response on the $(k + 1)$th trial in two different ways: Either he is in state C'_{k+1} and guesses correctly, or he learns and is in state C_{k+1}. Hence we find that

$$P[r_{k,1}] = (1 - g)\beta^{k-1}[(1 - c)g + c]$$
$$= (1 - g)\beta^{k-1}(1 - \beta).$$

If $n \geq 2$, the argument is similar, but the process starts with a correct guess in state C'_{n-1}, and the probability for this event is $g(1 - c)^{n-2}$. The probability that this event is followed by k wrong guesses and then a correct response is

$$\beta^k[(1 - c)g + c].$$

Hence for $n \geq 2$,

$$P[r_{k,n}] = g(1 - c)^{n-2}\beta^k[(1 - c)g + c]$$
$$= g(1 - c)^{n-2}\beta^k(1 - \beta). \quad \blacksquare$$

Theorem 7. The expected value of r_k, the number of runs of errors of length k, is given by

(29) $$E(r_k) = \beta^{k-1}(1 - \beta)^2E(T).$$

Proof. Each run of errors of length k must start on some particular trial. Hence

(30) $$E(r_k) = \sum_{n=1}^{\infty} P[r_{k,n}].$$

Using equation (27) when $n = 1$, and equation (28) for $n \geq 2$, we find that (30) gives

$$E(r_k) = \beta^{k-1}(1 - \beta)(1 - g) + \sum_{n=2}^{\infty} (1 - c)^{n-2} g \beta^k (1 - \beta)$$

$$= \beta^{k-1}(1 - \beta)(1 - g) + \beta^k(1 - \beta)g \sum_{n=0}^{\infty} (1 - c)^n$$

$$= \beta^{k-1}(1 - \beta)(1 - g) + \beta^k(1 - \beta)g \frac{1}{1 - (1 - c)}$$

$$(31) \qquad\qquad E(r_k) = \frac{\beta^{k-1}(1 - \beta)}{c}(c(1 - g) + \beta g).$$

Now $\beta = (1 - c)(1 - g)$, and hence $c(1 - g) = 1 - g - \beta$. So (31) becomes

$$E(r_k) = \frac{\beta^{k-1}(1 - \beta)}{c}(1 - g - \beta + \beta g)$$

$$= \frac{\beta^{k-1}(1 - \beta)^2(1 - g)}{c} = \beta^{k-1}(1 - \beta)^2 E(T). \quad \blacksquare$$

The experimental data gathered by Bower agreed quite well with the predictions given by the theory. However, with the passage of time, more stringent tests were applied, and some of the later experiments gave results that were contrary to the predictions. The next step in the development is to determine a *class* of experimental situations in which the all-or-none model gives accurate predictions and to devise a more general model that includes the all-or-none model as a special case but at the same time accounts for the discrepancies which arise when the experiments are outside of the class.

In the next exercise we give some problems based on some composed data. To keep the computations reasonably simple there is only one subject (a sales manager) and the set S has only six elements (his salesmen). Consequently one should not expect close agreement between the data and the theory.

Exercise 1

In a particular paired-associate learning experiment the newly appointed sales manager was presented with the names of six of his salesmen and asked to identify the territory that each one covered. The manager generated six sequences of the form $[x_1, x_2, \ldots, x_{10}]$:

Trial number 1, 2, 3, 4, 5, 6, 7, 8, 9, 10

$$[1, 0, 0, 0, 0, 0, 0, 0, 0, 0],$$
$$[1, 0, 1, 1, 0, 1, 0, 0, 0, 0],$$
$$[0, 0, 1, 0, 0, 0, 0, 0, 0, 0],$$
$$[1, 1, 0, 1, 0, 1, 1, 0, 0, 0],$$
$$[1, 1, 1, 0, 1, 1, 1, 1, 0, 0],$$
$$[1, 0, 0, 1, 0, 0, 0, 0, 0, 0].$$

Problems 1 through 8 refer to this experiment.

1. Does it make sense to regard these sequences as vectors (as the notation seems to indicate) and to add the vectors?
2. Does it make sense to multiply one of the "vectors" by an arbitrary number?
3. Find the experimental value of $E(T)$ and use this to compute c. *Hint:* Use equation (11).
4. Find the theoretical value of b.
5. Compute $P[T = k]$ for $k = 0, 1, 2, 3$ **(a)** from the theory and **(b)** from the data.
6. Compute $P[L = k]$ for $k = 1, 2, 3$ **(a)** from the theory and **(b)** from the data.
7. Compute each of $P[r_{1,1}]$, $P[r_{2,1}]$, $P[r_{3,1}]$, $P[r_{1,2}]$, and $P[r_{1,3}]$ **(a)** from the theory and **(b)** from the data.
8. Compute $E(r_k)$ for $k = 1, 2, 3, 4$ **(a)** from the theory and **(b)** from the data.
9. Prove that P^n is given by equation (2).
10. Let $S = 1 + a + a^2 + \cdots + a^{M-1}$. Prove that $S - aS = 1 - a^M$. From this prove formula (7).
11. Give a convincing argument for a proof of equation (23). *Hint:* Use formula (8) and multiply the two series on the right-hand side of

$$\frac{1}{1 - a} \cdot \frac{1}{1 - a} = (1 + a + a^2 + \cdots)(1 + a + a^2 + \cdots).$$

12. Check formulas (15) and (17) by using them to prove that

$$\sum_{k=0}^{\infty} P[T = k] = 1.$$

13. Check formulas (19) and (20) by using them to prove that

$$\sum_{k=0}^{\infty} P[L = k] = 1.$$

14. Explain why the sum must be 1 for the series given in problems 12 and 13.

15. Let R be the total number of runs of any length and let $E(R)$ be the expected value of R. Prove that $E(R) = E(T)(1 - \beta)$.

In problems 16 through 21 we indicate the development of another model for paired-associate learning. This model is sometimes called the *single-operator linear model*.

16. Let q_n be the probability that an error is made on the nth trial of a subject-item sequence, and let $p_n = 1 - q_n$ be the probability of a success on the nth trial. The basic assumption in this model is that $q_{n+1} = \alpha q_n$ for $n = 1, 2, 3, \ldots$, where α is some suitable constant. Prove that with this assumption

$$p_{n+1} = p_n + (1 - \alpha)(1 - p_n)$$

and

$$q_n = \alpha^{n-1} q_1.$$

17. Assume that q_1 is the probability of a wrong guess, so that $q_1 = 1 - g$, where g is the same here as for the all-or-none model. If we assume further that $\alpha = 1 - c$, then $P[x_n = 1] = (1 - c)^{n-1}(1 - g)$, and this is identical with the result for the all-or-none model [see equation (5)]. For this new model find $E(T)$.

18. Prove that the probability of a failure on every one of the first four trials is $\alpha^6 q_1^4 = (1 - c)^6(1 - g)^4$. Find a formula for the probability of a failure on every one of **(a)** the first five trials and **(b)** The first six trials.

19. Use the formula for $E(T)$ to find c for the sales manager data. With this value of c compute
   (a) $P[x_1 = 0, x_2 = 0]$.   **(b)** $P[x_1 = 0, x_2 = 1]$.
   (c) $P[x_1 = 1, x_2 = 0]$.   **(d)** $P[x_1 = 1, x_2 = 1]$.

★20. Let R be the number of runs and let $E(R)$ be the expected value of the number of runs. Prove that

$$E(R) = \frac{q_1}{1 - \alpha} - \frac{q_1^2 \alpha}{1 - \alpha^2}.$$

Hint: Each run ends with a sequence $x_n = 1$, $x_{n+1} = 0$, for some n. Hence

$$E(R) = \sum_{n=1}^{\infty} P[x_n = 1]P[x_{n+1} = 0].$$

21. Using the sales manager data, compute $E(R)$ **(a)** from the theory and **(b)** from the data.

3.   the Leontief input-output model

The economic behavior of any large portion of a civilized society is so complex that a fruitful mathematical analysis is almost impossible. One can simplify the problem by examining certain isolated parts of the economic system. Thus one might study "the theory of the firm." Or one might concentrate on "trade cycle theory" or on "utility theory." But even these isolated parts are so complicated that a wide gap still remains between the predictions obtained from the models and the real behavior of the systems represented by the models. Nevertheless, it is imperative that we continue to study such models in the hope that continued refinement will eventually lead to useful results. Consider, for example, the importance of knowing for certain that "an increase of 10 percent in the prime interest rate will definitely halt inflation."

In this section we develop a model for the study of the production and flow of goods in a system. It will soon be obvious that our assumptions (which may be called axioms) are drastic oversimplifications.

We consider n goods, $G_1, G_2, \ldots, G_n$, produced by n processes, $P_1, P_2, \ldots, P_n$ which use these goods as inputs. For simplicity we assume that each process produces just one of the goods and that each good is the result of some process. Then we may select the subscripts so that the process P_i produces the good G_i. The term *good* is used in a wide sense to include almost anything of use or interest to man. Thus law enforcement agencies and the entertainment industry both provide goods,[1] although the amount and value of the products may be difficult to measure.

The term *process* must also be understood in the wide sense. Thus the transportation process includes not only the use of buses, trains, trucks, and planes, but also the porters who carry bags by hand. Terms such as *activity, segment of the economy,* and *sector of the economy,* or *industry* are often used in place of *process.*

To keep n small, one may ignore minor processes such as the manufacture of false eyelashes and wigs, or one may subsume such items under the broader heading of the manufacture of drugs or the manufacture of clothing. Even with very gross subdivisions a recently published economic study[2] listed 87 different producing sectors in the U.S. economy.

In a first approach, we should not be too specific about the units used to measure G_i. In one application we may wish to use monetary units, while in a different context a different measure may be more appropriate. For example, we may speak of a dollar's worth of transportation or we may prefer to use ton-miles as the unit. In the first case we have a *monetary unit*, and in the second case we have a *physical unit.*

---

[1] Economists often distinguish between "goods" and "services" but for our objective such a separation is not necessary.

[2] *Survey of Current Business*, U.S. Dept. of Commerce, Washington, D.C., Sept. 1965.

Suppose now that a physical unit has been chosen for each good. Let

(32) a_{ij} = the amount of G_i used in the process
    P_j to produce one unit of G_j.

It is clear from the definition that $a_{ij} \geq 0$ and that if indeed the process P_j produces G_j, then for at least one i we must have $a_{ij} > 0$. The numbers a_{ii} are not always zero. For example, some electricity is required in the process of generating electricity, and automobiles are consumed (by both management and labor) in the production of automobiles.

The system consisting of these n processes is conveniently described by the matrix

(33)
$$A = \begin{bmatrix} a_{11} & a_{12} & \cdots & a_{1n} \\ a_{21} & a_{22} & \cdots & a_{2n} \\ \vdots & \vdots & & \vdots \\ a_{n1} & a_{n2} & \cdots & a_{nn} \end{bmatrix}$$

The matrix A is called the *technology input-output matrix* for the system, or simply the *technology* of the system. The jth column gives the various amounts of goods needed to produce one unit of G_j. The ith row indicates the distribution of G_i when each process is operating to produce one unit of its product.

Suppose now that we want the jth process to produce x_j units of the good G_j. The number x_j is called the *intensity* of P_j and we say that the process operates at *intensity* x_j. One of our basic assumptions is that the ratio of input to output is a constant. Thus if a_{ij} units of G_i are needed to produce one unit of G_j, then $a_{ij}x_j$ units of G_i are needed for x_j units of G_j. In some cases, such an assumption is quite reasonable. For example the amount of steel used to produce 1 million automobiles should be close to 1 million times the amount of steel needed to produce one automobile. On the other hand, if a night shift is needed to double production, then the amount of electricity used would more than double.

With the assumption of a fixed ratio of input to output we find that the sum

(34)
$$u_i = \sum_{j=1}^{n} a_{ij}x_j$$

gives the amount of G_i that is used for production when the system is operating to produce x_j units of G_j for $j = 1, 2, \ldots, n$. The form of (34) suggests a matrix product. Hence we introduce the three column vectors

$$(35) \qquad U = \begin{bmatrix} u_1 \\ u_2 \\ \vdots \\ u_n \end{bmatrix}, \qquad X = \begin{bmatrix} x_1 \\ x_2 \\ \vdots \\ x_n \end{bmatrix}, \qquad \text{and} \qquad D = \begin{bmatrix} d_1 \\ d_2 \\ \vdots \\ d_n \end{bmatrix}.$$

Naturally the vector U gives the amounts used when the system is in operation at the intensity described by X. The vector D is called the *demand vector* and represents the difference between the amount used in production and the amount produced. Consequently by the definition of D

$$(36) \qquad\qquad\qquad D = X - U.$$

The equation set (34) can be written as the matrix product $U = AX$, and when this is used in (36) we obtain $X = D + U = D + AX$ or

$$(37) \qquad\qquad\qquad X = AX + D.$$

We return to a discussion of D. Suppose that for some i, we have $d_i < 0$. In this case the system cannot operate for very long at the intensity X, for at this intensity the system uses more of G_i than it produces, and in due time any original stock of G_i will be exhausted. Consequently we assume henceforth that $d_i \geqq 0$ for $i = 1, 2, \ldots, n$. The vector D represents a surplus from the production. There are several ways of treating the surplus D. One may regard it as production that is available for export and attempt to adjust the system so that it produces just the surplus D that we wish. For this reason D is called the *demand*, or the *bill of goods*. On the other hand, if the cost of labor is not already included in the matrix A, then clearly some (or all) of the surplus D must be used to support the households that supply the labor. We can incorporate labor into the matrix A by creating a fictitious industry named *households* (or labor) that consumes the products of the other n industries and produces labor. The introduction of a household industry adds another variable and hence another row to each of the vectors and matrices already considered. The household (or labor) industry differs from the other industries in two important ways. First, there is a limited supply of labor, so that if labor is included as an industry, then the system cannot be operated at an arbitrary intensity. Second, the production of labor does not increase in direct proportion to consumption (doubling the supply of entertainment goods does not double the production of labor). Hence there are economically justifiable reasons for treating labor as a special type of product and not including it in the processes described by the matrix A.

Definition 1. A model is called a Leontief[1] open model, if

(I) It is an input-output model described by a technology matrix A.

(II) There is an extra input (usually labor) that is not included in the matrix A and is not the product of any of the processes of A.

(III) There is at least one demand vector $D \geqq \mathbf{0}$, but not identically zero, that the system can satisfy. (There is a $D \geqq \mathbf{0}$ such that the system of equations (37) has a solution $X \geqq \mathbf{0}$.)

In contrast, if labor is included in A we have

Definition 2. A model is called a Leontief closed model if

(I) It is an input-output model described by a technology matrix A that includes labor (households, consumers) both as an input and an output.

(II) There are no inputs other than those included in the matrix A, and there is no use for the outputs, except as inputs for the processes of A.

In either the closed or open model, we assume that $a_{ij} \geqq 0$ for all entries and that in each column there is at least one positive entry (we cannot produce something from nothing). Further, we rule out once and for all the trivial vector $X = \mathbf{0}$, where none of the processes are operating and the system is at a standstill. Thus we agree that if V is an economic vector (intensity, use, demand, etc.), then the notation $V \geqq \mathbf{0}$ means that at least one component of V is positive.

We now consider the open Leontief model and suppose that a given $D \geqq \mathbf{0}$ (demand, bill of goods) has been specified. To meet this demand, it is sufficient to solve the matrix equation (37) for X. Indeed, we can write (37) in the form

$$X - AX = D,$$

(38) $$(I - A)X = D,$$

---

[1] Named after Wassily W. Leontief because of his pioneer work in this field; see his article "Quantitative Input and Output Relations in the Economic System of the United States," *Review of Economic Statistics*, vol. 18 (1936), pp. 105–125, and his two books *The Structure of the American Economy 1919–1939*, 2nd ed., Oxford University Press, New York, 1951, and *Input-Output Analysis*, Oxford University Press, New York, 1966.

and if the inverse of $I - A$ exists, we can multiply both sides of (38) on the left-hand side by $(I - A)^{-1}$ and obtain

(39) $$X = (I - A)^{-1}D.$$

This gives the intensity X at which the system must operate to produce the demand D. Of course we hope that each component of X is nonnegative, for if some $x_i < 0$, equation (39) tells us that the corresponding process must operate in reverse and (by our understanding of the definition of an input-output matrix) this is impossible. We have

Theorem 8. If $(I - A)^{-1}$ exists and is a nonnegative matrix, then for any given demand $D \geq 0$, there is an intensity $X \geq 0$ such that the system produces D when operated at intensity X.

Proof. Since $D \geq 0$ and $(I - A)^{-1} \geq 0$, it is clear that equation (39) gives $X \geq 0$. ∎

EXAMPLE 1. If possible, find $(I - A)^{-1}$ for each of the following input-output matrices:

(a) $A = \begin{bmatrix} 0.4 & 0.6 \\ 0.6 & 0.4 \end{bmatrix}$ and (b) $A = \begin{bmatrix} 0.8 & 0.4 \\ 0.4 & 0.8 \end{bmatrix}$.

Solution. (a) We have

(40) $$I - A = \begin{bmatrix} 1 - 0.4 & -0.6 \\ -0.6 & 1 - 0.4 \end{bmatrix} = 0.6 \begin{bmatrix} 1 & -1 \\ -1 & 1 \end{bmatrix}.$$

In (40) we add the first row to the second row, and we obtain a row of zeros. Consequently there is no inverse for $I - A$ in this case.

(b) In this case we find that

$$I - A = \begin{bmatrix} 1 - 0.8 & -0.4 \\ -0.4 & 1 - 0.8 \end{bmatrix} = \begin{bmatrix} 0.2 & -0.4 \\ -0.4 & 0.2 \end{bmatrix}.$$

With a little computation one finds that

$$(I - A)^{-1} = \frac{5}{3} \begin{bmatrix} -1 & -2 \\ -2 & -1 \end{bmatrix}.$$

This shows that we may have an input-output matrix for which $(I - A)^{-1}$ has all entries negative. Naturally we are interested in those matrices A for which $(I - A)^{-1} \geq 0$.

Definition 3. Let $A \geqq 0$. If $(I - A)^{-1}$ exists, then we call this inverse the Leontief inverse of A. If $(I - A)^{-1} \geqq 0$, then we call A a Leontief matrix.

Theorem 8 states that if A is a Leontief matrix, then for any given demand $D \geqq 0$, one can find a level of operation for the system that will produce D. Theorems that identify Leontief matrices are always of interest. We shall not stop for proofs of the next three theorems but merely observe how they help to locate Leontief matrices.

Theorem 9. Let A be a Leontief matrix. Suppose that $A\star$ is the matrix that is obtained from A when the units for measuring the various goods G_i are changed. Then $A\star$ is a Leontief matrix.

Theorem 10. Suppose that $A \geqq 0$. If there is some $X_0 \geqq 0$ such that $AX_0 < X_0$, then A is a Leontief matrix.

Definition 4. Let $A \geqq 0$. Let

(41) $$s_j = \sum_{i=1}^{n} a_{ij}$$

be the sum of the elements in the jth column. Let N be the largest such sum; i.e., let

(42) $$N = \max\{s_1, s_2, \ldots, s_n\}.$$

Then N is called the norm of the matrix A and is denoted by $N(A)$.

Theorem 11. If $A \geqq 0$ and $N(A) < 1$, then A is a Leontief matrix.

EXAMPLE 2. Illustrate Theorem 10 for the technology matrix

(43) $$A = \begin{bmatrix} 0.5 & 0.1 & 0.1 \\ 0.2 & 0.6 & 0.2 \\ 0.1 & 0.2 & 0.6 \end{bmatrix}$$

Solution. As a first guess we set $x_i = 1.0$ for $i = 1, 2, 3$. Then we find

$$AX_0 = \begin{bmatrix} 0.5 & 0.1 & 0.1 \\ 0.2 & 0.6 & 0.2 \\ 0.1 & 0.2 & 0.6 \end{bmatrix} \begin{bmatrix} 1.0 \\ 1.0 \\ 1.0 \end{bmatrix} = \begin{bmatrix} 0.7 \\ 1.0 \\ 0.9 \end{bmatrix} \leqq X_0.$$

But since the equality sign holds for the second coordinate, this vector just misses. We change x_2 to 1.2 and then we find that

$$(44) \qquad AX_0 = \begin{bmatrix} 0.5 & 0.1 & 0.1 \\ 0.2 & 0.6 & 0.2 \\ 0.1 & 0.2 & 0.6 \end{bmatrix} \begin{bmatrix} 1.0 \\ 1.2 \\ 1.0 \end{bmatrix} = \begin{bmatrix} 0.72 \\ 1.12 \\ 0.94 \end{bmatrix} < \begin{bmatrix} 1.0 \\ 1.2 \\ 1.0 \end{bmatrix} = X_0.$$

Hence by Theorem 10, A is a Leontief matrix.

EXAMPLE 3. Use the matrix of Example 2 to illustrate Theorem 11.

Solution. The sums of the elements in the columns of A are

$$s_1 = 0.8, \qquad s_2 = 0.9, \qquad s_3 = 0.9.$$

Hence $N(A) = 0.9$ and since $0.9 < 1$, Theorem 11 tells us that A is a Leontief matrix.

To check these assertions we find the Leontief inverse. It is convenient to avoid fractions and put $I - A$ in the form

$$(45) \qquad I - A = \frac{1}{10} \begin{bmatrix} 5 & -1 & -1 \\ -2 & 4 & -2 \\ -1 & -2 & 4 \end{bmatrix}.$$

A little labor will then yield

$$(46) \qquad (I - A)^{-1} = \frac{1}{21} \begin{bmatrix} 60 & 30 & 30 \\ 50 & 95 & 60 \\ 40 & 55 & 90 \end{bmatrix}.$$

Just as the theorem predicted, all of the coefficients in $(I - A)^{-1}$ are positive.

EXAMPLE 4. Illustrate Theorem 9 for the matrix

$$(47) \qquad A^\star = \begin{bmatrix} \dfrac{1}{6} & 2 & 3 \\ \dfrac{1}{36} & \dfrac{1}{4} & \dfrac{1}{4} \\ \dfrac{1}{12} & \dfrac{1}{3} & \dfrac{1}{6} \end{bmatrix}.$$

Solution. For this matrix $N(A^\star) = 3\dfrac{5}{12}$. Since $N(A^\star) \geqq 1$, we cannot apply Theorem 11. Suppose now that we change the unit for measuring G_1 by reducing it by a factor of $1/6$. The reader should convince himself that the new technology matrix A is obtained from $A\star$ by dividing the first row by 6 and then multiplying the first column by 6. Hence

(48)
$$A = \begin{bmatrix} \dfrac{1}{6} & \dfrac{1}{3} & \dfrac{1}{2} \\[2mm] \dfrac{1}{6} & \dfrac{1}{4} & \dfrac{1}{4} \\[2mm] \dfrac{1}{2} & \dfrac{1}{3} & \dfrac{1}{6} \end{bmatrix}.$$

Now $N(A) = 11/12$, and Theorem 11 assures us that A is a Leontief matrix. By Theorem 9 it follows that $A\star$ is also a Leontief matrix.

The matrix $I - A$ can be put in the form

(49)
$$I - A = \frac{1}{12}\begin{bmatrix} 10 & -4 & -6 \\ -2 & 9 & -3 \\ -6 & -4 & 10 \end{bmatrix},$$

and a laborious computation will yield

(50)
$$(I - A)^{-1} = \frac{3}{32}\begin{bmatrix} 39 & 32 & 33 \\ 19 & 32 & 21 \\ 31 & 32 & 41 \end{bmatrix} > \mathbf{0}.$$

The reader may explain to himself why the inverse of $I - A\star$ is related to $(I - A)^{-1}$ in the same way that $A\star$ is related to A. This rule gives

(51)
$$(I - A\star)^{-1} = \frac{3}{32}\begin{bmatrix} 39 & 192 & 198 \\ \dfrac{19}{6} & 32 & 21 \\ \dfrac{31}{6} & 32 & 41 \end{bmatrix} > \mathbf{0}.$$

As indicated in our examples, the computation of the inverse of a matrix may be quite troublesome. Searching for alternative methods, the reader may recall that if $-1 < a < 1$, then

(52)
$$\frac{1}{1 - a} = 1 + a + a^2 + \cdots + a^n + \cdots = \sum_{n=0}^{\infty} a^n,$$

where the right-hand side is an infinite series (there are infinitely many terms). Now arguing by analogy one might guess

> **Theorem 12.** If A is a nonnegative matrix and $N(A) < 1$, then
>
> $$(53) \quad (I - A)^{-1} = I + A + A^2 + \cdots + A^n + \cdots = \sum_{n=0}^{\infty} A^n.$$

Although the right-hand side in Equation (53) is an infinite series of matrices, one can always obtain a good approximation to $(I - A)^{-1}$ by taking a suitable finite number of terms of the infinite series. The computation of A^2, A^3, etc., is still troublesome, but this can be turned over to a computer if a suitable one is available.

EXAMPLE 5. Use equation (53) to estimate $(I - A)^{-1}$ if

$$(54) \qquad A = \begin{bmatrix} 0.00 & 0.20 & 0.15 & 0.10 \\ 0.10 & 0.05 & 0.15 & 0.15 \\ 0.20 & 0.10 & 0.10 & 0.10 \\ 0.10 & 0.05 & 0.10 & 0.05 \end{bmatrix}.$$

Solution. The usual method gives

$$(I - A)^{-1} = \begin{bmatrix} 1.094 & 0.266 & 0.247 & 0.183 \\ 0.184 & 1.129 & 0.244 & 0.223 \\ 0.281 & 0.197 & 1.212 & 0.188 \\ 0.154 & 0.108 & 0.166 & 1.103 \end{bmatrix}$$

to the nearest thousandth. The computation of $(I - A)^{-1}$ by the infinite series (53) was given to a computer[1] and in a few minutes the computer returned the first 17 terms of the series:

$$\sum_{n=0}^{\infty} A^n = \begin{bmatrix} 1 & 0 & 0 & 0 \\ 0 & 1 & 0 & 0 \\ 0 & 0 & 1 & 0 \\ 0 & 0 & 0 & 1 \end{bmatrix} + \begin{bmatrix} 0.00 & 0.20 & 0.15 & 0.10 \\ 0.10 & 0.05 & 0.15 & 0.15 \\ 0.20 & 0.10 & 0.10 & 0.10 \\ 0.10 & 0.05 & 0.10 & 0.05 \end{bmatrix}$$

$$+ \begin{bmatrix} 0.0600 & 0.0300 & 0.0550 & 0.0500 \\ 0.0500 & 0.0450 & 0.0525 & 0.0400 \\ 0.0400 & 0.0600 & 0.0650 & 0.0500 \\ 0.0300 & 0.0350 & 0.0375 & 0.0300 \end{bmatrix}$$

$$+ \begin{bmatrix} 0.0190 & 0.0215 & 0.0240 & 0.0185 \\ 0.0190 & 0.0195 & 0.0235 & 0.0190 \\ 0.0240 & 0.0200 & 0.0265 & 0.0220 \\ 0.0140 & 0.0130 & 0.0165 & 0.0135 \end{bmatrix} +$$

[1] These computations were done by an IBM 1410 computer system at the Computer Research Center, University of South Florida.

The error that is made when one uses the first $K + 1$ terms of the series can be judged by examining M, the largest entry in the matrix:

$$(55) \qquad D = (I - A)^{-1} - (I + A + A^2 + \cdots + A^k).$$

For the matrix A given by (54), it was found that

If $K = 5$, then $M \approx 0.004$.

If $K = 10$, then $M \approx 0.000055$.

If $K = 15$, then $M \approx 0.000001$.

The rate of convergence of the series is strongly influenced by the norm of the matrix. In (54) we see that $N(A) = 0.50$. The same type of computation was performed for the matrix

$$A = \begin{bmatrix} 0.10 & 0.30 & 0.00 & 0.10 \\ 0.20 & 0.50 & 0.10 & 0.20 \\ 0.30 & 0.10 & 0.60 & 0.20 \\ 0.30 & 0.00 & 0.20 & 0.40 \end{bmatrix},$$

for which $N(A) = 0.90$. In this case the computer took five minutes to return the first 100 powers of A, and to evaluate the sum $\sum_{n=0}^{K} A^n$ for each K from 1 to 100. It was found that

If $K = 50$, then $M \approx 0.017$.

If $K = 75$, then $M \approx 0.0013$.

If $K = 100$, then $M \approx 0.00009$.

We now return to our model of an economy and we complicate it by considering the cost of labor, the rate of profit, and the price of each good produced. We keep the same physical units for G_j and introduce the following new symbols:

$p_j = $ the price of one unit of G_j (either to a consumer, or for use in a process),

$l_j = $ the cost of labor for producing one unit of G_j,

$r_j = $ the profit earned on one unit of G_j.

Here, of course, the units are dollars (or some other monetary unit), but we can drop all dollar signs without any harm. The fundamental equation

$$(56) \qquad p_j = p_1 a_{1j} + p_2 a_{2j} + \cdots + p_n a_{nj} + l_j + r_j$$

merely states that the price of one unit of G_j is the cost of producing one unit of G_j plus the profit ($j = 1, 2, \ldots, n$).

To put (56) into matrix form we introduce three new column vectors P, L, and R whose components are p_j, l_j, and r_j, respectively. However, we should note that in (56) multiplication is on the "wrong" side and that in order to use our column vectors we need A^T, the transpose of A, and not A. Then (56) has the form

(57) $$P = A^T P + L + R.$$

If we prefer to use A and not A^T, then we can introduce the row vectors P^T, L^T, and R^T, the transposes of P, L, and R. Then (57) takes the form

(58) $$P^T = P^T A + L^T + R^T.$$

Both (57) and (58) give the same relationship, only the first uses column vectors, and the second uses row vectors.

Suppose that we ask for a specific profit per unit of goods. Can we find a suitable set of prices to generate this profit? To answer this question we merely insert the identity matrix in (57) and write

(59) $$IP - A^T P = L + R.$$

Consequently we have

Theorem 13.   Let R be a preassigned nonnegative profit vector. If A is a Leontief matrix, then there is a nonnegative price vector P, given by

(60) $$P = (I - A^T)^{-1}(L + R)$$

that will yield the profit R.

Proof.   We merely multiply both sides of (59) by $(I - A^T)^{-1}$. If A is a Leontief matrix, then $(I - A)^{-1} \geqq 0$. Consequently the same is true for $(I - A^T)^{-1}$.   ■

Note carefully that Theorem 13 covers the case in which profit is per unit of good produced. In practice, this is not the standard procedure. One is more often interested in (a) profit per dollar invested or (b) profit as a fraction of price. To investigage (b) we introduce the rate of profit matrix

(61) $$[R\star] = \begin{bmatrix} r_1^\star & 0 & \cdots & 0 \\ 0 & r_2^\star & \cdots & 0 \\ \vdots & \vdots & & \vdots \\ 0 & 0 & \cdots & r_n^\star \end{bmatrix}$$

in place of the column vector R. Here r_i is the rate of profit as a fraction of the price of G_i. The fundamental equation (57) is now replaced by

(62) $$P = A^T P + L + [R\star]P,$$

and if $I - A^T - [R\star]$ has a nonnegative inverse, then the corresponding price vector is given by

(63) $$P = [I - A^T - [R\star]]^{-1}L.$$

Economists often simplify equation (63) by assuming that the rate of profit $r_j^\star$ is the same for all goods. The standard argument is that if one process (or good) shows a greater rate of profit than another, the investment dollars and production effort will flow from the less profitable good to the more profitable one until an equilibrium is reached. If we make this assumption and let $r\star$ be the common rate of profit ($r\star p_j$ is the profit on one unit of G_j), then (63) simplifies to

(64) $$P = [(1 - r\star)I - A^T]^{-1}L.$$

Both (63) and (64) state that price is a linear function of labor costs. If we grant that this model represents a reasonably true picture of the economic system, then we deduce that any increase in the cost of labor will cause a corresponding increase in the cost of goods. In fact, if[1] ΔL and ΔP denote the changes in the labor vector and the price vector, respectively, then equation (64) yields

(65) $$\Delta P = [(1 - r\star)I - A^T]^{-1}\Delta L.$$

It should be noted that this equation still holds if some of the components of ΔL are negative (decrease in cost of labor for certain goods), and in this case some of the components of ΔP may also be negative (decrease in the price of certain goods).

Exercise 2

1. Suppose that the matrix A, given by equation (43), represents a highly simplified economy in which the sectors are agriculture, manufactures, and transportation, respectively, and the units are selected suitably. Find

[1] The symbol Δ is the Greek D and is read "delta." It is often used in mathematics to denote the change in a quantity.

the intensity for the system if it is to have a surplus of 21 units of agriculture, 63 units of manufactures, and 0 units of transportation.

2. Find the intensity for the system in problem 1 if we wish a surplus of 63 units of agriculture, 21 units of manufactures, and 0 units of transportation.

3. For the matrices given by equations (43) and (46), use direct computation to show that $(I - A)(I - A)^{-1} = I$.

4. Explain why it is reasonable to form the sum of the elements in any one row of A. Why is the sum of the elements in any one column of A often physically meaningless? Describe a situation in which the sum of the elements in a column does have a meaning.

5. Suppose that in problem 1, the unit is "$1 worth" for each good. We hold this unit fixed for G_1 and G_3 but we increase the size of the unit to "$5 worth" for G_2. If $A\star$ is the technology matrix with this new unit, find $A\star$.

6. Find $(I - A\star)^{-1}$ for the matrix $A\star$ of problem 5. *Hint:* Use equation (46).

7. Let $A\star$ be the matrix given by equation (47). Illustrate Theorem 10 by finding a suitable X_0 for this matrix.

8. Explain why Theorem 9 is obviously true. *Hint:* You are not asked to prove Theorem 9.

9. Let

$$A = \begin{bmatrix} \dfrac{1}{2} & 2 \\ 0 & \dfrac{1}{2} \end{bmatrix}.$$

Find the norm of A and $(I - A)^{-1}$. Is A a Leontief matrix?

10. For the matrix given in problem 9, find A^2, A^3, A^4, and A^5.

11. For the matrix given in problem 9, guess at a formula for A^n and then prove it by mathematical induction.

12. Check equation (53) for the matrix given in problem 9 by computing $I + A + A^2 + A^3 + A^4 + A^5$. Compare your sum with $(I - A)^{-1}$ found in problem 9.

13. Does the matrix A given in problem 9 satisfy the conditions of Theorem 12? Does this example show that Theorem 12 is wrong?

14. Let $A \geqq \mathbf{0}$. Prove that for each element of A we have $0 \leqq a_{ij} \leqq N(A)$. Suppose that the elements of A^k are denoted by $a_{ij}^{(k)}$. Prove that for each positive integer k, we have

$$0 \leqq a_{ij}^{(k)} < [N(A)]^k.$$

15. Let $A \geqq \mathbf{0}$ and suppose that $N(A) < 1$. Prove that as k becomes large, the elements in A^k tend to zero.

16. Suppose that a technology matrix is

$$A = \begin{bmatrix} 0.8 & 0.2 & 0.1 \\ 0.4 & 0.8 & 0.2 \\ 0.0 & 0.2 & 0.8 \end{bmatrix}.$$

Find $N(A)$. Show that the positive demand vector D requires a negative intensity vector X, where

$$D = \begin{bmatrix} 4 \\ 2 \\ 4 \end{bmatrix} \geqq 0 \quad \text{and} \quad X = \begin{bmatrix} -15 \\ -30 \\ -10 \end{bmatrix} \leqq 0.$$

Is A a Leontief matrix?

17. Find $(I - A)^{-1}$ for the matrix A of problem 16.

18. Let D be a demand vector that represents a combination of the needs of labor and the export market for a period of one year. Suppose that we wish to build a stock of our goods for possible future use. Let $\alpha > 0$ be a constant, and suppose that for each i we wish to set aside αx_i amount of G_i for future use each year, where x_i is the amount of G_i produced that year. Show that if $[(1 - \alpha)I - A]^{-1}$ exists, then

$$X = [(1 - \alpha)I - A]^{-1}D.$$

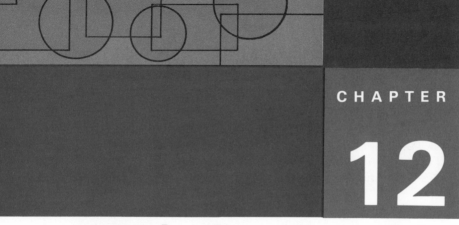

graph theory and its applications

The word *graph* has two different meanings in mathematics. The reader will recall that when we speak of the graph of a function such as $y = x^2$ we mean the set C of all pairs (x,y) for which the second number is the square of the first. The graph is usually pictured by marking those points in the plane whose coordinates are in the set C. When we do this for the particular function $y = x^2$, we obtain a smooth curve known as a parabola.

In this chapter the word *graph* has quite a different meaning. Intuitively a graph is a collection of points $A_1, A_2, \ldots$ together with curves which join some of these points in pairs. However, in graph theory we are not concerned with the shape of the curve joining the points A_i and A_j, but only with the existence of a direct path between A_i and A_j. Hence we are interested only in the presence or absence of the curve. For example, the two graphs G_1 and G_2 represented in Figs. 1 and 2 are the same, because in each case each pair of points that is joined by a curve in G_1 is also joined by a curve in G_2 and conversely.

It may seem unusual to regard G_1 and G_2 as equal, but, as we shall see in this chapter, there are many interesting theorems about graphs, and the theory has wide applications.

1. graphs and their representations

The representation of a graph in Figs. 1 and 2 helps us to grasp the concept of a graph but does not constitute a definition. In forming a definition we free ourselves from the geometric picture by replacing the word *point* by *vertex* and the word *curve* by *edge*. These abstractions lead to

> **Definition 1. Graph.** A graph is a set V of vertices $\{A_1, A_2, \ldots\}$ together with a second set E of edges. An edge is an unordered pair of distinct vertices A_iA_j and for each pair of vertices A_i and A_j either A_iA_j is in E or A_iA_j is not in E. The edge A_iA_j is regarded as the same as the edge A_jA_i. The order of a graph is the number of vertices in V.

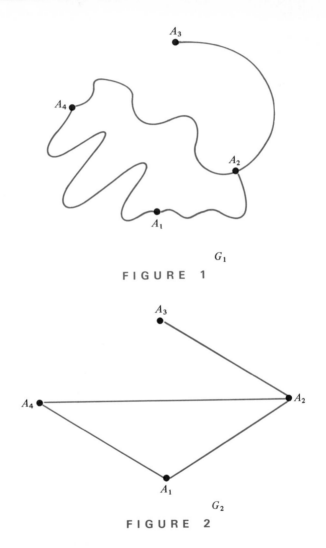

G_1

FIGURE 1

G_2

FIGURE 2

A geometric picture of a graph, such as that given in Fig. 1, is called a representation of the graph. Thus either Fig. 1 or Fig. 2 is a representation of the graph G defined by the two sets

(1) $$V = \{A_1, A_2, A_3, A_4\},$$

and

(2) $$E = \{A_1A_2, A_2A_3, A_1A_4, A_2A_4\}.$$

Once we grasp the abstract nature of Definition 1, we may relax our formal language and again refer to the vertices as points and the edges as lines or curves.

A graph is said to be *planar* if it can be represented by a plane diagram in which no two curves have common points except possibly at their end points.

EXAMPLE 1. Give two examples of graphs that are not planar.

Solution. Let K_5 be a graph consisting of five points such that every pair of points is joined by a line. This graph is pictured in Fig. 3.

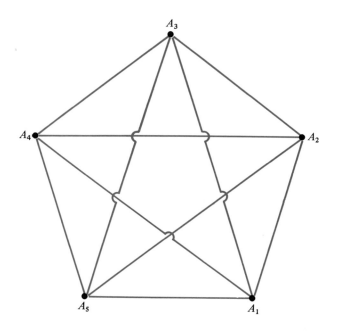

K_5

FIGURE 3

Let $K_{3,3}$ be a graph consisting of the points A_1, A_2, A_3, B_1, B_2, and B_3. Each A_i is joined by a line to every B_j, but no A_i is joined to an A_j, and no B_i is joined to a B_j. This graph is pictured in Fig. 4, using a tetrahedron in which there is one additional line joining one pair of midpoints of opposite sides.

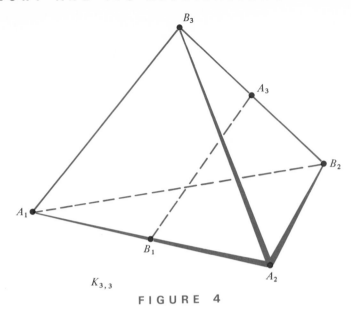

$K_{3,3}$

FIGURE 4

The proof that K_5 and $K_{3,3}$ are not planar is rather difficult and will be omitted. However, the student will easily convince himself of this fact if he tries to represent either of these graphs by a plane diagram without intersections.

A *subgraph H* of a graph G is merely a subset of the points and edges of G which also forms a graph. Thus if we select A_3A_5, an edge of G, to be in the subgraph H, we must also select the points A_3 and A_5 to be in H.

The special graphs K_5 and $K_{3,3}$ used in Example 1 are interesting because of the following remarkable theorem due to Kuratowski.[1]

Theorem 1. A graph G is planar if and only if it does not contain either K_5 or $K_{3,3}$ as a subgraph.

Graphs are extremely useful in the study of certain social structures. In a typical application we have a set of n persons and some symmetric relation among the persons. For example, the relation may be

(a) Two persons know each other.
(b) Two persons have yearly incomes within $100 of each other.
(c) Two persons belong to the same club.
(d) Two persons have sworn never to speak to each other again.

To simplify matters we concentrate our attention on the relation (a).

We relate the social structure described in (a) to the theory of graphs

---

[1] The curious reader may locate Kuratowski's proof in *Fundamenta Mathematica,* vol. 15 (1930), pp. 271–283, but he must be warned that the proof is difficult.

by using points to represent the people and lines to indicate the relation. Thus if A_i and A_j are points that represent Mrs. Brown and Mrs. Jones (also called A_i and A_j), we draw a line connecting the points A_i and A_j if and only if A_i and A_j are acquainted. At the moment we ignore the unsymmetric possibility that Mrs. Brown knows Mrs. Jones, while Mrs. Jones does not know Mrs. Brown. This unsymmetric situation will be considered in the next section. Further, at this moment we are not concerned with whether Mrs. Brown knows herself because in a proper graph an edge is not permitted to connect a point with itself. Of course, we can alter the definition of a graph to allow for *loops* (a curve that joins A_i to itself) or *multiple edges* (two or more curves joining the same pair of points), but we will have no need for these more general types of graphs. Multiple edges may be useful if one wants a diagram that represents the "strength" of an acquaintanceship.

EXAMPLE 2. Suppose that among five persons A_1, A_2, A_3, A_4, and A_5, no three are mutually acquainted and no three are mutually strangers. Is such a situation possible? If so, draw a graph representing this possibility.

Solution. The graph shown in Fig. 5 gives the desired relation. The reader should check that the conditions posed are satisfied.

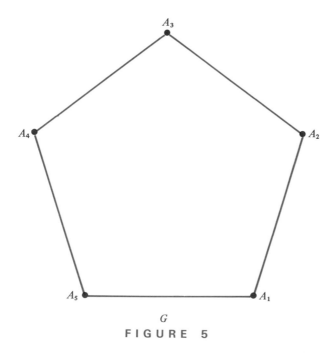

G

FIGURE 5

Each graph G has a *complementary* graph $G\star$ that is obtained from G as follows: The points A_i and A_j are connected in $G\star$ if and only if they are *not connected* in G. For example, the graph shown in Fig. 6 is the complement of the graph shown in Fig. 5.

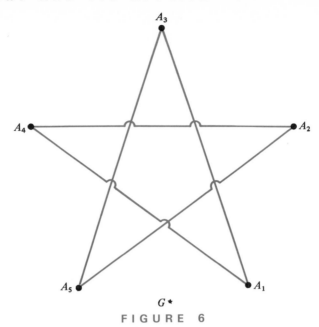

$G\star$

FIGURE 6

Let us now return to the question posed in Example 2. If there are three people who are mutually acquainted, this will be represented by a triangle in Fig. 5. Similarly three people who are mutually strangers will be represented by a triangle in $G\star$, shown in Fig. 6. Clearly neither of these graphs contains a triangle.

A graph G can also be represented by a matrix $A = [a_{ij}]$. Let $a_{ij} = 1$ if A_i and A_j are connected by a line, and let $a_{ij} = 0$ if A_i and A_j are not connected. Then A is called the *matrix representation* of graph G, or the *incidence matrix* for G.

For example, the incidence matrix for the graph shown in Fig. 5 is

(3)
$$A = \begin{bmatrix} 0 & 1 & 0 & 0 & 1 \\ 1 & 0 & 1 & 0 & 0 \\ 0 & 1 & 0 & 1 & 0 \\ 0 & 0 & 1 & 0 & 1 \\ 1 & 0 & 0 & 1 & 0 \end{bmatrix}.$$

Exercise 1

1. Give in abstract form the graphs shown in **(a)** Fig. 3, **(b)** Fig. 4, and **(c)** Fig. 6.
2. Is a tree always a graph?

3. Give the incidence matrix for each of the graphs shown in **(a)** Fig. 2, **(b)** Fig. 3, **(c)** Fig. 4, and **(d)** Fig. 6. Is the incidence matrix of a graph always symmetric?

4. A *complete* graph is a graph for which every pair of points is joined by a line. The symbol K_n denotes the complete graph of order n. Make a drawing for K_2, K_3, and K_4. In each case give the incidence matrix for the graph.

5. How many edges are in K_n?

6. How many edges are in $K_n^\star$, the complement of K_n?

7. Suppose that the vertices of a graph can be partitioned into two disjoint subsets U and V such that A_i and A_j are connected by an edge if and only if one of the vertices is in U and the other is in V. In this case the graph is called a *bipartite* graph and is denoted by the symbol $K_{m,n}$, where m is the number of vertices in U and n is the number of vertices in V. How many edges are in $K_{m,n}$? Make a drawing for the graph **(a)** $K_{1,5}$, **(b)** $K_{2,2}$, **(c)** $K_{3,4}$, and **(d)** $K_{2,8}$.

8. Give the incidence matrix for each of the graphs in problem 7.

9. Let U be an nth order square matrix in which every entry is 1 and let I be the usual identity matrix. Prove that if A is the incidence matrix for G, then $U - A - I$ is the incidence matrix for $G\star$.

10. Prove that for any graph $(G\star)\star = G$.

★11. Let G be the graph shown in Fig. 7. Give the incidence matrix for G. Is this graph (except for the lettering of the vertices) equal to any of the graphs already considered in this section?

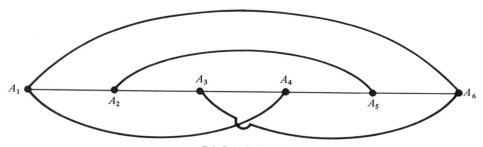

FIGURE 7

★12. Let G be an arbitrary graph with six vertices. Prove that either G or $G\star$ will contain a triangle. *Hint:* Select an arbitrary vertex A_1. Either three lines of G terminate at A_1, or this is true for $G\star$. (It might be helpful to consider the complete graph K_6 and color the lines red if they are in G and color them blue if they are in $G\star$. Then prove that K_6 contains either a red triangle or a blue triangle.)

13. The *degree* of a point A_i is the number of lines that have A_i as an end point. We let $d(A_i)$ denote this number. Prove that in any graph on n points $\sum_{i=1}^{n} d(A_i)$ is an even number.

2.   digraphs

We now consider a modification of the definition of a graph in which a line is assigned a direction. This direction is indicated by writing A_iA_j for the line from A_i to A_j and A_jA_i for the line in the reverse direction. We then speak of A_iA_j as a *directed line* (or edge or curve) and indicate the direction by an arrow as in Fig. 8. For the directed line A_iA_j, the point A_i is called the *initial point* (the *first point*, the *starting point*), and A_j is called the *terminal point* (the *second point,* the *final point*).

> **Definition 2. Digraph.**[1]   A digraph is a set V of vertices $\{A_1, A_2, \ldots\}$ together with a second set E of directed edges. The set E may contain both A_iA_j and A_jA_i, but E contains no loops (the initial and terminal points are different).

We remark at this point that either a graph or a digraph may have infinitely many vertices and infinitely many edges. However, in this text we shall consider only graphs and digraphs with finitely many vertices and edges.

EXAMPLE 1.   Draw all possible digraphs on two vertices.

Solution.   There are only four such and, as the reader can easily see, these are all included in Fig. 8. Of course, we must be clear about the objects being counted. If we allow a relabeling of the vertices, then in Fig. 8 the graph G_3 can be identified with G_2 merely by changing A_1 into A_2 and A_2 into A_1.

$$A_1 \qquad\qquad A_2 \qquad\qquad A_1 \qquad A_2 \qquad\qquad A_1 \qquad A_2 \qquad\qquad A_1 \qquad A_2$$
$$\quad G_1 \qquad\qquad\qquad G_2 \qquad\qquad\qquad G_3 \qquad\qquad\qquad G_4$$

FIGURE   8

> **Definition 3. Isomorphism.**   Two digraphs (or graphs) G and H are said to be isomorphic if we can relabel the vertices in H in such a way that the new graph obtained from H is identical with G.

Referring back to Fig. 8, we have seen that G_3 is isomorphic to G_2. Hence, up to an isomorphism, there are only three different digraphs with two points.

[1] The term *digraph* is merely a contraction of the longer term "directed graph," and as such it is a natural and welcome addition to our language.

There are a number of social structures that can be represented by digraphs, and this representation can be a valuable tool for the sociologist and those working in closely allied fields. We list a few such structures.

(I) Suppose that we have n persons and as in section 1 we wish to indicate that A_1 knows A_2. Then the corresponding digraph G contains the directed line A_1A_2. If, further, A_2 does not know A_1, then this is represented in G by the subgraph G_2 shown in Fig. 8. However, if A_1 and A_2 know each other, then G contains G_4, shown in Fig. 8. If everyone in a set of n persons knows everyone else in the set, then the digraph will have $n(n-1)$ lines. A similar type of structure arises if we replace *knows*, by *likes*, *respects*, or *trusts*. Thus A_iA_j is in G if and only if A_i likes (respects, trusts) A_j.

(II) In an army, or a dictatorship, it is usually clear who can give orders and who must obey them. The men in the army, or the citizens in a dictatorship, can be represented by the points of G and the directed line A_iA_j is in G if and only if A_i can give orders (which must be obeyed) to A_j. Such a digraph will usually be a directed tree but not always. In a more democratic organization the directed line A_iA_j may mean that A_i can alter the opinion of A_j by persuasion.

(III) A communication network may consist of n persons $A_1, A_2, \ldots, A_n$ together with some suitable means of communication between certain pairs of individuals. The individuals are represented by points in G, and the directed line A_iA_j is present if and only if A_i can communicate with A_j. It is by no means necessary that this relation be symmetric. It can happen that A_iA_j is in G while A_jA_i is not. For example, communication may be by radio, by a faulty telephone which works only in one direction, or by a carrier pigeon.

(IV) In a round robin tournament each contestant plays every other contestant one game. The directed line A_iA_j is in the digraph if and only if player A_i defeats player A_j. If there are no ties, then each pair of points in G is connected by exactly one directed line. Such a digraph is called a *dominance digraph*. It is also called a *tournament*. Clearly, if G is a dominance digraph, then for each pair i, j, with $i \neq j$, either A_iA_j is in G or A_jA_i is in G but not both.

(V) The points of G do not necessarily represent individuals. Suppose, for variety, that each point represents a statement. If we can prove that A_i implies A_j, then the directed line A_iA_j is in G. In a long mathematical work involving many theorems, a digraph G may be very helpful in bringing to light new results that might otherwise go unnoticed. To be specific, suppose that the lines A_2A_3, A_5A_9, A_6A_7, A_9A_1, A_7A_5, and A_3A_6 are in G. We see immediately that A_2A_1 must also be in G, and consequently (with almost no additional effort) we obtain another theorem, namely A_2 implies A_1.

There are still other interesting structures that can be represented by a digraph. By studying the theory of digraphs, we are sharpening some tools that can be used to study any one of these structures. Merely for the sake of simplicity and unity we shall devote most of our attention to the communication network described in (III).

EXAMPLE 2.   Give the incidence matrix for the digraph (communication network) shown in Fig. 9.

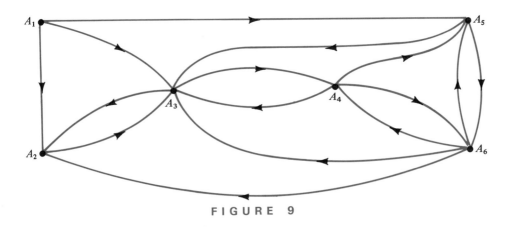

FIGURE 9

Solution.   By the definition of the incidence matrix of G, we have $a_{ij} = 1$ if and only if the directed line $A_i A_j$ is in G (A_i can communicate with A_j). For the digraph shown in Fig. 9,

$$(4) \qquad A = \begin{bmatrix} 0 & 1 & 1 & 0 & 1 & 0 \\ 0 & 0 & 1 & 0 & 0 & 0 \\ 0 & 1 & 0 & 1 & 0 & 0 \\ 0 & 0 & 1 & 0 & 1 & 1 \\ 0 & 0 & 1 & 0 & 0 & 1 \\ 0 & 1 & 1 & 1 & 1 & 0 \end{bmatrix}.$$

The matrix A does not contain any more information than the diagram. In some cases one can obtain the desired information more easily from the diagram; in other cases it is merely a matter of taste, while in still other cases the matrix A may be superior. However, the matrix A can be fed into a computer, and if the order of A is very large, this represents a distinct advantage over the diagram.

Example 2 serves to illustrate some interesting items, for example, no one communicates with A_1. This is clear from Fig. 9, but it also follows by inspection of the matrix A, because every element in the first column is zero.

The *outdegree* of a point A_i is the number of lines from A_i and is denoted by $od(A_i)$. Similarly, the *indegree* of A_i is the number of lines to A_i and is denoted by $id(A_i)$. The degree of A_i is denoted by $d(A_i)$ and by definition

$$d(A_i) = id(A_i) + od(A_i).$$

Clearly $od(A_i)$ is the sum of the elements in the ith row of A, and $id(A_i)$ is the sum of the elements in the ith column. Returning to our example, we observe from the matrix A [equation (4)] that A_6 has the largest outdegree and A_3 has the largest indegree.

In this example A_4 does not communicate directly with A_2 but can communicate with A_2 in two steps by the path A_4A_3, A_3A_2. Such a sequence is called a *two-stage communication*. In general we say that A_i has *k-stage communication* with A_j if there is in G a sequence of k directed lines $A_iA_q, A_qA_r, \ldots, A_tA_u, A_uA_j$ such that the terminal point of each line is the initial point of the next line. It is easy to compute the number of k-stage communications using the matrix A, as described in

Theorem 2. Let A be the incidence matrix for G, and set

$$(5) \qquad A^k = \begin{bmatrix} s_{11} & s_{12} & \cdots & s_{1n} \\ s_{21} & s_{22} & \cdots & s_{2n} \\ \vdots & \vdots & & \vdots \\ s_{n1} & s_{n2} & \cdots & s_{nn} \end{bmatrix}.$$

Then s_{ij} is the number of k-stage communications from A_i to A_j.

Proof. It is sufficient to prove this result for $k = 2$, because the general case will follow by mathematical induction. When $k = 2$,

$$(6) \qquad s_{ij} = a_{i1}a_{1j} + a_{i2}a_{2j} + \cdots + a_{in}a_{nj}.$$

In this sum, a term $a_{ik}a_{kj} = 1$ if and only if $a_{ik} = 1$ and $a_{kj} = 1$. This occurs if and only if A_i communicates with A_k and A_k communicates with A_j. Hence for each two-stage communication from A_i to A_j, there is a corresponding 1 on the right-hand side of (6). Conversely for each 1 in the sum s_{ij}, there is a two-stage communication from A_i to A_j. ∎

EXAMPLE 3. Find the number of two-stage and three-stage communications for the network given in Fig. 9.

Solution. By direct multiplication we find that

$$A^2 = \begin{bmatrix} 0 & 1 & 2 & 1 & 0 & 1 \\ 0 & 1 & 0 & 1 & 0 & 0 \\ 0 & 0 & 2 & 0 & 1 & 1 \\ 0 & 2 & 2 & 2 & 1 & 1 \\ 0 & 2 & 1 & 2 & 1 & 0 \\ 0 & 1 & 3 & 1 & 1 & 2 \end{bmatrix}$$

and

$$A^3 = \begin{bmatrix} 0 & 3 & 3 & 3 & 2 & 1 \\ 0 & 0 & 2 & 0 & 1 & 1 \\ 0 & 3 & 2 & 3 & 1 & 1 \\ 0 & 3 & 6 & 3 & 3 & 3 \\ 0 & 1 & 5 & 1 & 2 & 3 \\ 0 & 5 & 5 & 5 & 3 & 2 \end{bmatrix}.$$

Since the first column of A^3 contains only zeros, it follows that no one communicates with A_1 in three stages. But this was always obvious from Fig. 9, since no one communicates with A_1 at any stage. However the matrix A^3 also tells us that there are five different ways that A_6 can communicate with A_3 in three stages, and certainly this is not immediately obvious from Fig. 9. If we name the vertices in order, the five sequences are (1) $A_6A_3A_2A_3$, (2) $A_6A_3A_4A_3$, (3) $A_6A_4A_5A_3$, (4) $A_6A_4A_6A_3$, and (5) $A_6A_5A_6A_3$.

The sum of the entries in the ith row is greatest when $i = 6$, and this sum is 20. Consequently if one wishes to plant a rumor and have it spread most widely (and with the greatest number of variations) in three stages, it should be given to A_6. Looking at the sums of the entries in the jth column, we find that the maximum is 23 when $j = 3$. Consequently, one might pick up the maximum information (or misinformation) after three-stage transmission by listening to A_3.

A person can communicate with himself in two or more stages. This would be represented in the digraph by a sequence of directed lines A_iA_q, $A_qA_r, \ldots, A_uA_i$ that start and end with the same vertex A_i. Such a sequence is called a *cycle* and is denoted by merely naming the vertices in order. For example, the digraph shown in Fig. 9 contains the cycles $A_2A_3A_2$, $A_2A_3A_4A_6A_2$, and $A_2A_3A_4A_5A_6A_3A_2$. Thus if the person represented by A_2 started a rumor, he might later hear the same rumor (possibly distorted).

Exercise 2

1. Find A, A^2, and A^3 for the digraph shown in Fig. 10.

2. Let G be the digraph on the points $A_1, A_2, \ldots, A_n$ in which the line

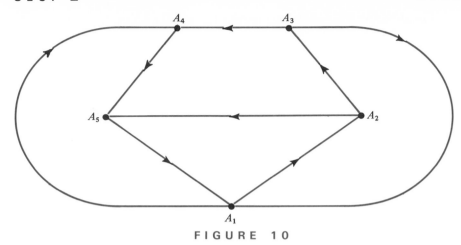

FIGURE 10

A_iA_j is present if and only if $i < j$. Make a drawing for this digraph and give the incidence matrix for $n = 3, 4,$ and 5.

3. For the matrix of problem 2, (with $n = 5$) find A^3.

4. For each of the following incidence matrices, draw the corresponding digraphs. Which of these matrices can represent a dominance digraph?

$$A = \begin{bmatrix} 0 & 0 & 0 \\ 0 & 0 & 0 \\ 0 & 0 & 0 \end{bmatrix}, \qquad B = \begin{bmatrix} 0 & 1 & 0 \\ 1 & 0 & 1 \\ 1 & 0 & 0 \end{bmatrix},$$

$$C = \begin{bmatrix} 0 & 0 & 0 & 1 \\ 0 & 0 & 1 & 1 \\ 0 & 1 & 0 & 0 \\ 0 & 1 & 1 & 0 \end{bmatrix}, \qquad D = \begin{bmatrix} 0 & 0 & 0 & 0 \\ 1 & 0 & 0 & 0 \\ 1 & 1 & 0 & 0 \\ 1 & 1 & 1 & 0 \end{bmatrix},$$

$$E = \begin{bmatrix} 0 & 1 & 1 & 0 \\ 1 & 0 & 0 & 1 \\ 1 & 0 & 0 & 1 \\ 0 & 1 & 1 & 0 \end{bmatrix}, \qquad F = \begin{bmatrix} 0 & 0 & 1 & 1 & 1 \\ 1 & 0 & 0 & 1 & 1 \\ 1 & 1 & 0 & 0 & 1 \\ 1 & 1 & 1 & 0 & 0 \\ 0 & 1 & 1 & 1 & 0 \end{bmatrix}.$$

5. For the communication network shown in Fig. 9, give all six of the three-stage communications of A_4 with A_3.

6. In Fig. 9, A_6 has five different three-stage communications with each of A_2, A_3, and A_4. Find them.

7. Let A be an incidence matrix for a digraph G on the points A_1, $A_2, \ldots, A_8$. Let B be the matrix obtained from A by interchanging the second row and the fifth row and then interchanging the second and fifth column. Prove that B is the incidence matrix of a digraph

H that can be obtained from G by interchanging the names of the points A_2 and A_5.

★8. Use the interchange of rows and columns in a matrix, described in problem 7, to give an alternative (and better) definition for isomorphic digraphs.

9. Set $A + A^2 = C = [c_{ij}]$. Prove that c_{ij} gives the total number of one-stage or two-stage communications from A_i to A_j.

10. Find $C = A + A^2$ for the communication system shown **(a)** in Fig. 9 and **(b)** in Fig. 10. In each case check some of the entries in C by tracing the proper communication sequences in the figure.

11. Draw a figure for the network corresponding to

$$A = \begin{bmatrix} 0 & 1 & 0 \\ 1 & 0 & 1 \\ 0 & 1 & 0 \end{bmatrix}.$$

By tracing all two-stage and three-stage communications, find A^2 and A^3. Then check your result by matrix multiplication. Can you find A^{2n+1} for each positive integer n?

12. Prove that for a dominance digraph, $A + A^T = U - I$ (see Exercise 1, problem 9).

★13. Complete the proof of Theorem 2, by carrying out the induction process.

3. distances in a graph

We shall measure distances by counting the edges in a path. Hence we need

> **Definition 4. Path.** A set of distinct vertices $A_i, A_q, A_r, \ldots, A_t, A_u, A_j$ in a graph G together with the edges $A_iA_q, A_qA_r, \ldots, A_tA_u, A_uA_j$ in G form a path joining A_i and A_j in G. The length of the path is the number of edges in the path.

If G is a digraph (instead of a graph) we naturally require that the edges have the direction specified by the order of the vertices, and in this case we refer to the path as a path from A_i to A_j.

It is worth emphasizing that in a path the vertices are distinct, and hence a path does not contain any "cycles." Thus a communication sequence from A_i to A_j that contains cycles (see Example 3 of the preceding section) is not a path.

If there is a path joining A_i and A_j in a graph G, then we say that A_i and A_j are *connected*. Otherwise A_i and A_j are not connected. A graph G is said to be *connected* if every pair of points in G are connected. In a digraph connectedness is more complicated because there may be a path from A_i to A_j but no path from A_j to A_i. In such a case A_i is connected to A_j but A_j is not connected to A_i.

EXAMPLE 1. For the digraph shown in Fig. 11, find the length of each path (a) from A_1 to A_9, (b) from A_9 to A_1, (c) from A_1 to A_5, (d) from A_5 to A_9, and (e) from A_9 to A_5.

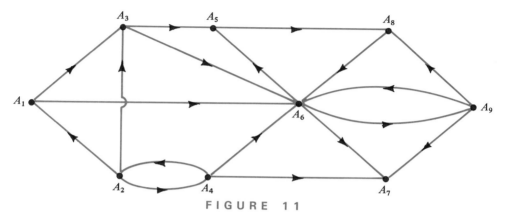

FIGURE 11

Solution. We use $L(XYZ)$ to denote the length of the path through the vertices X, Y, and Z in that order. Inspection yields

(a) $L(A_1A_6A_9) = 2$,
$L(A_1A_3A_6A_9) = 3$,
$L(A_1A_3A_5A_8A_6A_9) = 5$.

(b) There are no paths from A_9 to A_1.

(c) $L(A_1A_3A_5) = 2$,
$L(A_1A_6A_5) = 2$,
$L(A_1A_3A_6A_5) = 3$.

(d) $L(A_5A_8A_6A_9) = 3$.

(e) $L(A_9A_6A_5) = 2$,
$L(A_9A_8A_6A_5) = 3$.

Where there are several paths from A_i to A_j it is natural to call the length of the shortest path the *distance* from A_i to A_j.

Definition 5. Distance. If there is a path from A_i to A_j in the digraph G (or the graph G), then the length of the path that has the smallest length is called the distance from A_i to A_j and is denoted by $D(A_i, A_j)$. If no such path exists, then we say that the distance does not exist, and we write $D(A_i, A_j) = \infty$.

In the digraph shown in Fig. 11, we found three paths from A_1 to A_9. Clearly $D(A_1, A_9) = 2$. Similarly in that digraph $D(A_9, A_1)$ does not exist, $D(A_1, A_5) = 2$, $D(A_5, A_9) = 3$, and $D(A_9, A_5) = 2$.

We may recall from Euclidean geometry the *triangle inequality*, which states that for any three points A_1, A_2, and A_3,

(7) $$D(A_1, A_3) \leqq D(A_1, A_2) + D(A_2, A_3).$$

For either graphs or digraphs we have

Theorem 3. Let A_1, A_2, and A_3 be arbitrary points of a graph (or a digraph). If all of the distances involved exist, then inequality (7) holds.

One difference between graphs and digraphs is exposed in

Theorem 4. If there is a path joining A_i and A_j in a graph, then

(8) $$D(A_i, A_j) = D(A_j, A_i).$$

However, if we are concerned with a digraph, then (8) may not be true. To see this we return to Fig. 11 and observe that

$$D(A_5, A_9) = 3 \quad \text{and} \quad D(A_9, A_5) = 2.$$

This rather abstract concept of distance in a digraph can be used to measure the status of an individual in an organization. We examine the digraph G of an organization. The edges may be given various interpretations. For example, the directed edge $A_i A_j$ may mean that **(a)** A_i communicates with A_j or **(b)** A_i directly supervises the work of A_j, or **(c)** A_i gives orders to A_j. In any interpretation it is clear that A_i has some status over A_j, and this status is greater if A_j is farther from A_i in the chain. With these natural requirements, Harary[1] has introduced the following numerical measure for status.

Definition 6. Status. Let G be a digraph representing an organization. The status of A_i in G is denoted by $S(A_i)$ and is given by

(9) $$S(A_i) = \sum D(A_i, A_j),$$

where the sum is over all individuals A_j that are at finite distance from A_i (i.e., for which there is a path from A_i to A_j).

[1] F. Harary, "Status and Contrastatus," *Sociometry*, vol. 22 (1959), pp. 23–43. See also, *Structural Models*, F. Harary, R. Z. Norman, and D. Cartwright, Wiley, New York, 1965.

EXAMPLE 2. The structures of two organizations are shown in Fig. 12.
Compute the status for A_1, A_2, A_4, A_5, B_1, B_2, and B_6.

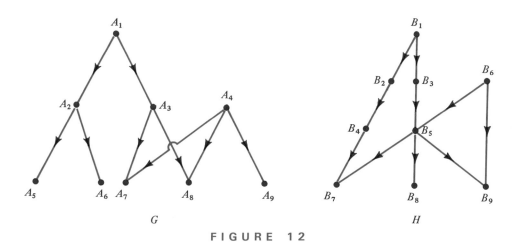

G   H

FIGURE 12

Solution. Direct computation gives the following:

$$S(A_1) = 10, \qquad S(A_2) = 2, \qquad S(A_4) = 3, \qquad S(A_5) = 0,$$
$$S(B_1) = 15, \qquad S(B_2) = 3, \qquad S(B_6) = 6.$$

It is interesting to observe that although A_1 and B_1 both head organizations with the same number of men, the Harary measure of status gives B_1 a greater status than A_1.

Exercise 3

1. Definition 5 gives the distance from A_i to A_j if these are distinct points. How would you define the distance from a point to itself?
2. For the digraph in Fig. 11, find **(a)** $D(A_1, A_6)$, **(b)** $D(A_2, A_9)$, **(c)** $D(A_9, A_2)$, **(d)** $D(A_5, A_7)$, and **(e)** $D(A_7, A_5)$.
3. Select three points from the digraph in Fig. 11, and check that for these points the triangle inequality is true.
4. Prove Theorem 3.
5. Is Theorem 4 obvious, or does it require a formal proof?
6. By citing a suitable example, explain the reason for the restriction on the sum in Definition 6.
7. In the famous sextet from the opera *Lucia di Lammermoor*, the members Alisa, Arturo, Edgardo, Enrico, Lucia, and Raimondo are all communicating with each other. Prove that in this organization each member has the same status. What is the status of Alisa?

8. Prove that if A_1 is not connected to any other person in an organization
G, then A_1 has zero status in G.

9. Suppose that the organizations G and H in Fig. 12 are merged by bringing
in an outside man C_1, who directly supervises A_1 and B_1, and there are
no other changes in G and H. Prove that the status of A_1 is unchanged.
Is it true that $S(C_1) = S(A_1) + S(B_1)$?

4.   tournaments

A round robin tournament gives a digraph G in a natural way. Each player
is represented by a point, and the directed line A_iA_j is in G if and only
if A_i beats A_j. It is assumed that each contestant plays every other contestant
and there are no ties. The digraph takes its name from the source.

> **Definition 7. Tournament.**   A digraph G is called a tournament if it
> has the property that for every pair of distinct points A_i and A_j, either
> A_iA_j is in G or A_jA_i is in G, but not both.

A tournament is also called a *dominance digraph*, because it also represents
a dominance relation, i.e., a relation in which among any two persons in
a set, one always dominates (gives orders to, influences) the other. The
incidence matrix for a tournament is called a *tournament matrix* or a *dominance matrix*.

Given three points, only two types of tournaments are possible. These
are shown in Fig. 13. The incidence matrices for these two digraphs are

$$(10) \qquad C = \begin{bmatrix} 0 & 1 & 0 \\ 0 & 0 & 1 \\ 1 & 0 & 0 \end{bmatrix} \quad \text{and} \quad D = \begin{bmatrix} 0 & 1 & 1 \\ 0 & 0 & 1 \\ 0 & 0 & 0 \end{bmatrix}.$$

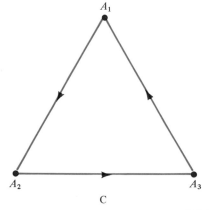

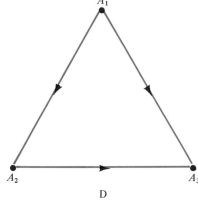

FIGURE   13

The digraph C is called a *cycle*. The digraph D is said to be *transitive*. To clarify the concept of transitivity it may be helpful to introduce the symbol $\gg$ (which resembles the inequality sign) to indicate dominance. Consequently $A_i \gg A_j$ means that A_i dominates A_j, or in a tournament A_i won over A_j. By definition, $A_i \gg A_j$ if and only if the directed line A_iA_j is in the digraph. At first one might expect that

(T)   If $A_i \gg A_j$ and $A_j \gg A_k$, then $A_i \gg A_k$.

But a little reflection shows that in nearly every tournament there are upsets; i.e., contestants A_1, A_2, and A_3 who form a cycle such as in Fig. 13.

Definition 8. Transitive. A tournament (or dominance digraph) is said to be transitive if for every triple of vertices the condition (T) is true.

EXAMPLE 1. The results of the tennis matches in Neuro City are given by the incidence matrix

(11)
$$
A = \begin{array}{c} \\ A_1 \\ A_2 \\ A_3 \\ A_4 \\ A_5 \\ A_6 \\ A_7 \end{array}
\begin{array}{c} \begin{array}{ccccccc} A_1 & A_2 & A_3 & A_4 & A_5 & A_6 & A_7 \end{array} \\
\left[\begin{array}{ccccccc}
0 & 0 & 0 & 1 & 1 & 1 & 1 \\
1 & 0 & 1 & 1 & 0 & 1 & 0 \\
1 & 0 & 0 & 0 & 1 & 1 & 1 \\
0 & 0 & 1 & 0 & 0 & 1 & 1 \\
0 & 1 & 0 & 1 & 0 & 0 & 1 \\
0 & 0 & 0 & 0 & 1 & 0 & 0 \\
0 & 1 & 0 & 0 & 0 & 1 & 0
\end{array} \right].
\end{array}
$$

Make a digraph for this tournament. How many cyclic triples does this tournament contain? How many transitive triples? Without any further play, how can we determine the winner?

Solution. We leave it for the reader to make the digraph. From the figure he can carefully and patiently count the number of cyclic triples and transitive triples. However, there are simple formulas for these quantities which we now develop.

Let S_i be the outdegree of the point A_i. By definition, this is exactly the number of games A_i won (or the number of people A_i dominates). The quantity S_i is also called the *score* for A_i. If we examine a particular vertex A_i, any pair of edges that leads from A_i automatically determines a transitive triple with A_i in the dominant position. If there are S_i edges leading from

A_i, then A_i is dominant in $S_i(S_i - 1)/2$ transitive triples. Consequently the formula

(12)
$$t = \sum_{i=1}^{n} \frac{S_i(S_i - 1)}{2}$$

gives the total number of transitive triples in any tournament.

The total number of triples is just the number of combinations of n things taken three at a time, and this is $n(n - 1)(n - 2)/6$. Since a triple is either transitive or cyclic, we find that

(13) $$c = \frac{n(n - 1)(n - 2)}{6} - t = \frac{n(n - 1)(n - 2)}{6} - \sum_{i=1}^{n} \frac{S_i(S_i - 1)}{2}$$

gives the number of cyclic triples.

Returning to Example 1, the score sequence $(S_1, S_2, S_3, S_4, S_5, S_6, S_7)$ can be found by looking at the rows of the matrix A. This gives (4, 4, 4, 3, 3, 1, 2), and using these values in equations (12) and (13) we find that

$$t = \frac{1}{2}(4 \cdot 3 + 4 \cdot 3 + 4 \cdot 3 + 3 \cdot 2 + 3 \cdot 2 + 1 \cdot 0 + 2 \cdot 1) = 25$$

and

$$c = \frac{7 \cdot 6 \cdot 5}{6} - t = 35 - 25 = 10.$$

In this particular tournament there is a three-way tie for first place. To settle on a winner without further play, we might decide to examine the number of two-stage dominances. If $A_i \gg A_j$ and $A_j \gg A_k$, we call this sequence a two-stage dominance for A_i. But (see Theorem 2) the number of two-stage dominances for A_i is the sum of the elements in the ith row of A^2. Using equation (11) we find that

Row sum

(14) $$A^2 = \begin{bmatrix} 0 & 2 & 1 & 1 & 1 & 2 & 2 \\ 1 & 0 & 1 & 1 & 3 & 3 & 3 \\ 0 & 2 & 0 & 2 & 2 & 2 & 2 \\ 1 & 1 & 0 & 0 & 2 & 2 & 1 \\ 1 & 1 & 2 & 1 & 0 & 3 & 1 \\ 0 & 1 & 0 & 1 & 0 & 0 & 1 \\ 1 & 0 & 1 & 1 & 1 & 1 & 0 \end{bmatrix} \quad \begin{matrix} 9 \\ 12 \\ 10 \\ 7. \\ 9 \\ 3 \\ 5 \end{matrix}$$

In view of this computation, we might award first prize to A_2, second prize to A_3, and third prize to A_1. Further, the tie for fourth and fifth place seems to be resolved in favor of A_5.

The total number of one-stage and two-stage dominances for A_i is called the *power* of A_i. In this example the power of A_2 is $4 + 12 = 16$.

If a tournament is transitive, then it has no cyclic triples. Consequently the number of cyclic triples given by equation (13) forms a good measure for the amount of deviation of a tournament from a transitive one. The maximum number of cyclic triples is given in

Theorem 5. In any tournament with n vertices

$$(15) \qquad c \leqq \begin{cases} \dfrac{n(n-1)(n+1)}{24}, & \text{if } n \text{ is odd,} \\[2mm] \dfrac{n(n-2)(n+2)}{24}, & \text{if } n \text{ is even,} \end{cases}$$

and if the tournament is selected properly, the equal sign can occur.

The proof is rather difficult and so we omit it.[1]

Contests and domination are not the only sources of tournaments. Sociologists and psychologists are frequently interested in "paired comparisons." Here the research worker selects a suitable set S such as

{ice cream, chocolate cake, apple pie, pecan pie, beer, pickles}.

The person who is being tested is then offered a pair of items from S, and he must indicate his preference. When all possible pairs have been considered (15 for this set) the indicated preferences determine a tournament, and the number of cyclic triples is a measure of the inconsistency of the person. If the set is large, it is almost certain that some cycles will appear.

Politicians are also vitally concerned with paired comparisons when they must select candidates. Frequently they devote a lot of time and effort to determine a composite digraph of voter preferences.

We close with an interesting theorem about round robin tournaments that is most easily phrased in terms of dominance.

Theorem 6. In a dominance digraph, the person who dominates the largest number of people in one stage will dominate everyone in either one stage or two stages.

[1] M. G. Kendall and B. B. Smith, "On the Method of Paired Comparisons," *Biometrika*, vol. 31 (1940), pp. 324–345.

Proof. Merely by a choice of subscripts we can let A_1 be the one who dominates the largest number of persons, and let the persons he dominates be $A_2, A_3, \ldots, A_k$. Let B be any other person that A_1 does not dominate (see Fig. 14). Since A_1 does not dominate B, then B dominates A_1 (in one stage). Assume now that A_1 does not dominate B in two stages. Then the edge joining B and A_2 must go from B to A_2; i.e., B dominates A_2. Similarly B dominates $A_3, A_4, \ldots$, and A_k. Thus B dominates in one stage more persons than A_1. This is a contradiction to the assumption about A_1. Hence at least one of the edges joining A_i and B must be from A_i to B, so that A_1 dominates B in two stages. ∎

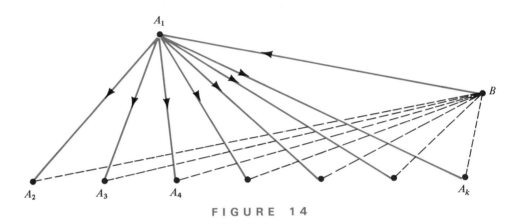

FIGURE 14

Exercise 4

1. Find the number of edges in a tournament with n vertices.
2. Find the sum of all the entries in an $n \times n$ dominance matrix.
3. In Uproar University, the intramural football contest between the fraternities gave the following dominance matrix:

$$A = \begin{bmatrix} 0 & 1 & 0 & 1 & 1 & 1 \\ 0 & 0 & 1 & 1 & 1 & 1 \\ 1 & 0 & 0 & 1 & 1 & 1 \\ 0 & 0 & 0 & 0 & 0 & 1 \\ 0 & 0 & 0 & 1 & 0 & 0 \\ 0 & 0 & 0 & 0 & 1 & 0 \end{bmatrix}.$$

Can the three-way tie for first place be resolved by the method used in Example 1? How about the tie for last place?
4. If C is the matrix given in equation (10), prove (in two different ways)

that $C^3 = I$, the identity matrix. Hence C is a cube root of I. Can you find a fourth-order matrix, different from I, that is a fourth root of I?

5. How many triples are there in a tournament of n players if **(a)** $n = 7$, **(b)** $n = 8$, **(c)** $n = 10$, and **(d)** $n = 13$? For each of these tournaments find the maximum number of cyclic triples that can occur.

★6. Give the incidence matrix for a tournament with five players in which the number of cycles is as large as possible. Draw the digraph.

7. Give an interpretation of Theorem 6 for a contest.

8. Let A be a dominance matrix and suppose that the first row has the largest row sum. Prove that in $A + A^2$ every element in the first row is positive except for the diagonal term (the element in the first row and the first column).

9. A *complete path* in a digraph is a path in which every vertex occurs (just once). Find a complete path **(a)** for C in Fig. 13, **(b)** for D in Fig. 13, **(c)** for the graph in Example 1, and **(d)** for the graph in problem 3.

★10. Prove the following interesting theorem. In every tournament there is at least one complete path. *Hint:* Use mathematical induction.

11. Give an interpretation of the theorem proved in problem 10 in terms of victories in a contest.

12. Find the number of transitive triples and the number of cyclic triples for the tournament given in problem 3.

13. In a certain tournament the score sequence was $(6, 5, 3, 3, 2, 2, 0)$. Find the number of transitive triples and the number of cyclic triples.

★14. The voting paradox: During the recent agitation at Uproar University, the students could not decide which building to seize: administration, botany, chemistry, or education. The organization selected B. Scraggly, L. Plumbum, and I. Zinnia as a three-man committee to make the final decision. Scraggly wanted to occupy the administration building, Plumbum hoped to hit chemistry, and Zinnia preferred to be planted in botany. Each was willing to consider other possibilities, and their

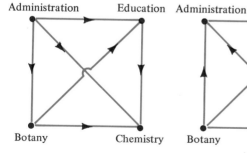

B. Scraggly

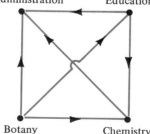

I. Zinnia

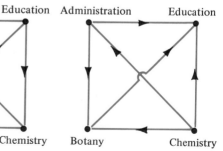

L. Plumbum

F I G U R E   1 5

preferences are given by the three transitive digraphs in Fig. 15. After much debate they decided to vote first with administration versus botany. The winner of that vote was to be pitted against chemistry, and the winner of that vote was to be pitted against education. Prove that with this arrangement, the committee recommended seizure of the education building, although no member of the committee really preferred this building.

★**15.** Suppose that in problem 14, L. Plumbum persuades the committee to alter the sequence of voting, so that chemistry is not considered until last. Prove that with this rearrangement the final balloting will always be between chemistry and administration, and chemistry will always win.

5.   the detection of cliques

We all have some intuitive idea of a clique. Formally we have

Definition 9. A clique in a communication network is any maximal collection of three or more individuals with the property that any two persons in the collection communicate with each other.

EXAMPLE 1. Suppose there is a set of six individuals $\{A_1, A_2, A_3, A_4, A_5, A_6\}$, such that **(a)** each of A_1, A_2, A_3, and A_4 communicates with everyone (except himself), **(b)** A_5 communicates only with A_1, and **(c)** A_6 communicates with A_2, A_4, and A_5. Find all possible cliques.

Solution. From the definition of a clique, we must look for the largest possible subsets (having at least three members) such that between every pair of individuals in the subset there is a two-way communication. We observe that there are only two cliques: $\{A_1, A_2, A_3, A_4\}$ and $\{A_2, A_4, A_6\}$. The subset $\{A_1, A_5\}$ is not a clique, since there are only two members in this subset. Also, the subset $\{A_1, A_2, A_3\}$ is not a clique, because it is not maximal (as large as possible).

If the number of individuals in a set is large, then it is difficult to determine the cliques by inspection. So, we look for the existence of cliques by matrix methods and use computers to do the calculations for us. To do this, suppose that $A = [a_{ij}]$ is the incidence matrix for a particular communication network with n individuals. We form a new matrix $B = [b_{ij}]$ as follows: If $a_{ij} = 1$ and $a_{ji} = 0$, then we set $b_{ij} = 0$; in all other cases we set $b_{ij} = a_{ij}$. The matrix B is called the *symmetric kernel* of A. For example, if A is the incidence matrix of Example 1, then

$$
\begin{array}{c}
 \begin{array}{cccccc} A_1 & A_2 & A_3 & A_4 & A_5 & A_6 \end{array} \\
(16) \qquad A = \begin{array}{c} A_1 \\ A_2 \\ A_3 \\ A_4 \\ A_5 \\ A_6 \end{array}
\begin{bmatrix}
0 & 1 & 1 & 1 & 1 & 1 \\
1 & 0 & 1 & 1 & 1 & 1 \\
1 & 1 & 0 & 1 & 1 & 1 \\
1 & 1 & 1 & 0 & 1 & 1 \\
1 & 0 & 0 & 0 & 0 & 0 \\
0 & 1 & 0 & 1 & 1 & 0
\end{bmatrix}.
\end{array}
$$

The symmetric kernel of this matrix is

$$
(17) \qquad B = \begin{bmatrix}
0 & 1 & 1 & 1 & 1 & 0 \\
1 & 0 & 1 & 1 & 0 & 1 \\
1 & 1 & 0 & 1 & 0 & 0 \\
1 & 1 & 1 & 0 & 0 & 1 \\
1 & 0 & 0 & 0 & 0 & 0 \\
0 & 1 & 0 & 1 & 0 & 0
\end{bmatrix}.
$$

From the definition of B it is clear that $b_{ij} = 1$ if and only if there is two-way communication between A_i and A_j.

To investigate the existence of cliques we examine B^3 instead of B.

Theorem 7. Let A be the incidence matrix for a communication network and let B be the symmetric kernel of A. Let $b_{ii}^{(3)}$ be the element in the ith row and the ith column (main diagonal) of B^3. If $b_{ii}^{(3)}$ is positive, then A_i belongs to some clique. Conversely if A_i belongs to some clique, then $b_{ii}^{(3)}$ is positive.

Proof. Suppose that $b_{ii}^{(3)}$ is positive. Then by Theorem 2 there is at least one three-stage communication from A_i to A_i. Since A_i does not communicate with himself, we must have two individuals A_j and A_k ($j \neq k$) so that

$$
A_i \longrightarrow A_j \longrightarrow A_k \longrightarrow A_i.
$$

(The arrow denotes the direction of communication.) Since B is the matrix having nonzero entries only if the communication is reciprocal, we must have

$$
A_i \longleftrightarrow A_j \longleftrightarrow A_k \longleftrightarrow A_i;
$$

hence A_i, A_j, and A_k are members of the same clique.

Conversely if A_i belongs to some clique, then there is a three-stage communication from A_i to A_i and hence $b_{ii}^{(3)}$ is positive. ∎

We remark that Theorem 7 does not provide any information regarding the number of individuals in a clique or the number of cliques to which A_i belongs.

EXAMPLE 2. Suppose the matrix A for a communication network is

(18)
$$A = \begin{bmatrix} 0 & 1 & 0 & 1 \\ 1 & 0 & 0 & 0 \\ 0 & 1 & 0 & 1 \\ 1 & 0 & 1 & 0 \end{bmatrix}.$$

What people belong to cliques?

Solution. The symmetric kernel of A is

(19)
$$B = \begin{bmatrix} 0 & 1 & 0 & 1 \\ 1 & 0 & 0 & 0 \\ 0 & 0 & 0 & 1 \\ 1 & 0 & 1 & 0 \end{bmatrix}.$$

Computing B^3, we get

(20)
$$B^3 = \begin{bmatrix} 0 & 2 & 0 & 3 \\ 2 & 0 & 1 & 0 \\ 0 & 1 & 0 & 2 \\ 3 & 0 & 2 & 0 \end{bmatrix}.$$

Since all the entries in the main diagonal of B^3 are zero, Theorem 7 tells us that there are no cliques.

The following theorem gives us some information about the size of the clique to which an individual A_i belongs.

Theorem 8. Suppose an individual A_i is a member of exactly one clique. Then this clique has m members if and only if

(21)
$$b_{ii}^{(3)} = (m - 1)(m - 2).$$

Proof. We shall prove our theorem for the case $m = 4$. The general case follows by a similar argument.

Suppose A_i belongs to exactly one clique and the size of the clique is four. Then we show that $b_{ii}^{(3)} = (4 - 1)(4 - 2) = 6$. Let $\{A_i, A_j, A_k, A_l\}$ be the four-person clique. By the definition of a clique, there is a two-way communication between these four individuals, as indicated in Fig. 16.

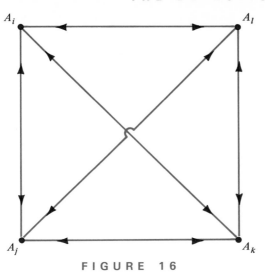

FIGURE 16

We observe from this digraph that a three-step connection from A_i to A_i can be established by going around any triangle in the graph starting from A_i. We thus obtain the following six connections:

Counterclockwise   Clockwise

$A_i \longrightarrow A_j \longrightarrow A_k \longrightarrow A_i$ $A_i \longrightarrow A_k \longrightarrow A_j \longrightarrow A_i$

$A_i \longrightarrow A_j \longrightarrow A_l \longrightarrow A_i$ $A_i \longrightarrow A_l \longrightarrow A_j \longrightarrow A_i$

$A_i \longrightarrow A_k \longrightarrow A_l \longrightarrow A_i$ $A_i \longrightarrow A_l \longrightarrow A_k \longrightarrow A_i$

Therefore $b_{ii}^{(3)} \geqq 6$.

Suppose now that $b_{ii}^{(3)} > 6$. Since B^3 counts the three-stage connections that are symmetric, there must exist A_p and A_q such that

$$A_i \longleftrightarrow A_p \longleftrightarrow A_q \longleftrightarrow A_i,$$

where either A_p or A_q is different from A_j and A_k. Consequently either A_i belongs to another clique or A_i belongs to a clique with more than four individuals. This contradicts our hypothesis that A_i belongs to exactly one clique of size four. Hence $b_{ii}^{(3)} = 6$.

To prove the converse, we assume that $b_{ii}^{(3)} = 6$. We must show that A_i belongs to a clique of size four. Since $b_{ii}^{(3)} \neq 0$, there must be a three-step connection from A_i to A_i involving two-way communication. Thus there are individuals A_j and A_k such that

$$A_i \longleftrightarrow A_j \longleftrightarrow A_k \longleftrightarrow A_i.$$

If these were the only two-way communications, we would have $b_{ii}^{(3)} = 2$. Since we are given $b_{ii}^{(3)} = 6$, there must be another three-stage connection from A_i to A_j. In other words, there are individuals A_l and A_p in the set such that

$$A_i \longleftrightarrow A_l \longleftrightarrow A_p \longleftrightarrow A_i$$

holds. Since A_i belongs to only one clique, A_i, A_j, A_k, A_l, and A_p must all belong to the same clique. This gives us a five-person clique, and by the first part of the theorem we should have $b_{ii}^{(3)} \geqq (5 - 1)(5 - 2) = 12$. Since $b_{ii}^{(3)} = 6 < 12$, A_l or A_p must be either A_j or A_k, and hence A_i belongs to one clique of size four. ∎

We remark that Theorem 8 cannot be applied directly if an individual A_i belongs to more than one clique. However, if we examine the matrix B^3 carefully, we shall see that the theorem can sometimes be useful even in this case. We illustrate this in

EXAMPLE 3. Find all cliques in a communication network whose matrix A is

$$(22) \qquad A = \begin{bmatrix} 0 & 1 & 1 & 0 & 1 \\ 1 & 0 & 1 & 1 & 1 \\ 1 & 1 & 0 & 1 & 1 \\ 0 & 1 & 1 & 0 & 1 \\ 1 & 0 & 1 & 1 & 0 \end{bmatrix}.$$

Solution. The symmetric kernel of A is

$$(23) \qquad B = \begin{bmatrix} 0 & 1 & 1 & 0 & 1 \\ 1 & 0 & 1 & 1 & 0 \\ 1 & 1 & 0 & 1 & 1 \\ 0 & 1 & 1 & 0 & 1 \\ 1 & 0 & 1 & 1 & 0 \end{bmatrix}.$$

A moderate amount of labor will yield

$$(24) \qquad B^3 = \begin{bmatrix} 4 & 8 & 8 & 4 & 8 \\ 8 & 4 & 8 & 8 & 4 \\ 8 & 8 & 8 & 8 & 8 \\ 4 & 8 & 8 & 4 & 8 \\ 8 & 4 & 8 & 8 & 4 \end{bmatrix}.$$

Before analyzing B^3, we tabulate the values of $b_{ii}^{(3)}(m)$ for various values of m. Indeed, equation (21) gives

$$(25) \qquad b_{ii}^{(3)}(3) = 2, \qquad b_{ii}^{(3)}(4) = 6, \qquad b_{ii}^{(3)}(5) = 12.$$

Now it is easy to see from B^3 that A_1 belongs to exactly two three-person cliques $(2 + 2 = 4)$. Similarly A_2, A_4, and A_5 each belong to two three-person cliques. Since $b_{33}^{(3)} = 8$, either A_3 belongs to a four-person clique and a three-person clique $(8 = 6 + 2)$ or A_3 belongs to four three-person cliques $(8 = 2 + 2 + 2 + 2)$. Because no other person belongs to a four-person clique, the first case is impossible. Hence A_3 belongs to four three-person cliques.

Exercise 5

1. Draw a communication network for the matrix B of Example 3.
2. A communication system with five persons has two four-person cliques. Draw a figure for such a system using the smallest number of lines possible.
3. Give the incidence matrix for the system described in problem 2. Let A_1 be a person who belongs to both cliques. Find $b_{11}^{(3)}$ in two different ways.
4. Suppose that A_1 is a member of an m-person clique and that he is not a member of any other clique. Prove that A_1 has three-stage communication with himself in $(m - 1)(m - 2)$ different ways.
5. Find all cliques for the communication network shown in Figure 9.
6. Find all cliques for the communication network defined by each of the following incidence matrices:

$$\begin{bmatrix} 0 & 1 & 1 & 0 & 1 \\ 1 & 0 & 1 & 1 & 0 \\ 1 & 1 & 0 & 1 & 1 \\ 1 & 1 & 1 & 0 & 1 \\ 1 & 1 & 1 & 1 & 0 \end{bmatrix} \qquad \begin{bmatrix} 0 & 1 & 1 & 1 & 1 & 1 \\ 1 & 0 & 1 & 1 & 1 & 1 \\ 1 & 1 & 0 & 1 & 1 & 0 \\ 1 & 1 & 1 & 0 & 1 & 1 \\ 1 & 0 & 1 & 1 & 0 & 1 \\ 1 & 1 & 0 & 1 & 1 & 0 \end{bmatrix}.$$

7. Find all cliques in the communication system for which the incidence matrix is

$$\begin{bmatrix} 0 & 1 & 0 & 0 & 0 & 1 & 0 \\ 1 & 0 & 1 & 1 & 0 & 1 & 1 \\ 0 & 1 & 0 & 1 & 0 & 0 & 0 \\ 0 & 1 & 1 & 0 & 1 & 1 & 1 \\ 0 & 0 & 0 & 1 & 0 & 1 & 0 \\ 1 & 1 & 0 & 1 & 1 & 0 & 1 \\ 0 & 1 & 0 & 1 & 0 & 1 & 0 \end{bmatrix}.$$

★6.   an open problem in graph theory

The student who reads only the text in his mathematics courses may well have the erroneous idea that all mathematics problems have been solved. What he sees are the well-polished portions of mathematics. The exercises are all carefully prepared so that (aside from misprints and oversights of the authors) all of the questions posed have reasonably easy answers.

The truth is quite otherwise. In mathematics there are many problems that are deep and fascinating, and yet, despite the best efforts of the strongest mathematicians, these problems remain unsolved. Most of the unsolved problems require an extensive background and cannot possibly be presented to the beginner. Among the ones that are rather easy to state,[1] most of them are in number theory (the theory of integers) and hence lie outside of the domain of this book. These include the celebrated "twin prime problem" and "Fermat's Last Theorem."

However, there is a nice open problem in graph theory, "the four-color problem," which we now discuss.

A plane graph divides the plane into pieces that are called *regions*. An example of such a graph is given in Fig. 17. It is natural to regard the regions as countries and the edges of the graph as boundaries of the countries. If we examine the graph in Fig. 17, it is clear that some boundaries are

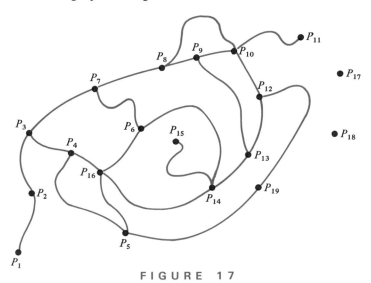

FIGURE 17

unnecessary because the same country lies on both sides. Such is the case for the edges P_1P_2, P_2P_3, $P_{10}P_{11}$, and $P_{14}P_{15}$. Further, some of the vertices

---

[1] Perhaps the reader has heard that no one has yet trisected an arbitrary angle using the straightedge and compass. This is *not* an unsolved problem. It has been solved. The *impossibility* of such a construction has been *proved*, and a proof can be found in almost any book on the theory of equations. See, for example, S. Borofsky, *Elementary Theory of Equations*, Macmillan, New York, 1950.

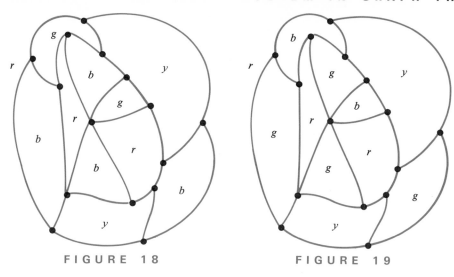

FIGURE 18   FIGURE 19

such as P_{17}, P_{18}, and P_{19} are also unnecessary. When all such unnecessary edges and vertices are removed, the remaining structure, consisting of vertices, edges, and regions, is called a *geographical map*. The regions are called the *countries of the map*, and the edges are called the *boundaries*. When coloring a geographical map, one naturally adopts the point of view described in

Definition 10. A map is said to be properly colored if each country has a definite color and if whenever two countries have a common boundary, these two countries have different colors.

For example, the map shown in Fig. 18 could be colored properly using 11 different colors, one for each of the 11 countries. But this would be wasteful, for we could also color this map properly using just 4 different colors. Let us denote the colors used by the letters *b*, *g*, *r*, and *y* (for blue, green, red, and yellow). With this notation a proper coloration for the map is shown in Fig. 18. Notice that a proper coloration is not necessarily unique. The coloring can be done in many ways and still be proper. Suppose that we have a proper coloring of a map. The interchange of any two of the colors will lead to a different coloring of the map which is also proper. This fact is illustrated by the coloring shown in Fig. 19, which is obtained from that used in Fig. 18 by interchanging the colors *b* and *g*.

It is an experimental fact that every geographical map that has ever been made on the plane or on the surface of a sphere can be properly colored using only four colors. But a proof that four colors will always suffice to color properly any such map has not yet been found.

This then is the four-color problem: to prove (or disprove) the conjecture that four colors will suffice for a proper coloring of any geographical map on the plane, or on the surface of a sphere.

The history of this problem has some interest in itself. There are claims that both Euler and Möbius were aware of the problem, but no written record has been found to support this claim. The first written source containing the problem is a letter from deMorgan to Hamilton in 1852. Interest in the problem grew gradually and about 27 years later both Kempe[1] and Tait[2] published "proofs" that four colors are sufficient. Actually Tait merely proved that the four-color problem could be solved if one could solve an equally difficult problem on the coloring of graphs. But needless to say, this graph problem is still unsolved to this day. The error in Kempe's proof is more delicate, and in fact for 10 years the error went undetected until Heawood[3] pointed out the mistake in Kempe's proof. In this very same paper Heawood proved that five colors are sufficient. Further, Heawood solved the map-coloring problem on the torus. This last result is remarkable, because the torus is a more complicated surface than the plane or sphere, and normally one would expect that the map-coloring problem would be more difficult on the torus.

The fact that this problem in the plane, now regarded as unsolved, was considered as solved during the years from 1880 to 1890, is rather disquieting. It suggests that perhaps there are today many theorems that we regard as proved that really have not been proved, because the "proofs" offered contain errors, as yet unnoticed.

Returning now to the problem, it may seem that the presence of regions removes the four-color problem from graph theory proper, because the latter is devoted to structures containing only vertices and edges. This difficulty is avoided by the following clever device. In each country we select a particular point A_i (the capital of the country). If A_i and A_j are the two points selected in two countries that have a common boundary, we join A_i and A_j by a curve that crosses the common boundary (we connect the two capitals with a railway line). The collection of points (capitals) and curves (railway lines) forms a new graph H that is called the *dual graph* of the original graph G. Clearly we can assign a color to each point (the color of the country in which it lies), and if the original geographical map is properly colored, then each curve of H will join two points that have different colors. The converse is also true. Hence the four-color problem is equivalent to the following problem on coloring the vertices of a planar graph.

[1] A. B. Kempe, "On the Geographical Problem of the Four Colours," *American Journal of Mathematics,* vol. 2 (1879), pp. 193–200.

[2] Peter Guthrie Tait, "Note on a Theorem of Position," *Transactions of the Royal Society of Edinburgh,* vol. 29 (1880), pp. 657–660.

[3] P. J. Heawood, "Map-colour Theorem," *Quarterly Journal of Mathematics,* vol. 24 (1890), pp. 332–338.

Let *H* be an arbitrary plane graph. *Prove (or disprove) that using only four colors it is possible to assign a unique color to each vertex of H so that each edge in H joins a pair of vertices that have different colors.*

It is not our intention to go further with this problem. There are a few books that contain material on the subject, and the casual reader might consult *The Pleasures of Math.*, Macmillan, New York, 1965. The most complete book on the subject in the English language is by O. Ore, *The Four-Color Problem,* Academic Press, New York, 1967.

The four-color problem is not the only open problem in graph theory, but it is certainly the most famous and the most tantalizing. The reader who wishes to examine other problems can find a multitude in *A Seminar on Graph Theory* edited by Frank Harary, Holt, Rinehart and Winston, New York, 1967.

functions and function notation

The concept of a function pervades most of mathematics. Consequently a clear understanding of functions and a facility with function notation is a must for any serious student. The reader who is already familiar with the topics covered in this appendix may either omit it or use it for review.

There are two different ways of defining a function: the classical and the modern one. We give the classical definition in Section 1. We shall discuss the modern definition in Section 4 after we become thoroughly familiar with the concept of a function.

1. the concept of a function

Briefly a function is a mapping from one set A to another set B. By a mapping we mean that to each element x in A, there is associated a uniquely determined element y in B.

Frequently a function can be expressed by a formula. Some rather elementary examples are

(1) $$y = 4x + 7,$$

(2) $$y = x^2 + 2x - 3,$$

and

(3) $$y = \sqrt{x(x - 1)(x - 2)}.$$

In these three examples the sets A and B are certain sets of real numbers, and each of these three formulas gives us a method (or rule) for finding the number y whenever the number x is given. We may think of the formula as taking a particular x into its corresponding y. For example, using equation (1), if $x = 1$, the corresponding y is 11, and if $x = 3$, the corresponding y is 19.

For the general concept of a function, the rule need not be given by a formula, but indeed may be quite wild. Further, it is not necessary for the rule to relate numbers. Indeed a function may relate any two sets. For example, one set might consist of automobiles, and the second set may consist of brand names. Then with each car there is associated some definite brand name: Chevrolet, Ford, Plymouth, etc. Thus the brand name is a function of the car. As another example we may consider a particular automobile race. With each car entered in that race, there is associated a particular person, the driver of the car in that race. Of course, these last two examples are not very interesting to a mathematician, but they illustrate the general nature of a function. Thus a function may relate elements from any two well-defined sets.

> **Definition 1. Function.**   Let A and B be any two nonempty sets, and let x represent an element from A, and let y represent an element from B. We say that a rule (method or procedure) f is a function from A to B if f associates with each x in A a unique y in B. A function is also called a mapping from A to B.

We can represent the function symbolically by writing

$$(4) \qquad\qquad f : A \longrightarrow B$$

to indicate that f carries elements from A to elements of B. However, in practical applications it is much more useful to write

$$(5) \qquad\qquad y = f(x)$$

(read "$y = f$ of x") to indicate that each x in A gives an associated y in B. The use of the notation $y = f(x)$ will be discussed in detail in the next section.

We can picture a function by representing the elements of the sets A and B by points in a plane, as indicated in Fig. 1. For each x in A, we imagine an arrow drawn from x to its associated y. Two arrows cannot begin at the same x, but two arrows may end at the same y. This is illustrated in Fig. 1, where x_4 and x_5 both go into the same element in B.

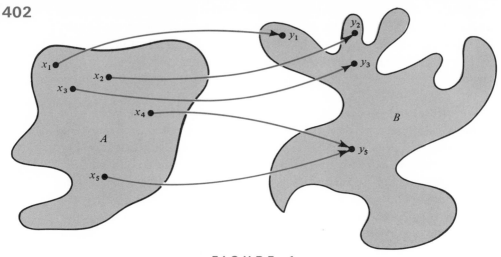

FIGURE 1

We naturally think of A as the *primitive* or *initial set* and observe that the function carries A to B, or *maps* A to B. If $y = f(x)$, the element x is called the *primitive*, and its associate y is called the *image* of x under the *mapping f*. As x runs through all elements of A, its image y may run through all elements of B. In this case we say that the mapping f is *onto B*. Otherwise, we say that the mapping is *into B*.

Definition 2. The set A is called the domain of the function f. The set C consisting of all y such that $y = f(x)$ for some x in A is called the range of the function f. If $C = B$, we say that the mapping is from A onto B. If $C \subset B$ and $C \neq B$, then the mapping is into B. If each y in C is the image of just one x in A, we say that the function f is a 1–1 (read "one-to-one") mapping from A onto C. When x represents an arbitrary element from A, we call x the independent variable. When y represents an arbitrary element from C, we call y the dependent variable. If A and C are sets of real numbers, then f is called a real-valued function of a real variable.

The function pictured in Fig. 1 is not 1–1 because both x_4 and x_5 have the same image element. The function that gives a brand name for each automobile is certainly not 1–1 because many cars carry the same brand name. However, the mapping generated by a particular auto race which associates a particular driver with each car in the race is a 1–1 function for the set of cars entered in the race.

EXAMPLE 1. Discuss the domain and range of each of the three functions defined by equations (1) through (3).

Solution. A function is not properly given unless the domain is stated in advance. Hence the question about the domain is an improper one, since this information has not been supplied. However, in any specific problem the domain is usually obvious from the nature of the problem, and then this minor detail may be omitted. This is especially the case when the function is given by a formula, for under these circumstances one naturally assumes that *the domain of the function is the largest domain for which the formula gives a real-valued function of a real variable.* With this agreement the functions defined by formulas $y = 4x + 7$ and $y = x^2 + 2x - 3$ have as their domain the set of all real numbers. But $y = \sqrt{x(x - 1)(x - 2)}$ is quite different. If $x < 0$ or if $1 < x < 2$, then $x(x - 1)(x - 2)$ is negative and y is not a real number. We let $D_1 = \{x | 0 \leq x \leq 1\}$ and $D_2 = \{x | 2 \leq x\}$. Then the domain of the function defined by (3) is $D_1 \cup D_2$.

The range R of the function $y = 4x + 7$ is the set of all real numbers. For $y = \sqrt{x(x - 1)(x - 2)}$, the range R is the set of all $y \geq 0$. For the function $y = x^2 + 2x - 3$, the determination of R is somewhat difficult without the calculus, but using the calculus it is easy to show that the range is the set $R = \{y | y \geq -4\}$.

In some cases we may want to specify a domain that is different from the one given by the agreement. In this case we write the domain to the right of the formula. Thus the notation

(6) $$y = 3x - 7, \qquad 2 \leq x \leq 5,$$

means that (6) defines a function on the domain $D = \{x | 2 \leq x \leq 5\}$. The reader should convince himself that $R = \{y | -1 \leq y \leq 8\}$ is the range for this function.

Finally, in a complicated situation, we may wish to define a function in pieces using several different formulas, each one valid in some specified subset of the domain. Thus the notation

(7) $$y = \begin{cases} 2 - x, & \text{if } 0 \leq x \leq 1, \\ x, & \text{if } 1 < x < 3, \\ (15 - x)/4, & \text{if } 3 \leq x \leq 7, \end{cases}$$

defines a function on the domain $0 \leq x \leq 7$. Given a particular x we examine it to see in which interval it lies. For example, if $x = 1/2$, we use the first line in (7) and find that $y = 2 - (1/2) = 3/2$. If $x = 2$, the second line in (7) gives $y = 2$. If $x = 6$, we use the third line in (7) and find that $y = (15 - 6)/4 = 9/4$.

The function defined by equation (7) may look artificial, but in truth many functions encountered in everyday affairs are defined in pieces just as the one given by (7). If y is the postage required for a first-class letter of weight x ounces, then y is a function of x for x in the interval $0 < x \leq 640$.

According to U.S. Postal Regulations we have

$$
(8) \qquad y = \begin{cases} 8, & \text{if } 0 < x \leqq 1, \\ 16, & \text{if } 1 < x \leqq 2, \\ 24, & \text{if } 2 < x \leqq 3, \\ 36, & \text{if } 3 < x \leqq 4, \qquad \text{etc.} \end{cases}
$$

The reader will recall other functions that are defined by different for-mulas for different intervals. The cost of a long-distance phone call as a function of the time is one such example. The income tax due as a function of the taxable income, or the value of a savings bond as a function of time are also natural examples of this type of function.

Exercise 1

In problems 1 through 6, a function is given by a formula. In each case find the domain and range in accordance with the agreement on page 403.

1. $y = x^2$.
3. $y = x^2 + 2$.
5. $y = 7 + \sqrt{x + 2}$.

2. $y = \sqrt{36 - x^2}$.
4. $y = 10/(x^2 + 5)$.
6. $y = 36/\sqrt{9 - x^2}$.

7. Let $n(A)$ denote the number of elements in a set. Is this a function? If so, give the domain and range.

In problems 8 through 12 a statement is given relating x and y. In each case state whether the relation defines a function.

8. y is the number of divisors of x, where x is a positive integer.
9. y is the son of x, where x is a U.S. male citizen.
10. y is the father of x.
11. y is the wife of x.
12. y is the weight of x at noon on Jan. 1, 1970.
13. Is it necessary to restrict ourselves to the letters x and y in the theory and problems of this section in order to define a function?

2. function notation

Letters such as $x, y, z, a, b, \ldots$ are customarily used to represent numbers or elements of a set. Whenever it is convenient or desirable we may use a letter with subscripts (for example, $x_1, x_2, x_3, \ldots$) to denote different num-bers or different elements of some set.

In the same way we also want to use a symbol to represent a function when the particular function is not explicitly known. We have already introduced the letter f to serve this purpose. Any letter or symbol will do, but the most popular ones for representing a function are f, g, h, ϕ, ψ, F, G, and H. We can also distinguish different functions by using letters with subscripts. Thus g_1, g_2, and g_3 may well represent three different functions.

As stated in Section 1, we shall use the notation $y = f(x)$ with the understanding that x is merely a representative element from the domain of f. If we need to specify particular elements for x, we can denote these by $x_0, x_1, a, \ldots$, and write $y_0 = f(x_0)$, $y_1 = f(x_1)$, $b = f(a)$, $\ldots$ for the images of $x_0, x_1, a, \ldots$, respectively. Before going further into the theory of functions, we need some exercise in the use of the notation $y = f(x)$.

EXAMPLE 1. Find $f(3)$ if $f(x)$ is the function defined by the formula

$$f(x) = \sqrt{x(x - 1)(x - 2)}.$$

Solution. Whenever a formula is given for $f(x)$ and we want to find $f(3)$, we merely replace x by 3 in the formula and compute in accordance with the formula. By definition

$$f(x) = \sqrt{x(x - 1)(x - 2)}; \quad \text{hence} \quad f(3) = \sqrt{3 \cdot 2 \cdot 1} = \sqrt{6}.$$

EXAMPLE 2. Let $F(x) = 4x + 5$. Is it true that

(9) $$F(x_1 + x_2) = F(x_1) + F(x_2)$$

for all pairs of real numbers x_1, x_2?

Solution. The left-hand side of (9) looks like a multiplication problem in which F multiplies the sum of x_1 and x_2. Since the distributive law of multiplication

(10) $$A(B + C) = AB + AC$$

is true for all real numbers A, B, C, a beginner might easily believe that (9) is also true for any function F. But in general (9) is *false*. To settle this we need only one *counterexample*. We need to find one pair of numbers for which (9) is false. Selecting at random, let $x_1 = -1$ and $x_2 = 3$. Since $F(x) = 4x + 5$, the left-hand side of equation (9) gives

$$F(x_1 + x_2) = F(-1 + 3) = F(2) = 4 \cdot 2 + 5 = 13.$$

For the right-hand side of (9) we find

$$F(x_1) + F(x_2) = F(-1) + F(3) = \left(4(-1) + 5\right) + \left(4 \cdot 3 + 5\right)$$
$$= 1 + 17 = 18.$$

Since $13 \neq 18$, the assertion (9) is in general false.

EXAMPLE 3. If $g(x) = 4^x$, prove that for all real numbers t, u, and v,

$$\textbf{(a)}\ g(t + 2) = 16g(t) \quad \text{and} \quad \textbf{(b)}\ g(u + v) = g(u)g(v).$$

Solution. Using the definition of the function g,

$$\textbf{(a)}\ g(t + 2) = 4^{t+2} = 4^t 4^2 = 16 \cdot 4^t = 16g(t).$$
$$\textbf{(b)}\ g(u + v) = 4^{u+v} = 4^u 4^v = g(u)g(v).$$

Exercise 2

1. Let $y = x^3 - 2x^2 + 3x - 4$. Find the corresponding y when **(a)** $x = 1$, **(b)** $x = 2$, **(c)** $x = 3$, **(d)** $x = 0$, and **(e)** $x = -2$.
2. Let $f(x) = x^3 - 2x^2 + 3x - 4$. Find **(a)** $f(1)$, **(b)** $f(2)$, **(c)** $f(3)$, **(d)** $f(0)$, and **(e)** $f(-2)$.
3. If $f(x) = 2^x$, show that $f(1) = 2, f(5) = 32, f(0) = 1$, and $f(-2) = 1/4$.
4. If $f(x) = 11x$, prove that $f(x + y) = f(x) + f(y)$ for all x, y.
5. If $f(x) = 3x + 5$, prove that $f(4x) = 4f(x) - 15$ for all x.
6. If $f(x) = 7x - 11$, prove that $f(3x) = 3f(x) + 22$ for all x.
★7. If $f(x) = x^2$, prove that $f(x + h) - f(x) = h(2x + h)$ for all x, h.
★8. If $F(x) = x^2 - 3x - 5$, find all values of x such that $F(x) = x$. These are called *fixed points* for the function.
★9. Find the fixed points for the functions

$$\textbf{(a)}\ G(x) = 5x + 8, \quad \textbf{(b)}\ H(x) = x^2, \quad \text{and} \quad \textbf{(c)}\ J(x) = x^2 - 12x.$$

★10. Define the function $\pi(x)$ to be the number of primes less than or equal to x. By definition, 1 is not a prime. Find $\pi(9)$, $\pi(18)$, $\pi(27)$, $\pi(36)$, $\pi(\sqrt{97})$, and $\pi(\pi^2)$.
11. Let x be the length of one edge of a cube. The surface area of the cube is then a function of x. Find a formula for this function.
12. The surface area of a cube is a function of the volume. Find a formula for this function.

3. the graph of a function

We recall that a function is any mapping from a set A to another set B. Because of the general nature of the sets A and B, it is not always possible to make a picture or diagram of the function. But when A and B are subsets of the real numbers, then a picture of the function is a simple matter. In this case we use a rectangular coordinate system and mark in the plane those points whose coordinates are related by the function. In most cases (but not in all) we shall obtain a smooth curve, and this curve gives us a good picture of the function.

> **Definition 3.** Let $y = f(x)$ be a real-valued function of a real variable. The graph of f is the collection of all points (x,y) in the plane for which x is in the domain of f and y is the image of x under f.

EXAMPLE 1. Sketch the graph of the function $y = f(x)$ if

(11) $$f(x) = \frac{1}{3}x - 1.$$

Solution. It is easy to find pairs (x,y) such that $y = f(x)$ as demanded by Definition 3. For example, if $x = 15$, we find that

$$f(15) = \frac{1}{3} \times 15 - 1 = 4 = y$$

and hence (15,4) is one point on the graph of equation (11).

It is desirable to introduce some system into the computations by arranging the work in a little table, selecting values for x for which the computation is easy and finding the corresponding y. Such a table is shown below, and the corresponding points are marked in Fig. 2.

x	-6	-3	0	3	6	9	12	15	18
$y = \frac{1}{3}x - 1$	-3	-2	-1	0	1	2	3	4	5

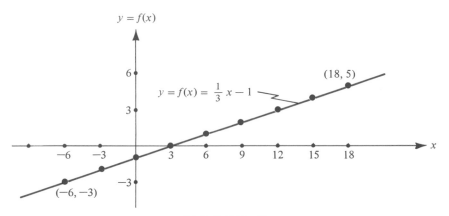

FIGURE 2

The points all seem to lie on a straight line and we feel reasonably confident that if we continue to select values for x, compute the corresponding y, and plot the point (x,y), then these new points will also fall on the line determined by the points already plotted in Fig. 2. This is indeed true, and follows as a special case of

> **Theorem 1.** Let m and b be real numbers. Then the graph of the function defined by
>
> (12) $$y = mx + b$$
>
> is a straight line.

We shall not linger for a formal proof of this theorem, because the proof belongs properly in the domain of analytic geometry. However, the student may observe this theorem in action as he proceeds with further examples and exercises.

The constant m in equation (12) is called the *slope* of the line and measures the rate of rise of the line (or fall if m is negative). For example, $m = 1/3$

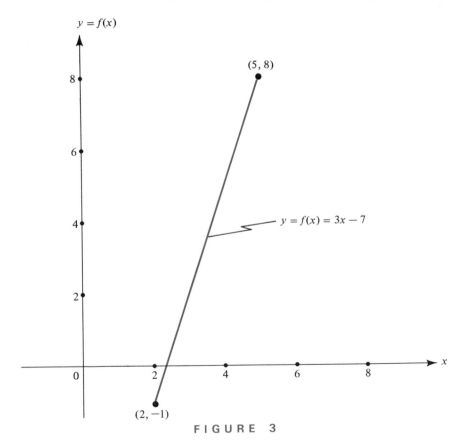

FIGURE 3

in equation (11) tells us that for each change of 3 units in x, there is a change of 1 unit in y, and this is easy to observe in Fig. 2. Indeed, if x increases from 9 to 12, then y increases from 2 to 3.

EXAMPLE 2. Sketch the graph of the function defined by equation (6) (see Section 1).

Solution. In equation (11) it is assumed that the domain of f is the set of all real numbers. Hence we assume that in Fig. 2, the line extends indefinitely in both directions and that we have drawn only a (very small) portion of the graph. In equation (6) the domain has been specified as the interval $2 \leqq x \leqq 5$. Hence the graph is only a segment of a straight line. This segment is shown in Fig. 3.

EXAMPLE 3. Sketch the graph of the function defined by equation set (7) (see Section 1).

Solution. Since the function is defined by several formulas, we may expect the graph to look a little unusual. This graph is shown in Fig. 4, and the student is invited to check several points of the graph.

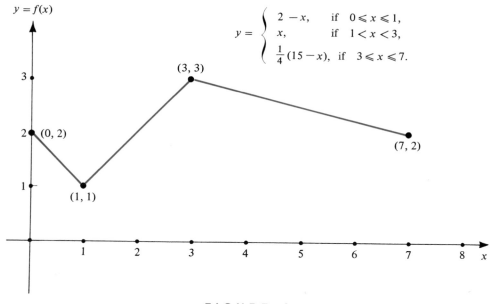

$$y = \begin{cases} 2 - x, & \text{if } 0 \leqslant x \leqslant 1, \\ x, & \text{if } 1 < x < 3, \\ \frac{1}{4}(15 - x), & \text{if } 3 \leqslant x \leqslant 7. \end{cases}$$

FIGURE 4

Exercise 3

In problems 1 through 26, a function is defined by a formula (or several formulas). In each case sketch a graph of the function.

1. $y = 2x + 5$.   2. $y = -3x + 7$.

3. $y = x - 4$.   4. $y = -2x - 5$.

5. $3y = x + 6$.   6. $4y = -x - 8$.

7. $y = x^2$.   8. $y = x^2/4$.

9. $y = x^3$.   10. $y = x^2 - 6x$.

11. $y = |x|$.   ★12. $y = x(x - 2)(x - 4)$.

13. $y = -|x|$.   14. $y = 2 - |x|$.

15. $y = 2 + |x|$.   ★16. $y = |x + 2|$.

★17. $y = 3 + |x - 2|$.   ★18. $y = 1 + x + |x|$.

★19. $y = 1 + x - |x|$.   ★20. $y = |x^2 - 1|$.

★21. $y = \begin{cases} 4, & \text{if } x \geq 2, \\ 1, & \text{if } 1 \leq x < 2, \\ -1, & \text{if } x < 1. \end{cases}$

★22. $y = \begin{cases} \sqrt{4 - x^2} + 2, & \text{if } -2 \leq x \leq 2, \\ 2, & \text{for all other } x. \end{cases}$

★23. $y = \begin{cases} 1, & \text{if } x \geq 1, \\ x, & \text{if } -1 < x < 1, \\ -1 & \text{if } x \leq -1 \end{cases}$

★24. $y = \begin{cases} 2, & \text{if } x \geq 2, \\ 3x - 4, & \text{if } 1 < x < 2, \\ -1, & \text{if } x \leq 1. \end{cases}$

★25. $y = \begin{cases} 4 - x^2, & \text{if } -1 \leq x \leq 1, \\ 2, & \text{for all other } x. \end{cases}$

★26. $y = \begin{cases} x^2, & \text{if } -2 < x < 2, \\ 2x, & \text{for all other } x. \end{cases}$

4. the modern definition of a function

We recall from Section 1 that f is called a function if f is a rule (method or procedure) which associates with each element x in A a uniquely determined y in B. This definition seems to be natural because in each example studied so far the function was presented by a rule. Thus the function $y = x^2$ [or $f(x) = x^2$] is defined by the rule: Select a real number x; then its associate (or image) y is obtained by squaring x.

There is, however, some logical basis for objecting to this definition. The modern attitude is that the function stands apart from the mechanism by which it is constructed. We may create a function in any way we wish (rule, method, and so on), but the definition of a function should stand by itself.

What then is a function? We can form an ordered pair (x,y), where x

is an element of the domain of the function f and y is its image, $f(x)$. Since the image is unique, we see that for two distinct ordered pairs (x, y) and $(x\star, y\star)$ we must have $x \neq x\star$. The modern point of view is that the function is just this set of ordered pairs.

Definition 4. A function f is a set S of ordered pairs (x, y) such that if (x, y) and $(x\star, y\star)$ are distinct elements of S, then $x \neq x\star$. The set of all x such that (x, y) is in S is called the domain of f. The set of all y such that (x, y) is in S is called the range of f.

All other concepts relating to functions can be defined via ordered pairs. From a logical point of view Definition 4 is somewhat better then Definition 1. However, it is a *trivial* matter to prove that both definitions are equivalent: *Any f that can be defined by ordered pairs can also be defined by a rule and conversely.*

Since the rule definition is natural, easy to understand, and is the one that is *always used* when building a function, we have selected Definition 1 in preference to Definition 4. Further, the notation $y = f(x)$ has many advantages. The reader need only compare $y = x^2$ or $f(x) = x^2$ with

$$f = S = \{(x, x^2)\},$$

which gives the same function.

We should mention one source of confusion that is present with either notation. The reader must realize that the letters x and y are merely convenient symbols that represent elements from A and B. Any letter may be used. For example, $f(t) = t^2$, $f(y) = y^2$, $q = r^2$, $\zeta = \xi^2$, $S = \{(w, w^2)\}$, and $S = \{(\alpha, \alpha^2)\}$ all represent the same function. One merely follows custom in writing $y = f(x)$.

inequalities

Anyone using numbers (or formulas involving numbers) should have some feeling for the relative magnitudes of the quantities involved. The theory of inequalities is a systematic exploration of the problem of determining the larger of two given numbers. But the theory itself goes far beyond this first elementary problem, and indeed forms a domain of mathematics that is fascinating, deep, and beautiful. The ambitious reader may consult the two books by N. Kazarinoff, *Analytic Inequalities,* Holt, Rinehart and Winston, New York, 1961 and *Geometric Inequalities,* Random House, New York, 1961.

1. the elementary theorems

The set R of all real numbers can be divided into three mutually disjoint sets in a natural way: R^+, the set of all positive numbers; R^-, the set of all negative numbers; and $\{0\}$, the set consisting of the single number 0. Thus

$$R = R^+ \cup R^- \cup \{0\}$$

and

$$R^+ \cap R^- = R^+ \cap \{0\} = R^- \cap \{0\} = \varnothing.$$

We assume that the reader is familiar with the following properties of the real numbers.

(I) A real number x is a negative number if and only if $-x$ is a positive number. In symbols

$$x \in R^- \iff -x \in R^+.$$

(II) The sum of any two positive numbers is a positive number. In symbols

$$x \in R^+, y \in R^+ \implies x + y \in R^+.$$

(III) The product of any two positive numbers is a positive number. In symbols

$$x \in R^+, y \in R^+ \implies xy \in R^+.$$

The inequality relation for any two real numbers is defined in terms of the set R^+ of positive numbers.

Definition 1. The number a is said to be less than b, and we write

(1) $\qquad\qquad a < b \qquad$ (read "a is less than b")

if and only if the difference $b - a$ is positive. Under these conditions we also write

(2) $\qquad\qquad b > a \qquad$ (read "b is greater than a").

The relations (1) and (2) are called *inequalities*. By definition, the two inequalities (1) and (2) are equivalent. It also follows from this definition that $0 < b$ if and only if b is a positive number. As trivial examples we have

$$86 < 99. \qquad \text{because } 99 - 86 = 13 \in R^+,$$
$$-1000 < 1, \qquad \text{because } 1 - (-1000) = 1001 \in R^+,$$
$$\frac{21}{34} < \frac{55}{89}, \qquad \text{because } \frac{55}{89} - \frac{21}{34} = \frac{1}{3026} \in R^+.$$

On the other hand, it is not at all obvious that

$$\sqrt{83} + \sqrt{117} < \sqrt{99} + \sqrt{101},$$

and it will take some effort to prove this (see problem 5 of Exercise 1).

Since (1) and (2) are equivalent, one of the symbols $<$ and $>$ is really unnecessary and could be dropped. However, it is convenient to have both of them available for use. It is also convenient to have a compound symbol $a \leqq b$, which means that either $a < b$ or $a = b$. Similar, $a \geqq b$ means that either $a > b$ or $a = b$.

We now prove a number of theorems about inequalities.

Theorem 1. If a and b are any two real numbers, then exactly one of the three relations

$$\textbf{(A) } a < b, \qquad \textbf{(B) } a = b, \qquad \textbf{(C) } b < a$$

holds.

Proof. For any two numbers a and b we have either $b - a$ is positive, or $b - a = 0$, or $b - a$ is negative, and these three cases are mutually exclusive. In the first case $a < b$, in the second case $a = b$, and in the third case if $b - a$ is negative, then $-(b - a) = a - b$ is positive, and hence $b < a$. ∎

Theorem 2. If $a < b$ and $b < c$, then $a < c$.

Proof. By hypothesis $b - a$ and $c - b$ are positive numbers. Then the sum $(b - a) + (c - b)$ is a positive number. But this sum is $c - a$. Since $c - a$ is positive, we have by Definition 1 that $a < c$. ∎

Theorem 3. If $a < c$ and $b < d$, then $a + b < c + d$.

In other words, two inequalities can be added termwise to give a true inequality. Of course, the inequality sign must be in the same direction in all three of the inequalities.

Proof. By hypothesis $c - a$ and $d - b$ are positive numbers. Consequently the sum $c - a + d - b$ is a positive number. Hence $c + d - (a + b)$ is positive. ∎

Theorem 4. If $a < b$ and c is any number, then $a + c < b + c$.

Proof. Clearly $(b + c) - (a + c) = b - a$ is positive. ∎

Thus a true inequality remains true when the same number is added to both sides. Note that c can be a negative number, so that this theorem includes subtraction.

> **Theorem 5.** If $a < b$ and c is any positive number, then $ca < cb$.

In other words, an inequality remains true when multiplied on both sides by the same positive number.

Proof. Since $a < b$, then $b - a$ is a positive number. Since the product of two positive numbers is again positive we have that

$$c(b - a) = cb - ca$$

is positive. By Definition 1 this gives $ca < cb$. ∎

We leave to the student the proof of

> **Theorem 6.** If $a < b$ and c is any negative number, then $ca > cb$.

In other words, when an inequality is multiplied on both sides by the same negative number, the inequality sign is reversed.

> **Theorem 7.** If $0 < a < b$, then
>
> (3) $$\frac{1}{a} > \frac{1}{b} > 0.$$

Thus reciprocation reverses the inequality sign when both members are positive.

Proof. Multiply both sides of $a < b$ by the positive number $1/ab$ and use Theorem 5. ∎

> **Theorem 8.** If $0 < a < b$ and $0 < c < d$, then $ac < bd$.

Thus multiplication of the corresponding terms of an inequality preserves the inequality. Of course all terms should be positive, and the inequality sign must be in the same direction in all three inequalities.

Proof. Using Theorem 5, the inequality $a < b$ yields $ac < bc$. Similarly $c < d$ yields $bc < bd$. Since $ac < bc$ and $bc < bd$, Theorem 2 gives $ac < bd$. ∎

> **Theorem 9.** If $0 < a < b$ and n is any positive integer, then
>
> (4) $$a^n < b^n.$$

Proof. Apply Theorem 8, $n - 1$ times with $c = a$ and $d = b$. ∎

Theorem 10. If $0 < a < b$ and n is any positive integer, then

$$\sqrt[n]{a} < \sqrt[n]{b},$$

where, if n is even, the symbol $\sqrt[n]{}$ means the positive nth root.

Proof. The proof is a little complicated because it uses the method of contradiction. By Theorem 1 there are only three possibilities:

(A) $\sqrt[n]{a} < \sqrt[n]{b}$,   **(B)** $\sqrt[n]{a} = \sqrt[n]{b}$,   **(C)** $\sqrt[n]{b} < \sqrt[n]{a}$.

In each of the last two cases we take the nth power of both sides. In case **(B)** we find obviously that $a = b$. But this is impossible, because by hypothesis $a < b$. In case **(C)** we apply Theorem 9 to $\sqrt[n]{b} < \sqrt[n]{a}$ and find that $b < a$. Again this is contrary to the hypothesis that $a < b$. Since each of the cases **(B)** and **(C)** leads to a contradiction, the only case that can occur is **(A)**. ∎

In most of these theorems we can allow the equality sign to occur in the hypotheses as long as we make suitable modifications in the conclusions. In this way we obtain a large number of new theorems that vary only slightly from the ones already proved. There is no need to list them all, because they are really obvious. (The student who is in doubt on this point should investigate the matter on his own.) Merely as an illustration of the type of theorem to be expected, we give the following variations on Theorems 4 and 8.

Theorem 4′. If $a \leqq b$ and c is any number, then $a + c \leqq b + c$.

Theorem 8′. If $0 < a \leqq b$ and $0 < c < d$, then $ac < bd$.

There is one other important tool in proving inequalities, namely the innocent remark that the square of any number is either positive or zero; i.e., $c^2 \geqq 0$ for any number c. Of course if c is a complex number, then c^2 may be negative, but in this text all of the numbers used are real numbers.

EXAMPLE 1. Prove that for any two numbers

(5) $$2ab \leqq a^2 + b^2$$

and that the equality sign occurs if and only if $a = b$.

Solution. By our remark above

(6) $$(a - b)^2 \geqq 0$$

and equality in (6) occurs if and only if $a = b$. Expanding (6) we have

$$a^2 - 2ab + b^2 \geqq 0$$

or

$$0 \leqq a^2 - 2ab + b^2.$$

Then by Theorem 4′ (adding $2ab$ to both sides)

$$2ab \leqq a^2 + b^2.$$

Since the equality sign occurs in (6) if and only if $a = b$, it also occurs in (5) under the same conditions.

EXAMPLE 2. Prove that if a, b, c, and d are any four positive numbers, then

(7) $$ab + cd \leqq \sqrt{a^2 + c^2} \sqrt{b^2 + d^2}.$$

It is not easy to see the proper starting place for this problem, so we work backward. That is, we start with the inequality (7) and see if we can deduce one that we know to be true. This operation is called the *analysis* of the problem.

Analysis. If (7) is true, we can square both sides and obtain

(8) $$a^2b^2 + 2abcd + c^2d^2 \leqq (a^2 + c^2)(b^2 + d^2),$$

or

(9) $$a^2b^2 + 2abcd + c^2d^2 \leqq a^2b^2 + c^2b^2 + a^2d^2 + c^2d^2,$$

or, on transposing (Theorem 4′),

(10) $$0 \leqq c^2b^2 - 2abcd + a^2d^2,$$

or

(11) $$0 \leqq (cb - ad)^2.$$

But we know that this last inequality is always true. Hence if we can reverse our steps, we can prove that the given inequality is also true.

Solution. We begin with the known inequality (11) and on expanding we find that (10) is also true. Then adding $a^2b^2 + 2abcd + c^2d^2$ to both sides of (10) we obtain (9). Factoring the right-hand side of (9) gives (8). Finally, taking the positive square root of both sides of (8) gives (7).

It is customary to do the analysis on scratch paper and then write the solution in the proper order; i.e., in the reverse order of the analysis. The student should write out in detail the correct solution of this example, following the outline just given.

EXAMPLE 3. Without using tables, prove that

(12) $$\sqrt{2} + \sqrt{6} < \sqrt{3} + \sqrt{5}.$$

Solution. We give the analysis. Squaring both sides of (12) yields

(13) $$2 + 2\sqrt{2}\,\sqrt{6} + 6 < 3 + 2\sqrt{3}\,\sqrt{5} + 5$$

or, on subtracting 8 from both sides and dividing by 2,

(14) $$\sqrt{12} < \sqrt{15}.$$

But since $12 < 15$, the inequality (14) is obviously true (Theorem 10). To prove the inequality (12) we start with the remark that $12 < 15$ and reverse the above steps.

The absolute value of x (or numerical value of x), denoted by $|x|$, is often useful in the study of inequalities. We recall from Chapter 0

Definition 2. Absolute Value. For each real number x, the absolute value of x is defined by

(15) $$|x| = \begin{cases} x, & \text{if } x \geqq 0, \\ -x, & \text{if } x < 0. \end{cases}$$

For example, if $x = -7$, then it falls in the second case of (15) and hence $|x| = |-7| = -(-7) = 7$. By the nature of the definition, we have the obvious

Theorem 11. For all x

(16) $$-|x| \leqq x \leqq |x| \qquad \text{and} \qquad |x| \geqq 0.$$

Further, $|x| = 0$ if and only if $x = 0$.

Slightly less obvious properties are given in

Theorem 12. If x and y are any pair of real numbers, then

(17) $$|xy| = |x|\,|y|,$$

(18) $$|x - y| = |y - x|,$$

(19) $$\sqrt{x^2} = |x|,$$

and, if $y \neq 0$, then

(20) $$\left|\frac{x}{y}\right| = \frac{|x|}{|y|}.$$

The proofs of Theorems 11 and 12 are trivial and my be omitted.

Theorem 13. If x and y are any pair of real numbers, then

(21) $$|x + y| \leqq |x| + |y|$$

and

(22) $$|x| - |y| \leqq |x - y|.$$

Proof. For (21) we consider various cases. If $x \geqq 0$ and $y \geqq 0$, then (21) merely asserts that $x + y \leqq x + y$ and this is obviously true when the equal sign is selected.

Suppose that $x \geqq 0$ and $y \leqq 0$, but $|y| \leqq x$. Then $x + y \geqq 0$ and

$$|x + y| = x + y \leqq x + |y| = |x| + |y|.$$

We leave the other cases of (21) for the reader.

If we apply (21) to the identity $x = y + (x - y)$, we find that

$$|x| = |y + (x - y)| \leqq |y| + |x - y|,$$

and subtracting $|y|$ from both sides we obtain (22).

We leave it for the reader to show that from (22) one can easily obtain the more complicated looking inequality

$$\big||x| - |y|\big| \leqq |x - y|.$$

Exercise 1

In problems 1 through 4, determine which of the two given numbers is the larger without using tables.

1. $\sqrt{19} + \sqrt{21}$, $\sqrt{17} + \sqrt{23}$.   2. $\sqrt{11} - \sqrt{8}$, $\sqrt{17} - \sqrt{15}$.
3. $\sqrt{17} + 4\sqrt{5}$, $5\sqrt{7}$.   4. $2\sqrt{2}$, $\sqrt[3]{23}$.
5. Prove that if $1 < k < n$, then

$$\sqrt{n - k} + \sqrt{n + k} < \sqrt{n - 1} + \sqrt{n + 1}.$$

6. Prove that the inequality of Example 2 [equation (7)] is true even if some or all of the numbers are negative.

In problems 7 through 18, prove the given inequality under the assumption that all the quantities involved are positive. Determine the conditions under which the equality sign occurs.

7. $a + \dfrac{1}{a} \geqq 2.$ 8. $\dfrac{a}{5b} + \dfrac{5b}{4a} \geqq 1.$

9. $\sqrt{\dfrac{c}{d}} + \sqrt{\dfrac{d}{c}} \geqq 2.$ 10. $(c + d)^2 \geqq 4cd.$

11. $\dfrac{a + b}{2} \geqq \sqrt{ab} \geqq \dfrac{2ab}{a + b}.$ 12. $(a + 5b)(a + 2b) \geqq 9b(a + b).$

13. $x^2 + 4y^2 \geqq 4xy.$ ★14. $x^2 + y^2 + z^2 \geqq xy + yz + zx.$

★15. $\dfrac{c^2}{d^2} + \dfrac{d^2}{c^2} + 6 \geqq \dfrac{4c}{d} + \dfrac{4d}{c}.$ 16. $\dfrac{a + 3b}{3b} \geqq \dfrac{4a}{a + 3b}.$

★17. $cd(c + d) \leqq c^3 + d^3.$

★18. $4ABCD \leqq (AB + CD)(AC + BD).$

19. Which of the above inequalities are still meaningful and true if the letters are permitted to represent negative numbers?
20. Complete the proof of Theorem 13.
21. Prove that if x, y, and z are any real numbers, then
   (a) $|xyz| = |x|\,|y|\,|z|.$
   (b) $|x + y + z| \leqq |x| + |y| + |z|.$
   (c) $|x - y| \leqq |x - z| + |y - z|.$
22. Prove that if $|x| < |y|$, then $-|y| < x < |y|.$
23. Let $r > 0$. Prove that $|x - a| < r$ if and only if $a - r < x < a + r.$
24. Sometimes we are told that the absolute value sign means "You drop the minus sign." Explain why this is bad. *Hint:* Consider such expressions as $|x - y|$, $|x - y + z - w|$, and $|(x^3 - y^3)/(x - y)|.$
★25. For each integer $n \geqq 2$ the quantity $n!$ (n factorial) is usually defined as the product of all positive integers less than or equal to n. This definition is extended to include $n = 0$ and $n = 1$ by the definitions $0! = 1$ and $1! = 1$. It is customary to write $n! = 1 \cdot 2 \cdot 3 \cdots n$ [or $n(n - 1) \cdots 3 \cdot 2 \cdot 1$] with suitable agreements if n is small. Prove that if $n \geqq 1$, then $n^n \leqq (n!)^2.$

★**26.** Prove that if $n \geqq 1$, then

$$1 \cdot 3 \cdot 5 \cdots (2n - 1) \leqq n^n.$$

Hint: $(n + k)(n - k) \leqq n^2$.

2.   conditional inequalities

An inequality that involves a variable may be true for some values of the variable and false for others. Such inequalities are called conditional inequalities. To solve a conditional inequality one must find the truth set for the inequality; i.e., the set of all values of the variable for which the asserted inequality is indeed true.

EXAMPLE 1.   For what values of x is

(23) $$x^2 - x - 30 > 0?$$

Solution.   Factoring this expression we have

(24) $$x^2 - x - 30 = (x - 6)(x + 5).$$

This is certainly positive if both factors are positive. This happens only if $x > 6$. The product is also positive if both factors are negative. This happens only for $x < -5$. If $-5 < x$ and $x < 6$, the first factor is negative and the second factor is positive, so that the product is negative. Whence we conclude that (23) is true if and only if either $x < -5$ or $x > 6$. By introducing special letters for these sets, such as

(25) $$A = \{x \mid x < -5\} \quad \text{and} \quad B = \{x \mid x > 6\},$$

we can write that (23) is true if and only if $x \in A \cup B$. Consequently $A \cup B$ is the truth set for the inequality (23).

EXAMPLE 2.   Solve the conditional inequality

(26) $$x^3 - 30x \leqq 4x^2 - 9x.$$

Solution.   The inequality (26) is equivalent to

$$x^3 - 4x^2 - 21x \leqq 0,$$

and this is equivalent to

(27) $$x(x - 7)(x + 3) \leqq 0.$$

Now the left-hand side of (27) is zero if $x = 0$, or $x = 7$, or $x = -3$. These three numbers divide the real numbers naturally into four sets and we examine the sign of each factor in (27) in each of these four sets. The results are arranged systematically in Table 1. From this table it is clear that (27) is true if and only if $x \in A \cup C$, where $A = \{x \mid x \leq -3\}$ and $C = \{x \mid 0 \leq x \leq 7\}$.

TABLE   1

Value of x	Sign of the factor			Sign of $x(x-7)(x+3)$
	$x + 3$	x	$x - 7$	
$x < -3$	$-$	$-$	$-$	$-$
$-3 < x < 0$	$+$	$-$	$-$	$+$
$0 < x < 7$	$+$	$+$	$-$	$-$
$7 < x$	$+$	$+$	$+$	$+$

EXAMPLE 3.   Solve the conditional inequality

(28)
$$\frac{3x}{x-1} + \frac{x}{x-4} < 5.$$

Solution.   Our first impulse is to multiply both sides of (28) by $(x - 1)(x - 4)$ and obtain the new inequality

(29)
$$3x(x - 4) + x(x - 1) < 5(x - 1)(x - 4).$$

But we are now in trouble, because the inequality (29) is *not* equivalent to (28). The source of the trouble is the multiplier $(x - 1)(x - 4)$, which is *positive* for some values of x and *negative* for others. When $(x - 1)(x - 4) < 0$, the process of multiplication reverses the sign of the inequality in going from (28) to (29).

To avoid this difficulty, we transpose the left-hand side of (28) to the right-hand side obtaining the equivalent inequality

(30)
$$0 < 5 - \frac{3x}{x-1} - \frac{x}{x-4}.$$

Clearly (30) is equivalent to

$$0 < \frac{5(x - 1)(x - 4) - 3x(x - 4) - x(x - 1)}{(x - 1)(x - 4)},$$

$$0 < \frac{5x^2 - 25x + 20 - 3x^2 + 12x - x^2 + x}{(x - 1)(x - 4)},$$

(31)
$$0 < \frac{x^2 - 12x + 20}{(x - 1)(x - 4)} = \frac{(x - 2)(x - 10)}{(x - 1)(x - 4)}.$$

Let $A = \{x \mid x < 1\}$, $B = \{x \mid 2 < x < 4\}$, and $C = \{x \mid 10 < x\}$. By analyzing the signs of the factors on the right-hand side of (31) we see that (31) is true if and only if $x \in A \cup B \cup C$. Hence (28) is true if and only if $x \in A \cup B \cup C$.

Exercise 2

In problems 1 through 10, find the set of values of x for which the given inequality is true.

1. $x(x - 1) > 0$.
2. $(x - 8)(x + 1) < 0$.
3. $10x - 15 < 7(x - 3)$.
4. $2x - 19 \leq 11(x - 2)$.
5. $x^2 - 8x + 24 \leq 9$.
6. $4x^2 - 13x + 4 < 1$.
★7. $x^3 - 16x \geq 0$.
★8. $x^3 + 3x^2 \geq 13x + 15$.
★9. $x^4 + 36 \geq 13x^2$.
★10. $x^2 + 22 \leq 10x$.

11. Prove that if $a < b$, then $a < (a + b)/2 < b$.
12. Prove that if $a_1 < a_2 < a_3 < \cdots < a_n$, then

$$a_1 < \frac{a_1 + a_2 + a_3 + \cdots + a_n}{n} < a_n.$$

In problems 13 through 17, find the set of values of x for which the given inequality is true.

★13. $x - \dfrac{8}{x} < 7$; compare with problem 2.

★14. $x + \dfrac{15}{x} \leq 8$; compare with problem 5.

★15. $x^2 - 13 \geq 3\left(\dfrac{5}{x} - x\right)$; compare with problem 8.

★16. $\dfrac{4}{x} + \dfrac{3}{x - 2} \leq 1$.

★17. $\dfrac{3}{x - 1} + \dfrac{1}{x - 5} \leq 1$.

mathematical induction

One of the most important techniques for proving theorems bears the title mathematical induction. The reader who is already familiar with this technique may omit this appendix entirely or use it for review.

1. the nature of mathematical induction

Let us try to find the sum of the first n odd positive integers, where n is any positive integer. By direct computation we find that

$$
\begin{array}{lll}
\text{if } n = 1, & 1 = 1 & = 1^2, \\
\text{if } n = 2, & 1 + 3 = 4 & = 2^2, \\
\text{if } n = 3, & 1 + 3 + 5 = 9 & = 3^2, \\
\text{if } n = 4, & 1 + 3 + 5 + 7 = 16 & = 4^2, \\
\text{if } n = 5, & 1 + 3 + 5 + 7 + 9 = 25 & = 5^2, \\
\text{if } n = 6, & 1 + 3 + 5 + 7 + 9 + 11 = 36 & = 6^2.
\end{array}
$$

An examination of these cases leads us to believe that the sum is always the square of the number of terms in the sum. To express this symbolically we should observe that if n is the number of terms, it seems as though $2n - 1$ is the last term in the sum. To check this we notice that

$$\text{if } n = 1, \quad 2n - 1 = 2 - 1 = 1,$$
$$\text{if } n = 2, \quad 2n - 1 = 4 - 1 = 3,$$
$$\text{if } n = 3, \quad 2n - 1 = 6 - 1 = 5,$$
$$\text{if } n = 4, \quad 2n - 1 = 8 - 1 = 7,$$

and so on. Thus we can express our assertion for general n by the equation

(1) $$1 + 3 + 5 + 7 + \cdots + (2n - 1) = n^2,$$

because the nth odd integer is $2n - 1$.

We have seen by direct calculation that equation (1) holds when $2n - 1$ has any one of the values 1, 3, 5, 7, 9, and 11. Does this mean that equation (1) is true for any positive odd integer $2n - 1$? Can we settle this by continuing our numerical work? We might try the case when $2n - 1 = 23$, so that $n = 12$. Direct computation shows that

$$1 + 3 + 5 + 7 + 9 + 11 + 13 + 15 + 17 + 19 + 21 + 23 = 144 = 12^2,$$

so that again our formula (1) seems to hold. One might be tempted to say that since the terminal odd number 23 was selected at random this proves that (1) is true for every possible choice of the terminal number. Actually, no matter how many cases we check, we can never prove that (1) is always true, because there are infinitely many cases, and no amount of pure computation can settle them all. What is needed is some *logical* argument that will prove that (1) is always true. Before we give the details of this logical argument, we shall give some examples of assertions which can be checked experimentally for small values of n, but which after careful investigation turn out to be *false* for certain other values of n.

Let us first examine the numbers A of the form

(2) $$A = 2^{2^n} + 1,$$

where n is a positive integer. We find by direct computation that

$$\text{if } n = 0, \quad A = 2^1 + 1 = 3,$$
$$\text{if } n = 1, \quad A = 2^2 + 1 = 5,$$
$$\text{if } n = 2, \quad A = 2^4 + 1 = 17,$$
$$\text{if } n = 3, \quad A = 2^8 + 1 = 257,$$
$$\text{if } n = 4, \quad A = 2^{16} + 1 = 65{,}537.$$

Each of these values for A is a prime; i.e., an integer (whole number) which has no divisor other than itself and one (with the quotient also an integer). Can we assert on the basis of these five examples that A is always prime for every positive integer n? Of course not. We might conjecture (guess) that this is true, but we should not make a positive assertion unless we can supply a proof valid for all n. In fact Fermat (1601–1665), the great French mathematician, did conjecture exactly that, but about a hundred years later Euler (1707–1783), the leading Swiss mathematician, showed that for $n = 5$, $A = 2^{2^5} + 1 = 4{,}294{,}967{,}297 = 641 \times 6{,}700{,}417$, and hence A is *not* a prime for $n = 5$.

Next consider the inequality

(3)
$$2^n < n^{10} + 2.$$

If we compute both sides of (3) for the first four values of n, we find that

if $n = 1$,	we have	$2 < 1 + 2 = 3$,
if $n = 2$,	we have	$4 < 1024 + 2 = 1026$,
if $n = 3$,	we have	$8 < 59{,}051$,
if $n = 4$,	we have	$16 < 1{,}048{,}578$.

It certainly looks as though the inequality (3) is true for every positive integer n. If we try a large value of n, say $n = 20$, then the inequality (3) asserts that

$$126{,}976 < 10{,}240{,}000{,}000{,}002,$$

which is certainly true. But even this computation does not prove that (3) is always true, and in fact such an assertion is *false*. For if we set $n = 59$, we find that (approximately)

$$2^{59} = 5.764 \times 10^{17} \quad \text{and} \quad 59^{10} + 2 = 5.111 \times 10^{17}.$$

In fact, it is not hard to prove that for every $n \geq 59$ *the inequality (3) is false*.

We give one more example, which, although trivial, still illustrates our central point. We make the assertion that

"Every positive integer n is less than 1,000,001."

This is obviously false, and yet if we begin by setting first $n = 1$, then $n = 2$, then $n = 3$, $\ldots$, it is clear that for the first million cases that we check the assertion is true. The falsity is not revealed until we set $n = 1{,}000{,}001$. Although this example may seem artificial, it is not difficult to give quite natural examples of the same phenomenon, although such examples would necessarily be more complicated.

We cannot conclude that an assertion involving an integer n is true for all positive values of the integer n merely by checking specific values of n, no matter how many we check.

2.   the general principle

We use the symbol $p(n)$ (read "p of n") to denote some proposition which depends on the positive integer n. For example, $p(n)$ might denote the proposition that

(4) $$1 + 3 + 5 + 7 + \cdots + (2n - 1) = n^2.$$

We let T be the set of positive integers for which $p(n)$ is true, and let F be the set of positive integers for which $p(n)$ is false. To prove that $p(n)$ is always true, it is sufficient to show that the set F is empty. We concentrate our attention on F.

Now it is obvious that if F is a nonempty set, then it has a smallest positive integer. Two cases can occur.

1. The smallest integer in F is 1.
2. If the smallest integer in F is not 1, we denote the smallest integer by $k + 1$. Then by the definition of $k + 1$, $p(k)$ is true and $p(k + 1)$ is false.

To show that the set F is empty, if suffices to show that neither of the two cases listed above can occur. First, we must prove that $p(1)$ is true, so that 1 is in T and not in F. Second, we must prove that for each positive integer k, if $p(k)$ is true, then $p(k + 1)$ is true. If we have proved this, then the second case cannot occur. If we have proved these two statements, then F must be empty and T must contain all of the positive integers.

Thus the proof that $p(n)$ is true for all positive integers can be given in two steps and these two steps form the

Principle of Mathematical Induction. If for a given assertion $p(n)$ we can prove that

1. The assertion is true for $n = 1$,
2. If it is true for index $n = k$, then it is also true for index $n = k + 1$,

then the assertion is true for every positive integer n.

We now illustrate the use of this principle by proving the assertion embodied in equation (4).

1. We have already seen that equation (4) is true when $n = 1$.
2. We now assume that $p(n)$ is true for the index $n = k$; i.e., we assume that indeed

(5) $$1 + 3 + 5 + \cdots + (2k - 1) = k^2.$$

To obtain the sum of the first $k + 1$ odd integers we merely add the next odd one, $2k + 1$, to both sides of (5). This gives

(6) $$1 + 3 + 5 + \cdots + (2k - 1) + (2k + 1) = k^2 + 2k + 1$$
$$= (k + 1)^2.$$

But this equation is precisely equation (4) when the index n is $k + 1$, and hence we have shown that if the assertion is true for index k, it is also true for index $k + 1$.

By the principle of mathematical induction this completes the proof that equation (4) is true for every positive integer n.

3.   further examples

We next prove that for every positive integer n

(7) $$\frac{1}{1 \cdot 2} + \frac{1}{2 \cdot 3} + \frac{1}{3 \cdot 4} + \cdots + \frac{1}{n(n + 1)} = \frac{n}{n + 1}.$$

1. For $n = 1$ the assertion of equation (7) is that

$$\frac{1}{1 \cdot 2} = \frac{1}{1 + 1},$$

and this is certainly the case.
2. We assume that (7) is true for index k. Thus we assume that

(8) $$\frac{1}{1 \cdot 2} + \frac{1}{2 \cdot 3} + \frac{1}{3 \cdot 4} + \cdots + \frac{1}{k(k + 1)} = \frac{k}{k + 1}.$$

To obtain the left-hand side of equation (7) for the index $n = k + 1$, we must add $1/(k + 1)(k + 2)$ to the left-hand side of equation (8).

This gives

$$(9) \quad \frac{1}{1 \cdot 2} + \frac{1}{2 \cdot 3} + \frac{1}{3 \cdot 4} + \cdots + \frac{1}{k(k+1)} + \frac{1}{(k+1)(k+2)}$$

$$= \frac{k}{k+1} + \frac{1}{(k+1)(k+2)}$$

$$= \frac{k(k+2) + 1}{(k+1)(k+2)}$$

$$= \frac{k^2 + 2k + 1}{(k+1)(k+2)} = \frac{(k+1)^2}{(k+1)(k+2)}$$

$$= \frac{k+1}{(k+1)+1}.$$

But this is just equation (7) when the index n is $k + 1$. Hence by the principle of mathematical induction (7) is true for every positive integer n.

An assertion $p(n)$ may be false or meaningless for certain small values of n. In this case the assertion would necessarily be modified to state only what is actually true. Then the principle of mathematical induction would also be altered to meet the situation. Thus in step 1 of the process we would not set $n = 1$, but we would use instead some integer n_0 such that $p(n)$ is true for every $n \geq n_0$. For example, let us investigate the inequality

$$(10) \qquad\qquad 2^n > 2n + 1.$$

For $n = 1$ this states that $2 > 3$, and this is false.
For $n = 2$ this states that $4 > 5$, and this is also false.
But for $n = 3$, the inequality asserts that $8 > 7$, and this is true.
Now assume that (10) is true for the index $n = k$, where $k > 2$. Since $2^k > 2$, we may add these terms to the assumed inequality $2^k > 2k + 1$ without disturbing the inequality. We find that

$$2^k + 2^k > 2k + 1 + 2,$$
$$2 \cdot 2^k > 2(k + 1) + 1,$$
$$2^{k+1} > 2(k + 1) + 1.$$

But this is just the inequality (10) when the index n is $k + 1$. Hence, using the principle of mathematical induction we have proved that *the inequality (10) is true for every integer $n \geq 3$.*

Exercise 1

In problems 1 through 14, prove that the given assertion is true for all positive integers n.

1. The nth positive even integer is $2n$.

2. The nth positive odd integer is $2n - 1$.

3. $1 + 2 + 3 + 4 + \cdots + n = \dfrac{n(n + 1)}{2}$.

4. $1^2 + 2^2 + 3^2 + 4^2 + \cdots + n^2 = \dfrac{n(n + 1)(2n + 1)}{6}$.

5. $1 \cdot 2 + 2 \cdot 3 + 3 \cdot 4 + \cdots + n(n + 1) = \dfrac{n(n + 1)(n + 2)}{3}$.

6. $1^2 + 3^2 + 5^2 + \cdots + (2n - 1)^2 = \dfrac{n(2n - 1)(2n + 1)}{3}$.

7. $1^3 + 2^3 + 3^3 + \cdots + n^3 = \dfrac{n^2(n + 1)^2}{4}$.

8. $\dfrac{1}{1 \cdot 3} + \dfrac{1}{3 \cdot 5} + \dfrac{1}{5 \cdot 7} + \cdots + \dfrac{1}{(2n - 1)(2n + 1)} = \dfrac{n}{2n + 1}$.

9. $\dfrac{1}{1 \cdot 4} + \dfrac{1}{4 \cdot 7} + \dfrac{1}{7 \cdot 10} + \cdots + \dfrac{1}{(3n - 2)(3n + 1)} = \dfrac{n}{3n + 1}$.

10. $1 + x + x^2 + \cdots + x^n = \dfrac{x^{n+1} - 1}{x - 1}$, if $x \neq 1$.

11. $\dfrac{1}{x(x + 1)} + \dfrac{1}{(x + 1)(x + 2)} + \cdots + \dfrac{1}{(x + n - 1)(x + n)}$

$$= \dfrac{n}{x(x + n)}, \text{ if } x > 0.$$

★12. $\dfrac{1}{n} + \dfrac{1}{n + 1} + \dfrac{1}{n + 2} + \cdots + \dfrac{1}{2n - 1}$

$$= 1 - \dfrac{1}{2} + \dfrac{1}{3} - \dfrac{1}{4} + \cdots + \dfrac{1}{2n - 1}.$$

★13. $1^3 + 2^3 + 3^3 + \cdots + n^3 = (1 + 2 + 3 + \cdots + n)^2$.

14. $1 \cdot 3 + 3 \cdot 5 + 5 \cdot 7 + (2n - 1)(2n + 1) = \dfrac{n(4n^2 + 6n - 1)}{3}$.

15. For which positive integers n is $2^n > n^2$?

16. For which positive integers n is $2^n > n^3$?

17. Prove that if $x > -1$ and n is an integer greater than 1, then we have the inequality $(1 + x)^n > 1 + nx$.

18. Prove that for $n \geqq 2$,

$$\dfrac{1}{n + 1} + \dfrac{1}{n + 2} + \dfrac{1}{n + 3} + \cdots + \dfrac{1}{2n} > \dfrac{13}{24}.$$

19. Prove that $1 + 5 + 9 + \cdots + (4n - 3) = n(2n - 1)$.

20. Prove that

$$3 \cdot 4 + 4 \cdot 7 + 5 \cdot 10 + \cdots + (n + 2)(3n + 1) = n(n + 2)(n + 3).$$

21. Prove the formula for the sum of the terms of an arithmetic progression,

$$a + (a + d) + (a + 2d) + \cdots + [a + (n - 1)d] = n\left[a + \frac{(n - 1)d}{2}\right].$$

Apply this formula to obtain the formulas of problems 3 and 19.

22. Prove that $2 \cdot 2 + 3 \cdot 2^2 + 4 \cdot 2^3 + \cdots + (n + 1)2^n = n2^{n+1}$.

★**23.** Prove that $x + y$ divides $x^{2n-1} + y^{2n-1}$ for each positive integer n.

★**24.** Prove that if $x \neq \pm 1$, then

$$\frac{1}{1 + x} + \frac{2}{1 + x^2} + \frac{4}{1 + x^4} + \cdots + \frac{2^n}{1 + x^{2^n}} = \frac{1}{x - 1} + \frac{2^{n+1}}{1 - x^{2^{n+1}}}.$$

tables

TABLE A. BINOMIAL COEFFICIENTS $\binom{n}{k}$

n \ k	0	1	2	3	4	5	6	7	8
2	1	2	1						
3	1	3	3	1					
4	1	4	6	4	1				
5	1	5	10	10	5	1			
6	1	6	15	20	15	6	1		
7	1	7	21	35	35	21	7	1	
8	1	8	28	56	70	56	28	8	1
9	1	9	36	84	126	126	84	36	9
10	1	10	45	120	210	252	210	120	45
11	1	11	55	165	330	462	462	330	165
12	1	12	66	220	495	792	924	792	495
13	1	13	78	286	715	1,287	1,716	1,716	1,287
14	1	14	91	364	1,001	2,002	3,003	3,432	3,003
15	1	15	105	455	1,365	3,003	5,005	6,435	6,435
16	1	16	120	560	1,820	4,368	8,008	11,440	12,870
17	1	17	136	680	2,380	6,188	12,376	19,448	24,310
18	1	18	153	816	3,060	8,568	18,564	31,824	43,758
19	1	19	171	969	3,876	11,628	27,132	50,388	75,582
20	1	20	190	1,140	4,845	15,504	38,760	77,520	125,970
21	1	21	210	1,330	5,985	20,349	54,264	116,280	203,490
22	1	22	231	1,540	7,315	26,334	74,613	170,544	319,770
23	1	23	253	1,771	8,855	33,649	100,947	245,157	490,314
24	1	24	276	2,024	10,626	42,504	134,596	346,104	735,471
25	1	25	300	2,300	12,650	53,130	177,100	480,700	1,081,575
26	1	26	325	2,600	14,950	65,780	230,230	657,800	1,562,275

Note: $\binom{n}{k} = C(n, k) = \dfrac{n!}{k!(n-k)!}$ $C(n, k) = C(n, n-k)$.

k n	9	10	11	12	13
16	11,440	8,008	4,368	1,820	560
17	24,310	19,448	12,376	6,188	2,380
18	48,620	43,758	31,824	18,564	8,568
19	92,378	92,378	75,582	50,388	27,132
20	167,960	184,756	167,960	125,970	77,520
21	293,930	352,716	352,716	293,930	203,490
22	497,420	646,646	705,432	646,646	497,420
23	817,190	1,144,066	1,352,078	1,352,078	1,144,066
24	1,307,504	1,961,256	2,496,144	2,704,156	2,496,144
25	2,042,975	3,268,760	4,457,400	5,200,300	5,200,300
26	3,124,550	5,311,735	7,726,160	9,657,700	10,400,600

TABLE B. VALUES OF $b(k, n, p)$

n	$k \diagdown p$	.05	.10	.20	.30	.40	.50
10	0	.599	.349	.107	.028	.006	.001
	1	.315	.387	.268	.121	.040	.010
	2	.075	.194	.302	.233	.121	.044
	3	.010	.057	.201	.267	.215	.117
	4	.001	.011	.088	.200	.251	.205
	5	.000	.001	.026	.103	.201	.246
	6	.000	.000	.006	.037	.111	.205
	7	.000	.000	.001	.009	.042	.117
	8	.000	.000	.000	.001	.011	.044
	9	.000	.000	.000	.000	.002	.010
	10	.000	.000	.000	.000	.000	.001
15	0	.463	.206	.035	.005	.000	.000
	1	.366	.343	.132	.031	.005	.000
	2	.135	.267	.231	.092	.022	.003
	3	.031	.129	.250	.170	.063	.014
	4	.005	.043	.188	.219	.127	.042
	5	.001	.010	.103	.206	.186	.092
	6	.000	.002	.043	.147	.207	.153
	7	.000	.000	.014	.081	.177	.196
	8	.000	.000	.003	.035	.118	.196
	9	.000	.000	.001	.012	.061	.153
	10	.000	.000	.000	.003	.024	.092
	11	.000	.000	.000	.001	.007	.042
	12	.000	.000	.000	.000	.002	.014
	13	.000	.000	.000	.000	.000	.003
	14	.000	.000	.000	.000	.000	.000
	15	.000	.000	.000	.000	.000	.000

Note: $b(k, n, p) = b(n - k, n, 1 - p)$.

n	k \ p	.05	.10	.20	.30	.40	.50
	0	.358	.122	.012	.001	.000	.000
	1	.377	.270	.058	.007	.000	.000
	2	.189	.285	.137	.028	.003	.000
	3	.060	.190	.205	.072	.012	.001
	4	.013	.090	.218	.130	.035	.005
	5	.002	.032	.175	.179	.075	.015
	6	.000	.009	.109	.192	.124	.037
	7	.000	.002	.055	.164	.166	.074
	8	.000	.000	.022	.114	.180	.120
	9	.000	.000	.007	.065	.160	.160
	10	.000	.000	.002	.031	.117	.176
20	11	.000	.000	.000	.012	.071	.160
	12	.000	.000	.000	.004	.035	.120
	13	.000	.000	.000	.001	.015	.074
	14	.000	.000	.000	.000	.005	.037
	15	.000	.000	.000	.000	.001	.015
	16	.000	.000	.000	.000	.000	.005
	17	.000	.000	.000	.000	.000	.001
	18	.000	.000	.000	.000	.000	.000
	19	.000	.000	.000	.000	.000	.000
	20	.000	.000	.000	.000	.000	.000

Note: $b(k, n, p) = b(n - k, n, 1 - p)$.

TABLE C. AREAS UNDER THE UNIT NORMAL
CURVE

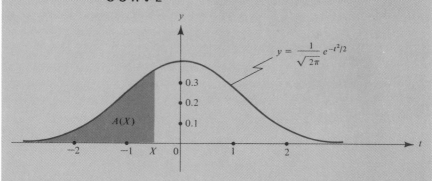

X	.00	.01	.02	.03	.04	.05	.06	.07	.08	.09
−3.8	.0001	.0001	.0001	.0001	.0001	.0001	.0001	.0001	.0001	.0000
−3.7	.0001	.0001	.0001	.0001	.0001	.0001	.0001	.0001	.0001	.0001
−3.6	.0002	.0002	.0001	.0001	.0001	.0001	.0001	.0001	.0001	.0001
−3.5	.0002	.0002	.0002	.0002	.0002	.0002	.0002	.0002	.0002	.0002
−3.4	.0003	.0003	.0003	.0003	.0003	.0003	.0003	.0003	.0003	.0002
−3.3	.0005	.0005	.0004	.0004	.0004	.0004	.0004	.0004	.0004	.0003
−3.2	.0007	.0007	.0006	.0006	.0006	.0006	.0006	.0005	.0005	.0005
−3.1	.0010	.0009	.0009	.0009	.0008	.0008	.0008	.0008	.0007	.0007
−3.0	.0013	.0013	.0013	.0012	.0012	.0011	.0011	.0011	.0010	.0010
−2.9	.0019	.0018	.0017	.0017	.0016	.0016	.0015	.0015	.0014	.0014
−2.8	.0026	.0025	.0024	.0023	.0023	.0022	.0021	.0020	.0020	.0019
−2.7	.0035	.0034	.0033	.0032	.0031	.0030	.0029	.0028	.0027	.0026
−2.6	.0047	.0045	.0044	.0043	.0041	.0040	.0039	.0038	.0037	.0036
−2.5	.0062	.0060	.0059	.0057	.0055	.0054	.0052	.0051	.0049	.0048
−2.4	.0082	.0080	.0078	.0075	.0073	.0071	.0069	.0068	.0066	.0064
−2.3	.0107	.0104	.0102	.0099	.0096	.0094	.0091	.0089	.0087	.0084
−2.2	.0139	.0136	.0132	.0129	.0126	.0122	.0119	.0116	.0113	.0110
−2.1	.0179	.0174	.0170	.0166	.0162	.0158	.0154	.0150	.0146	.0143
−2.0	.0228	.0222	.0217	.0212	.0207	.0202	.0197	.0192	.0188	.0183
−1.9	.0287	.0281	.0274	.0268	.0262	.0256	.0250	.0244	.0238	.0233
−1.8	.0359	.0352	.0344	.0336	.0329	.0322	.0314	.0307	.0300	.0294
−1.7	.0446	.0436	.0427	.0418	.0409	.0401	.0392	.0384	.0375	.0367
−1.6	.0548	.0537	.0526	.0516	.0505	.0495	.0485	.0475	.0465	.0455
−1.5	.0668	.0655	.0643	.0630	.0618	.0606	.0594	.0582	.0570	.0559
−1.4	.0808	.0793	.0778	.0764	.0749	.0735	.0722	.0708	.0694	.0681
−1.3	.0968	.0951	.0934	.0918	.0901	.0885	.0869	.0853	.0838	.0823
−1.2	.1151	.1131	.1112	.1093	.1075	.1056	.1038	.1020	.1003	.0985
−1.1	.1357	.1335	.1314	.1292	.1271	.1251	.1230	.1210	.1190	.1170
−1.0	.1587	.1562	.1539	.1515	.1492	.1469	.1446	.1423	.1401	.1379
−0.9	.1841	.1814	.1788	.1762	.1736	.1711	.1685	.1660	.1635	.1611
−0.8	.2119	.2090	.2061	.2033	.2005	.1977	.1949	.1922	.1894	.1867
−0.7	.2420	.2389	.2358	.2327	.2297	.2266	.2236	.2206	.2177	.2148
−0.6	.2743	.2709	.2676	.2643	.2611	.2578	.2546	.2514	.2483	.2451
−0.5	.3085	.3050	.3015	.2981	.2946	.2912	.2877	.2843	.2810	.2776

TABLE C. AREAS UNDER THE UNIT NORMAL CURVE (Cont.)

X	.00	.01	.02	.03	.04	.05	.06	.07	.08	.09
−0.4	.3446	.3409	.3372	.3336	.3300	.3264	.3228	.3192	.3516	.3121
−0.3	.3821	.3783	.3745	.3707	.3669	.3632	.3594	.3557	.3520	.3483
−0.2	.4207	.4168	.4129	.4090	.4052	.4013	.3974	.3936	.3897	.3859
−0.1	.4602	.4562	.4522	.4483	.4443	.4404	.4364	.4325	.4286	.4247
−0.0	.5000	.4960	.4920	.4880	.4840	.4801	.4761	.4721	.4681	.4641
0.0	.5000	.5040	.5080	.5120	.5160	.5199	.5239	.5279	.5319	.5359
0.1	.5398	.5438	.5478	.5517	.5557	.5596	.5363	.5675	.5714	.5753
0.2	.5793	.5832	.5871	.5910	.5948	.5987	.6026	.6064	.6103	.6141
0.3	.6179	.6217	.6255	.6293	.6331	.6368	.6406	.6443	.6480	.6517
0.4	.6554	.6591	.6628	.6664	.6700	.6736	.6772	.6808	.6844	.6879
0.5	.6915	.6950	.6985	.7019	.7054	.7088	.7123	.7157	.7190	.7224
0.6	.7257	.7291	.7324	.7357	.7389	.7422	.7454	.7486	.7517	.7549
0.7	.7580	.7611	.7642	.7673	.7703	.7734	.7764	.7794	.7823	.7852
0.8	.7881	.7910	.7939	.7967	.7995	.8023	.8051	.8078	.8106	.8133
0.9	.8159	.8186	.8212	.8238	.8264	.8289	.8315	.8340	.8365	.8389
1.0	.8413	.8438	.8461	.8485	.8508	.8531	.8554	.8577	.8599	.8621
1.1	.8643	.8665	.8686	.8708	.8729	.8749	.8770	.8790	.8810	.8830
1.2	.8849	.8869	.8888	.8907	.8925	.8944	.8962	.8980	.8997	.9015
1.3	.9032	.9049	.9066	.9082	.9099	.9115	.9131	.9147	.9162	.9177
1.4	.9192	.9207	.9222	.9236	.9251	.9265	.9278	.9292	.9306	.9319
1.5	.9332	.9345	.9357	.9370	.9382	.9394	.9406	.9418	.9430	.9441
1.6	.9452	.9463	.9474	.9484	.9495	.9505	.9515	.9525	.9535	.9545
1.7	.9554	.9564	.9573	.9582	.9591	.9599	.9608	.9616	.9625	.9633
1.8	.9641	.9648	.9656	.9664	.9671	.9678	.9686	.9693	.9700	.9706
1.9	.9713	.9719	.9726	.9732	.9738	.9744	.9750	.9756	.9762	.9767
2.0	.9772	.9778	.9783	.9788	.9793	.9798	.9803	.9808	.9812	.9817
2.1	.9821	.9826	.9830	.9834	.9838	.9842	.9846	.9850	.9854	.9857
2.2	.9861	.9864	.9868	.9871	.9874	.9878	.9881	.9884	.9887	.9890
2.3	.9893	.9896	.9898	.9901	.9904	.9906	.9909	.9911	.9913	.9916
2.4	.9918	.9920	.9922	.9925	.9927	.9929	.9931	.9932	.9934	.9936
2.5	.9938	.9940	.9941	.9943	.9945	.9946	.9948	.9949	.9951	.9952
2.6	.9953	.9955	.9956	.9957	.9959	.9960	.9961	.9962	.9963	.9964
2.7	.9965	.9966	.9967	.9968	.9969	.9970	.9971	.9972	.9973	.9974
2.8	.9974	.9975	.9976	.9977	.9977	.9978	.9979	.9979	.9980	.9981
2.9	.9981	.9982	.9982	.9983	.9984	.9984	.9985	.9985	.9986	.9986
3.0	.9987	.9987	.9987	.9988	.9988	.9989	.9989	.9989	.9990	.9990
3.1	.9990	.9991	.9991	.9991	.9992	.9992	.9992	.9992	.9993	.9993
3.2	.9993	.9993	.9994	.9994	.9994	.9994	.9994	.9995	.9995	.9995
3.3	.9995	.9995	.9996	.9996	.9996	.9996	.9996	.9996	.9996	.9997
3.4	.9997	.9997	.9997	.9997	.9997	.9997	.9997	.9997	.9997	.9998
3.5	.9998	.9998	.9998	.9998	.9998	.9998	.9998	.9998	.9998	.9998
3.6	.9998	.9998	.9999	.9999	.9999	.9999	.9999	.9999	.9999	.9999
3.7	.9999	.9999	.9999	.9999	.9999	.9999	.9999	.9999	.9999	.9999
3.8	.9999	.9999	.9999	.9999	.9999	.9999	.9999	.9999	.9999	1.000

answers to exercises

Chapter 1. Exercise 1, page 9

1. (a) Nonstatement, (b) compound, (c) nonstatement, (d) compound, (g) compound, (h) nonstatement.
3. (b) And, (d) but, (f) if..., then, (g) not, (i) if..., then, and.

Chapter 1. Exercise 2, page 11

1. (a) 2, (b) 5, (c) 1, (d) 6, (e) 4, (f) 3.
3. (a) $p \wedge \sim q$, (c) $\sim(\sim p \wedge q)$, (e) $\sim(\sim p \vee \sim q)$.
5. (a) All cats are not pets, (c) some cats are not pets, (e) some pets are cats.
6. (a) The girl is neither blonde nor under twenty-one, (c) the girl is not blonde or she is not under twenty-one, (e) the girl is blonde or she is under twenty-one.
7. (a) $p \wedge \sim q$, (c) $p \vee \sim q$, (e) $\sim p \vee q$.

Chapter 1. Exercise 3, page 15

1. F T T T. 3. F T T T. 5. F T T F. 7. T T T F T F T F.
9. F F F F T F T F. 11. T T T T T T T F. 15. $S = \sim(p \vee q)$, $T = \sim p \wedge q$, $U = \sim p$.

Chapter 1. Exercise 4, page 18

1. (a) F T T T, **(b)** T T F F, **(c)** F F T T, **(d)** T F F T, **(e)** F T T F.
3. (a) F T F F, **(b)** F F F F, **(c)** F T F T, **(d)** F T T T.
5. (a) T F T F, **(b)** F F F T, . **(c)** F F T T, **(d)** F F T T.
7. (a) $\sim p \wedge q$, **(b)** $\sim p \wedge \sim q$, **(c)** $\sim (p \longrightarrow q)$,
  (d) $\sim (p \longleftrightarrow q)$.

Chapter 1. Exercise 5, page 20

1. T. **3.** N. **5.** N. **7.** T. **9.** T. **11.** C. **13.** T.

Chapter 1. Exercise 6, page 22

1. $\sim p \wedge q \Longrightarrow p \longrightarrow q$.
3. $(p \longleftrightarrow q) \Longleftrightarrow (p \longrightarrow q) \wedge (\sim p \longrightarrow \sim q)$.
5. Neither.
7. $\sim (p \text{ - } \rightarrow \sim q) \Longleftrightarrow p \wedge q$.
9. (a) $\sim (\sim p \vee \sim q)$, **(b)** $\sim p \vee q$, **(c)** $\sim [\sim (\sim p \vee q) \vee \sim (\sim q \vee p)]$,
  (d) $\sim p \vee q$.
11. $(a) \Longrightarrow (b)$, $(c) \Longrightarrow (b)$, $(c) \Longrightarrow (a)$.

Chapter 1. Exercise 7, page 24

1. (a) $\sim p \wedge (\sim p \vee q)$, **(b)** $p \wedge (p \vee q)$, **(c)** $\sim p \vee \sim q$, **(d)** $\sim p$.

Chapter 1. Exercise 8, page 27

3. (a) If the sun is shining, then it is warm, **(c)** if you are a man, then
  you are mortal.
5. (a) Converse: If a baby is happy, then he does not cry.
  Contrapositive: If a baby is sad, then he cries.
  Inverse: If a baby cries, then he is sad.
7. All. **9. (a)** and **(d)**, **(b)** and **(c)**.

Chapter 1. Exercise 9, page 32

1. V. **3.** V. **5.** F. **7.** V. **9.** F. **11.** V.
13. No student in this school gets drunk.

14. Cab drivers never wrestle alligators.

15. A major in the 99th regiment does not push peanuts with his nose.

16. Snakes do not vote in the presidential elections.

17. Kuratowski is not a classical mathematician.

18. Isaac Newton was not a first-class mathematician. (Here the argument is valid, but the conclusion is false!)

19. Concrete workers are happy.

Chapter 1. Exercise 10, page 37

1.

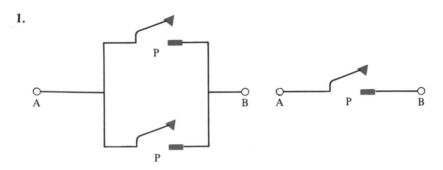

3.

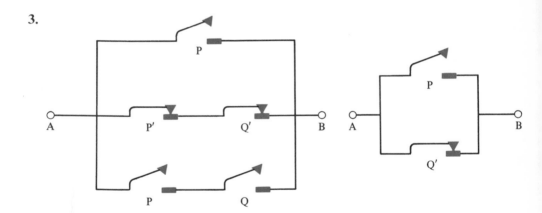

5.

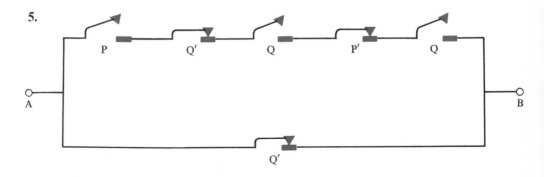

7.

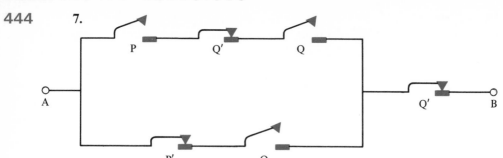

9.

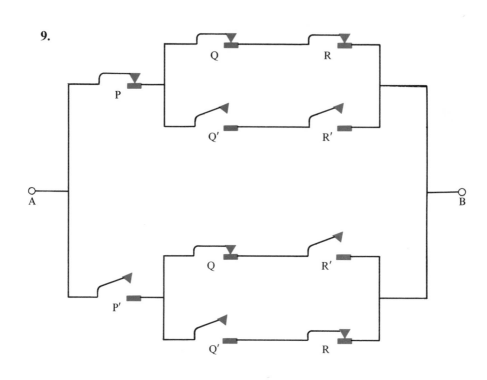

10.

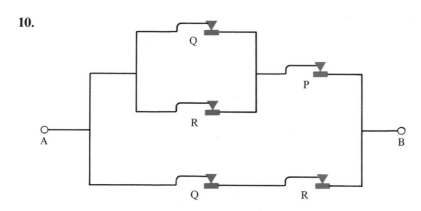

11.

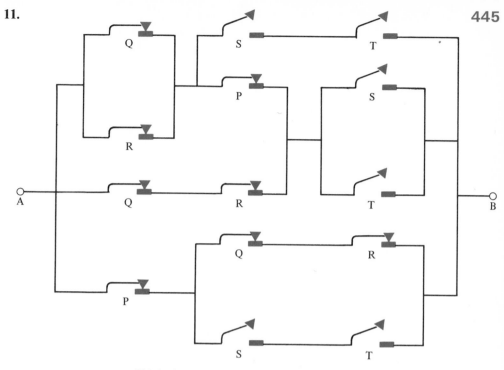

This is the circuit when P, Q, and R favor the measure.

Chapter 2. Exercise 1, page 42

1. N. 3. W; this set is probably empty.
5. W; this set probably includes all men in the U.S. 7. W. 9. W. 11. W.
   12. N. 13. W.
15. $\{4,5,6,7,8\}$. 17. $\{-4,-3,-2,-1,0,1,2\}$.
19. $\{0,1,16,81\}$. 21. $\{0,\pm1,\pm2,\pm3,\pm4\}$.
23. (a) $\{2n \mid n$ is an integer$\}$, (b) $\{2n+1 \mid n$ is an integer$\}$,
   (c) $\{5n+2 \mid n$ is a nonnegative integer$\}$,
   (d) $\{p^n q^m \mid n,m$ are positive integers, $p,q \geq 2$ are distinct primes$\}$.
24. 6. 25. *mn*. 26. 24.

Chapter 2. Exercise 2, page 50

1. (a) $\{1,2,3,4,5,6\}$, (b) $\{1,2,3,4\}$, (c) $\{2,4,5,6,7\}$,
   (d) $\{5,7\}$, (e) $\varnothing$, (f) $\{1,2,3,4,5,6,7\}$,
   (g) $\{5,7\}$, (h) $\{2,4\}$, (i) $\{1,2,3,4,5,6,7\}$.
3. (a), (c), (d), and (f) are true.
5. $A \supset B \not\supset C$. 6. Let C_n be the set of all integers that are divisible
   by 2^n. Then $C_1 \supset C_2 \supset \cdots$.

7. $A \neq B$ in **(c)** and **(e).**   **9.** **(a)** 4, **(b)** 8, **(c)** 2^n.
10. (a) All even integers, **(b)** all odd integers, **(c)** all integers.
11. (a) $2^6 = 64$, **(b)** 2^{2n} subsets with an odd number of elements.

Chapter 2. Exercise 3, page 53

1. Theorem 10, $\{1,2,3,5\}$; Theorem 11, $\{1,2,3,4,5,6,7\}$.

3.

A	B	C	D	Region
Y	Y	Y	Y	1
Y	Y	Y	N	2
Y	Y	N	Y	3
Y	N	Y	Y	4
Y	Y	N	N	5
Y	N	Y	N	6
Y	N	N	Y	7
Y	N	N	N	8
N	Y	Y	Y	9
N	Y	Y	N	10
N	Y	N	Y	11
N	N	Y	Y	12
N	Y	N	N	13
N	N	Y	N	14
N	N	N	Y	15
N	N	N	N	16

5.

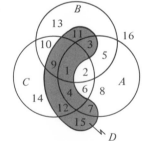

6.

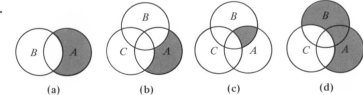

(a)   (b)   (c)   (d)

9.

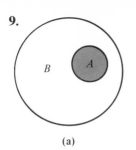

(a)

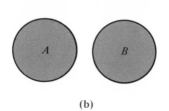

(b)

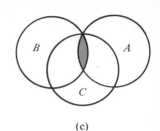

(c)

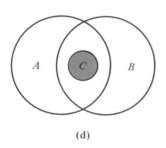

(d)

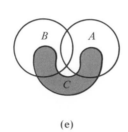

(e)

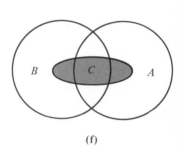

(f)

Chapter 2. Exercise 4, page 57

1. $A' = \{4,5,6,7,8,9\}$, $B' = \{1,3,5,7,9\}$, $C' = \{0,4,8\}$.
16. (a) U, (b) $C \cap D'$, (c) U,
 (d) $(A' \cap B') \cup (A \cap B)$.
17. $X = \varnothing$.
18. (a) $P \cup (Q \cap R')$, (b) $(P \cap Q' \cap R') \cup (Q \cap R)$,
 (c) $[(P \cap Q) \cup (Q \cap R) \cup (R \cap P)] \cap (P' \cup Q' \cup R')$,
 (d) $(P \cup Q \cup R) \cap (P' \cup Q' \cup R')$.
19. There is no essential gain or loss. We do lose some simplicity.
20. (a) $C \cap R' \cup T'$, regions 2 and 4; (b) $C' \cap R \cap T'$, regions 5 and 9; (c) $C' \cap R' \cap T$, regions 1 and 11.

Chapter 2. Exercise 5, page 65

1. $\{2,3,5,7\}$, 4, 8, 10, 12, 15.
3. (a) 12, (b) 144. 5. 42.
7. $n(A \times B \times C \times D) = n(A)n(B)n(C)n(D)$. $(p,q,r,s) = (t,u,v,w)$ if and only if $p = t$, $q = u$, $r = v$, and $s = w$.
9. (a) 125, (b) 50. 11. 200,000.
13. (a) 27, (b) 61.
15. (a) 43 percent, (b) 7 percent, (c) 73 percent.

447

16. (a) 97, (c) 17, (e) 64, (g) 21.
17. (a) 52, (b) 6. **19.** (a) 100, (b) 120, (c) 160.
20. 80.

Chapter 2. Exercise 6, page 71

1. (a) 2, (b) 6, (c) 4.
3. 144, no. **5.** (a) 35, (b) 5.
6. (a) 8, (b) 4. **7.** (a) 52, (b) 35, (c) 17.
8. (a) 8, (b) 3, (c) 16, (d) 2^n.
9. (a) 8, (b) 18. **10.** 10.
11. $m + mn + mnp = m(1 + n + np)$, if we exclude the head.

Chapter 2. Exercise 7, page 74

1. x is a prime between 10 and 30.
2. x is a positive number less than 25 that has a remainder 2 when divided by 3.
3. x is the product of two different primes between 1 and 10.
4. x was vice president of the U.S. between 1900 and 1930.
5. x is the symbol for a chess piece (king, queen, bishop, knight, rook, pawn).
6. x is the first name of a man who invented the airplane or the electric light.
7. x is the type of figure that can be obtained by cutting a cone (conic section).
8. x is a letter of the English alphabet that is also in the Russian alphabet, but represents a different sound: v, ye, n, r, ch, oo.
9. (a) $P \cup Q$, (b) $P' \cap Q$, (c) $P \cap Q \cap R$,
  (d) $(P \cap Q) \cup R$, (e) $(P \cup Q) \cap R'$,
  (f) $Q \cup (P \cap R)$.
11. $Q \subset R \cap P$.
13. $P \cup Q = (P \cap Q) \cup R$.
15. $P \cap Q = (P' \cup Q')'$.
17. $P' \cup Q = (Q')' \cup P'$.

Chapter 2. Exercise 8, page 76

1. If $n(A) = m + 1$, A is a M.W.C. If $n(A) = m$, A is a M.B.C.
3. The minimal winning and blocking coalitions are not changed.
5. $\{p\}$ is a M.W.C. There are no M.B.C's.

7. M.W.C: {all 5 of the large nations} ∪ {any 4 of the other 10 nations}.
   M.B.C: {any one of the large nations} or {any 7 of the other 10 nations}.
12. Let a, b, and c have 4, 2, and 1 votes, respectively, and the majority
   rules ($p = 1/2$). Then $\{a\}$ is a M.W.C. Add 4 votes to each so that the
   votes are 8, 6, and 5. Then $\{a\}$ is a losing coalition.

Chapter 3. Exercise 1, page 86

1. 1. 3. $m = n = 1$.
5. If $n = 2$, then $n^2! = 24$ and $(n!)^2 = 4$. If $n = 3$, we have 362,880 and
   36, respectively.
7. The empty set can be considered as a permutation; hence $P(n,0) = 1$.
   Theorem 1 does not make sense, but from equation (12) we find that
   $P(n,0) = n!/n! = 1$.
9. 720. 11. 720. 13. 42 minutes.
15. 41 scores from 0 to 40.
16. Yes. Suppose that A votes (1,10,9) for "right," "abstention," and
   "wrong," while B votes (0,14,6). Using $x = 2$, $y = 1$, and $z = 0$ as in
   problem 14, B scores 14, and A scores 12, so that B is more liberal than
   A. Using $x = 10$, $y = 3$, and $z = 1$, A scores 49 and B scores 48, so
   that B is less liberal than A.
17. $P(10,10) = 3,628,800$. In 1964, there were only eight states with a voting
   population larger than $P(10,10)$.
19. 8.
22. From equation (6), $P(k,k) = k!$.
23. We should make the definition $0! = 1$. In fact this is the universally
   adopted definition for $0!$.

Chapter 3. Exercise 2, page 89

1. 10,10.
5. (a) $C(8,4) = 70$, (b) $C(8,4) + 2C(8,3) + 2C(8,2) + 2C(8,1) = 254$
7. (a) $C(10,3) = 120$, (b) $C(n,3)$.
9. The maximum number of folders is $C(10,5) = 252$.
11. $C(20,3) = 1140$. 13. (a) Yes, (b) yes. 14. Yes.

Chapter 3. Exercise 3, page 92

1. $2 \cdot 5! - 4 \cdot 4! = 144$. 3. $2(4!)^2 = 1152$.
5. $P(6,3) = 120$. 6. (a) 3360, (c) 630.
7. $C(6,1) + C(6,2) + C(6,3) + C(6,4) + C(6,5) + C(6,6) = 63$.

9. 9. **10. (a)** 3, **(b)** $(n-1)!/2$. **11.** $C(8,4) \times 3 \times 3/2 = 315$.
12. (a) $7 \times 5 \times 3 \times 1 = 105$, **(b)** $105 \times 2^4 = 1680$.
13. $p(p-1)q(q-1)/4$. **14.** $C(n,4)$.
15. $3C(n,4)$. **16.** $C(n,2) - C(k,2) + 1$.
17. $C(p+2,2)C(q+2,2)$. **19.** $3^{10} = 59{,}049$.
21. (a) $2^3 C(10,7) = 960$, **(b)** 1161.
22. (a) 848, **(b)** 57,888.

Chapter 4. Exercise 1, page 99

3. T. **5.** F. **7.** F. **9.** F. **11.** T.
13. $\pm\sqrt{11}$. **15.** $A^2B^5 + A^3B^4 + A^4B^3 + A^5B^2 + A^6B$.

16. $n! = \displaystyle\prod_{k=1}^{n} k$. **17. (a)** $2^6 = 64$, **(c)** $5! = 120$,

    (e) $(9!)^2$. **19.** 12. **20.** $(n+3)(n+1)!/3$.
21. 1. **22.** $n!$. **23.** 2700. **25. (a)** T, **(b)** T, **(c)** T.
26. The standard definition is 1.

Chapter 4. Exercise 2, page 106

1. (a) $1, 8, 28, 56, 70, 56, 28, 8, 1$;
    (b) $1, 9, 36, 84, 126, 126, 84, 36, 9, 1$.
2. 1080. **3.** $-7/324$. **5.** 252. **7.** 55. **9.** 1.
11. 30. **13.** $8/15$. **15.** 1. **21. (a)** 24, **(b)** 0.
22. (a) 0, **(b)** 15, **(c)** 270. **23. (a)** 0, **(b)** 0, **(c)** 11. **24. (a)** 35,
    (b) 35.
25. (a) 84, **(b)** 0, **(c)** 84.

Chapter 4. Exercise 3, page 111

1. 210. **3.** 360. **5.** 72. **7.** 15. **9.** 15.
13. $\dbinom{7}{1,1,1,1,1,1,1} = 7! = 5040$. **14.** 560

15. 1260. **17. (a)** $\dbinom{20}{6,7,7}$, **(b)** $\dbinom{17}{5,6,6}$.

18. (a) 4^{12}, **(c)** $\dbinom{12}{3,6,3} = 18{,}480$, **(e)** 138,600.

19. 945. **21.** 2100. **23.** $\dfrac{1}{24}\dbinom{40}{10,10,10,10}$. **25.** $\dbinom{52}{13,13,13,13}$

26. (a) $5\binom{31}{6,6,6,6,7}$, **(b)** $5\binom{32}{6,6,6,6,8} + 10\binom{32}{6,6,6,7,7}$.

28. (a) 30, **(c)** 3780.

29. $5040 - 1 = 5039$.

Chapter 4. Exercise 4, page 115

3. (6,0,0,0), (5,1,0,0), (4,2,0,0), (4,1,1,0), (3,3,0,0), (3,2,1,0), (3,1,1,1), (2,2,2,0), (2,2,1,1).

5. $-6720x^4y^3z$. **7.** $(4y + 6z^2)x^6$.

9. $-2500x^3z^3(x + 2z)^3y^3$.

11. $(4y^3 + 12yz + 12y^2zw + 12z^2w)x^7$.

17. (a) 36, **(b)** 165, **(c)** 252.

Chapter 4. Exercise 5, page 120

1. 8. **2.** 70. **3. (a)** 4, **(b)** 8, **(c)** 16, **(d)** 2^r.

6. r. **7.** Yes, but in some cases the strict inequality may hold.

8. Use equation (48) with $n = 5$ and $n(C) = 44$.

9. 1854.

Chapter 5. Exercise 1, page 129

1. (a) $1/4$, **(b)** $1/2$, **(c)** $3/4$, **(d)** $1/4$.

3. (a) $1/4$, **(b)** $1/13$, **(c)** $1/52$, **(d)** $12/13$.

5. $1/9$. **7. (a)** $67/135$, **(b)** $68/135$.

8. $(d_1e_2 + e_1d_2)/(d_1 + e_1)(d_2 + e_2)$.

9. $(e_1 - d_1)(e_2 - d_2) > 0$.

11. $3/4$. **12.** $2/5$, $1/4$.

13. (a) $1/216$, **(b)** $1/36$, **(c)** $5/72$, **(d)** 0. **15.** $1/8$.

Chapter 5. Exercise 2, page 132

1. 2^n. **3.** $P(n,3) = n(n - 1)(n - 2)$.

5. $C(52,13)$. **7. (a)** A does not occur, **(b)** either A or B must occur, **(c)** either A and B occur or C occurs, **(d)** A occurs and either B or C occurs.

9. Let $U = \{A,B,C\}$, $P[A] = P[B] = 1/2$, $P[C] = 0$. This is a probability function. Hence the statement is false.

19. No. **20. (a)** $1/4$, **(b)** $3/4$, **(c)** $7/24$, **(d)** $11/12$.

21. 12 balls (#4), 3 balls each (#2 and #6), 2 balls each (#1, #3, and #5), total 24 balls.

22. (a) 2/9, (b) 4/9, (c) 1/3.

23. (a) 7/60, (b) 11/20.

24. (a) 1/8, 3/8, 1/4, 1/4, (b) 5/8.

25. 3/20. **26.** 1/3.

Chapter 5. Exercise 3, page 138

3. $C(4,2)C(4,1) \times 44 \times 40/2C(52,5)$.

5. $C(24,13)/C(52,13)$.

7. (a) $4C(13,4)C^3(13,3)/C(52,13) \approx 0.105$.
   (b) $12C^2(13,5)C(13,2)C(13,1)/C(52,13) \approx 0.032$.
   (c) $24C(13,6)C(13,4)C(13,2)C(13,1)/C(52,13) \approx 0.047$.
   (d) $24C(13,7)C(13,5)C(13,1)/C(52,13) \approx 0.001$.

8. (a) W. 4/17, T. 1/17, (b) W. 16/51, T. 1/17,
   (c) W. 28/51, T. 1/17, (d) W. 44/51, T. 1/17.

9. $17/45 \approx 0.378$.

10. (a) $72/425 \approx 0.169$, (b) $12/5525 \approx 0.002$, (c) $274/5525 \approx 0.0496$.

11. 7 to 5. **12.** 1 to 1. **13.** 8 to 5. **14.** 158 to 63.

15. 13 to 3.

16. $P[A] = 3/5$, $P[B] = 1/5$, $P[C] = 1/10$, $P[D] = 1/25$.
   $\sum P_i = 0.940 < 1$, because the track pays back only a fraction of the amount bet.

18. 3 to 1. **19.** 9 to 11.

Chapter 5. Exercise 4, page 142

1. (a) 0.083, (b) 0.236, (c) 0.427, (d) 0.618.

2. Use equation (15) with $n = 52$, for $P[C] \approx 1/e \approx 0.368$. The probability of a match is 0.632. The approximate odds for a match are 5 to 3. Yes.

3. $P = \dfrac{1}{2!} - \dfrac{1}{3!} + \dfrac{1}{4!} - \cdots + (-1)^{n-1}\dfrac{1}{(n-1)!}, n \geqq 3$.

Chapter 5. Exercise 5, page 145

1. (a) 1/221, (b) 1/17, (c) 0.

3. (a) 1/17, (b) 1/17.

4. (a) 3/10, (b) 3/7.

5. (a) 8/9, (b) 6/13.

6. $C(35,13)/C(39,13) \approx 0.182$.

7. 20/21.

9. In "practical" situations $P[B] = 0$ means that B cannot occur. Hence $A \cap B$ cannot occur and the question of finding $P[A \mid B]$ is not really meaningful.

10. If $A \cap B = \emptyset$, then $P[A \mid B] = 0$.

Chapter 5. Exercise 6, page 150

1. $P[R,R,R] = 3/10$, $P[R,G,R] = 9/20$, $P[G,R,R] = 1/10$, $P[G,G,R] = 3/20$, $P[X,Y,G] = 0$, where X and Y denote either a red or green ball.

3. $C(4,2)/2^4 = 3/8$.

5. (a) 3/10, (b) 1/3, (c) 2/9.

7. Yes. 9. 7/13. 11. (a) 0, (b) 3/4.

13. (a) $P[A \cap B] = P[A]P[B] = 1/4$. (b) $P[A \cap B \cap C] = 0$. $P[A]P[B]P[C] = (1/2)^3 \neq 0$.

Chapter 5. Exercise 7, page 155

1. $P[H_1 \mid B] = 18/23$, $P[H_2 \mid B] = 4/23$, $P[H_3 \mid B] = 1/23$.

2. $P[H_1 \mid C] = 6/19$, $P[H_2 \mid C] = 12/19$, $P[H_3 \mid C] = 1/19$.

4. 2/3. 5. $13/27 \approx 0.48$. 6. $17/29 \approx 0.59$.

7. $15/23 \approx 0.65$. 8. 100/199. 9. $50/67 \approx 0.75$.

10. Approximately $2/9.886 \approx 0.202$. 11. (a) 1/172, (b) 9/28.

12. $45/49 \approx 0.92$. 13. (1) 4/5, (2) 1/5, (3) 0.

15. It is not always true.

Chapter 5. Exercise 8, page 160

2. 1/32, 5/32, 10/32, 10/32, 5/32, 1/32.

3. $1/D$, $10/D$, $40/D$, $80/D$, $80/D$, $32/D$, where $D = 243$.

5. 1/8. 7. 763/3888.

9. (a) 4/9, (b) $80/243 \approx 0.329$, (c) $1792/6561 \approx 0.273$.

11. 0.317. 12. If $n = 8$, $P[\text{hit}] \approx 0.961$.

13. 0.168.

17. See problem 15, $p \approx 0.95$.

18. If $p(n + 1)$ is an integer, then for maximum $b(k,n,p)$, set $k = p(n + 1)$ or $p(n + 1) - 1$. Otherwise let k be the largest integer $\leqq p(n + 1)$.

1. $3.50. **3.** $-\$1/37$.
5. $-\$5/52$, the dealer. **6.** $E = \$199/216 \approx \$0.92$. He will lose on the average 8 cents per game. **7.** 7/2.
9. (a) $135, **(b)** $135 + operating costs + profit + reserve for unusually bad years.
11. $2. **12. (a)** $729/737 \approx 0.99$, **(b)** $729/769 \approx 0.948$. **13.** $-\$1/12$.
15. (a) 2, **(b)** 2, **(c)** 4, **(d)** 3. **16.** $np = 200$. No. The probability of this event ≈ 0.03.

Chapter 6. Exercise 1, page 174

1. $p = (5 + m - 2n)/10$. We must have $-5 \leqq m - 2n \leqq 5$; otherwise p will be negative or exceed 1.
3. $P[A] = P[B] = 5/16$, $P[C] = 1/4$, $P[5 \text{ or more rounds}] = 1/8$.
4. $P[A] = P[B] = 11/64$, $P[C] = 9/16$, $P[5 \text{ or more rounds}] = 3/32$.
5. (a) $P[A] = P[B] = 5/16$, $P[C] = 1/8$, $P[\text{tie}] = 1/4$.
(b) $P[A] = P[B] = 13/64$, $P[C] = 27/64$, $P[\text{tie}] = 11/64$.
6. (a) 13/27, **(b)** 79/216, $P[\text{tie}] = 1123/5832$.
7. (a) 1/30, **(b)** 1/45, **(c)** 8/405.
9. 0.
10. (a) $5/12 < 1/2$, **(b)** 3/11.
11. (a) $P[(7,1)] = P[(1,7)] = 5/16$, $P[(5,3)] = P[(3,5)] = 3/16$,
(b) 231/1024.
12. (a) 4/7, **(b)** 16/31, **(c)** 64/127.
13. After drawing $N - 1$ green balls in a row, the urn will contain 1 red ball and $2^{N+1} - 2$ green balls. $P[N \text{ green balls in a row}] = 2^N/(2^{N+1} - 1)$.

Chapter 6. Exercise 2, page 180

1. (a) $\begin{bmatrix} \dfrac{2}{3} & \dfrac{1}{3} \\ \dfrac{1}{4} & \dfrac{3}{4} \end{bmatrix}$,
(b) $\begin{bmatrix} \dfrac{1}{6} & \dfrac{1}{2} & \dfrac{1}{3} \\ \dfrac{1}{2} & \dfrac{1}{4} & \dfrac{1}{4} \\ \dfrac{5}{12} & \dfrac{1}{4} & \dfrac{1}{3} \end{bmatrix}$,

$$(c) \begin{bmatrix} \frac{1}{2} & \frac{3}{10} & \frac{1}{5} & 0 \\ 0 & 0 & \frac{1}{2} & \frac{1}{2} \\ \frac{1}{3} & 0 & \frac{1}{3} & \frac{1}{3} \\ \frac{1}{4} & \frac{3}{4} & 0 & 0 \end{bmatrix}, \quad (d) \begin{bmatrix} 0 & 1 & 0 \\ 0 & 0 & 1 \\ 1 & 0 & 0 \end{bmatrix}.$$

$$4. \begin{bmatrix} \frac{13}{36} & \frac{13}{36} & \frac{5}{18} \\ \frac{19}{48} & \frac{19}{48} & \frac{5}{24} \\ \frac{7}{24} & \frac{7}{24} & \frac{5}{12} \end{bmatrix}.$$

$$5. \begin{bmatrix} 0.1175 & 0.3375 & 0.4450 & 0.1000 \\ 0.0625 & 0.3775 & 0.3950 & 0.1650 \\ 0.1000 & 0.3300 & 0.4900 & 0.0800 \\ 0.0525 & 0.3825 & 0.3900 & 0.1750 \end{bmatrix}.$$

$$6. \begin{bmatrix} \frac{19}{36} & \frac{17}{36} \\ \frac{17}{48} & \frac{31}{48} \end{bmatrix}.$$

$$7. \quad P^{(3m)} = \begin{bmatrix} 1 & 0 & 0 \\ 0 & 1 & 0 \\ 0 & 0 & 1 \end{bmatrix}, \quad P^{(3m+1)} = \begin{bmatrix} 0 & 1 & 0 \\ 0 & 0 & 1 \\ 1 & 0 & 0 \end{bmatrix}, \quad P^{(3m+2)} = \begin{bmatrix} 0 & 0 & 1 \\ 1 & 0 & 0 \\ 0 & 1 & 0 \end{bmatrix}.$$

$$10. \quad P^{(2)} = \begin{bmatrix} 0.675 & 0.185 & 0.140 \\ 0.185 & 0.675 & 0.140 \\ 0.350 & 0.350 & 0.300 \end{bmatrix}, \quad (a) \; 0.675 \quad (b) \; 0.140.$$

$$11. \quad P = \begin{bmatrix} 0.90 & 0.05 & 0.05 \\ 0.50 & 0.30 & 0.20 \\ 0.90 & 0.00 & 0.10 \end{bmatrix}.$$

12. $p_{ij} = 0$ for all i,j except as follows: $p_{i,i+1} = (N - i)^2/N^2$, $p_{ii} = 2i(N - i)/N^2$, $p_{i,i-1} = i^2/N^2$,

$$P = \begin{bmatrix} 0 & 1 & 0 & \cdots & 0 & 0 \\ \frac{1}{N^2} & \frac{2(N-1)}{N^2} & \frac{(N-1)^2}{N^2} & \cdots & 0 & 0 \\ 0 & \frac{4}{N^2} & \frac{4(N-2)}{N^2} & \cdots & 0 & 0 \\ \vdots & \vdots & \vdots & & \vdots & \vdots \\ 0 & 0 & 0 & \cdots & \frac{2(N-1)}{N^2} & \frac{1}{N^2} \\ 0 & 0 & 0 & \cdots & 1 & 0 \end{bmatrix}.$$

1. 0.198
2. (a) 0.261, (b) 0.296, (c) 0.315
3. 4/5.
7. $p = 38/65$.
8. $(2^B - 1)/(2^{B+1} - 1) < 1/2$.
13. No. The expected value of the number of games is 36. The corresponding time of departure is 2:00 A.M.
14. 700 years.
15. (a) 20 years, (b) 1.000.
16. (a) 0.922, (b) 7.28.

Chapter 6. Exercise 4, page 193

1. (a) 0.194, (b) 0.000, (c) 0.135.
2. Let H_0 be $p \geqq 0.95$ for production of a good microwidget. (a) If there are 4 or more defective pieces in 20, reject H_0 with a significance level 1.5 percent, (b) If there are 5 or more defective pieces reject H_0 with a significance level 0.2 percent ($\alpha = 0.002$).
3. $\alpha = 0.012$.
4. 11.3 percent.
5. 0.252.
6. (a) If $p = 0.6$, then $P[X \geqq 15] = 0.125$. We cannot reject H_0. (b) If $p = 0.9$, then $P[X \leqq 15] = 0.043$. We can reject H_0.
7. $P[X \geqq 14 \mid p = 0.5] = 0.058$. (a) Reject H_0 at 10 percent, (b) accept H_0 at 5 percent.
8. 13.
9. $P[T \mid 0.9] = 0.930$, $P[D \mid 0.5] = 0.890$, $P[H \mid 0.1] = 0.930$.

Chapter 6. Exercise 5, page 196

1. (a) 0.656, (b) 0.698, (c) 0.736.

Chapter 6. Exercise 6, page 204

1. (a) 5000, (b) 50.
2. (a) 120, (b) 10.
3. (a) 600, (b) 10.
4. (a) 75, (b) 7.5.

5. $61 \leqq s \leqq 89$.

6. 58.3. Design conditions for the airplane may require that the number of seats be divisible by 4. Then one would provide 60 seats.

7. 62.

8. (a) 0.390, (b) 0.956.

9. 0.906.

10. 0.0062.

12. From problem 11, $n = 2500$.

13. 0.0035.

14. The critical region is $x \geqq 180$. Reject H_0.

15. $P[h \leqq 20.5 | p = 0.3] = 0.131$. If H_0 is $p \geqq 0.3$, then his performance does not permit him to reject H_0 at the 5 percent significance level.

16. Let s be the number of odd numbers. Then at the 5 percent significance ievel the hypotheses $p = 1/2$ for odd numbers cannot be rejected if $12{,}644 \leqq x \leqq 12{,}956$. However, the fact that odd and even numbers occur with equal probability is only one of the many requirements for a set of numbers to be random. For example, the probability that a number selected at random from the book is divisible by 3 should be 1/3, etc.

17.

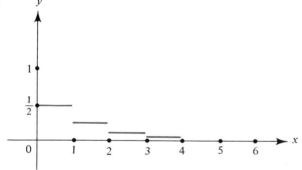

19. 1.

Chapter 7. Exercise 1, page 211

1. (a) $[3, -3, 7, 7]$, (c) $[6, 6, 0, 15]$.

3. (a) Not possible, (c) $\begin{bmatrix} 21 \\ 0 \\ 76 \end{bmatrix}$

7. $\left[\dfrac{1}{4}, \dfrac{1}{4}, \dfrac{1}{8}, \dfrac{1}{9} \right]$

9. $[800, 16, 12]$, $[1200, 10, 15]$, $[2000, 26, 27]$.

5. (a) $\dfrac{\sqrt{2}}{10}A.$

7. $(-4,9)$, $(-1,6)$, $(0,4)$.

Chapter 7. Exercise 3, page 220

1. (a) -29,   **(c)** not possible,   **(e)** 17,   **(g)** not possible.
3. (a) No,   **(b)** no.
5. $A \cdot C + A \cdot D + B \cdot C + B \cdot D.$
7. $A \cdot B = \$444.$ This is the total amount received for the sale of all the shirts.
9. (a) Demand $D = [22,\ 19,\ 16]$,   **(b)** profit $P = [86,\ 55,\ 68]$.

Chapter 7. Exercise 4, page 223

1. $a_{13} = 3$, $a_{33} = 11$.   **3.** No.   **5.** $x = 4$.
7. All real numbers.
9. No solution.

$$
\begin{array}{c}
 \\
1 \\
2 \\
\textbf{11. } 3 \\
4 \\
5
\end{array}
\begin{array}{ccccc}
1 & 2 & 3 & 4 & 5 \\
\left[\begin{array}{ccccc}
0 & 1 & 0 & 0 & 1 \\
0 & 0 & 1 & 0 & 0 \\
0 & 0 & 0 & 1 & 0 \\
1 & 0 & 0 & 0 & 0 \\
0 & 0 & 0 & 1 & 0
\end{array}\right].
\end{array}
$$

13.

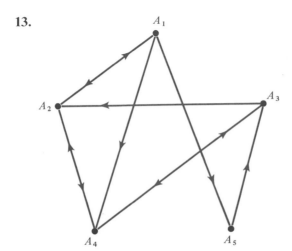

Chapter 7. Exercise 5, page 227

1. (a) $\begin{bmatrix} 3 & 6 & -9 \\ 9 & 12 & 15 \\ 6 & -3 & 21 \end{bmatrix}$, **(c)** $\begin{bmatrix} -2 & 2 & -9 \\ -3 & 1 & 5 \\ 2 & -1 & 4 \end{bmatrix}$,

(e) $\begin{bmatrix} 0 & 0 & 0 \\ 0 & 0 & 0 \\ 0 & 0 & 0 \end{bmatrix}$.

3. $E = \dfrac{1}{6} \begin{bmatrix} 31 & 8 & 57 \\ 11 & -42 & 65 \\ 18 & 11 & 25 \end{bmatrix}$

Chapter 7. Exercise 6, page 232

1. $AB = \begin{bmatrix} 8 & 0 \\ 24 & 0 \end{bmatrix}$, $BA = \begin{bmatrix} 8 & 12 & 16 \\ 0 & 8 & 16 \\ 0 & -4 & -8 \end{bmatrix}$.

2. (a) $\begin{bmatrix} 0 & 7 \\ 2 & 11 \end{bmatrix}$, **(c)** $\begin{bmatrix} -13 & 5 \\ 24 & 1 \end{bmatrix}$,

(e) $\begin{bmatrix} 35 & 7 \\ 57 & 19 \end{bmatrix}$, **(g)** $\begin{bmatrix} 37 & 54 \\ 81 & 118 \end{bmatrix}$,

(i) $\begin{bmatrix} 11 & 13 \\ 25 & 27 \end{bmatrix}$, **(k)** $\begin{bmatrix} 7 & -12 \\ 24 & 79 \end{bmatrix}$.

3. (a) $A^n = A$ if n is odd, and $A^n = I$ if n is even.
5. $m = p$. **7.** $AX = B$.
17. $C(AB)$ or $(CA)B$.
19. No such matrix B exists.
21. m rows and q columns.

Chapter 7. Exercise 7, page 236

1. (a) $\begin{bmatrix} 2 \\ 5 \\ 7 \end{bmatrix}$, **(b)** $\begin{bmatrix} 2 & 3 \\ -1 & 5 \\ 5 & -2 \end{bmatrix}$.

9. $A = \begin{bmatrix} 1 & 2 \\ 2 & 0 \end{bmatrix}$, $B = \begin{bmatrix} 2 & 1 \\ 1 & 1 \end{bmatrix}$.

1. $x_1 = -13$, $x_2 = 10$.

3. $x_1 = 19/2$, $x_2 = -(3/2)$, $x_3 = 1/2$.

5. No solution.

7. $x_1 = 4$, $x_2 = 3$, $x_3 = 2$.

9. $x_1 = 3$, $x_2 = -2$, $x_3 = 1$.

11. $k = 7$.

13. $x_1 + 2x_2 - 3x_3 = 4$, and $x_1 + 2x_2 - 3x_3 = -13$. This system represents two parallel planes.

15. 594.

Chapter 8. Exercise 2, page 255

1. $\begin{bmatrix} 28 & -9 \\ -3 & 1 \end{bmatrix}$.

3. $\begin{bmatrix} -1 & 2 & -1 \\ 0 & 1 & -1 \\ -2 & 5 & -4 \end{bmatrix}$.

5. $\begin{bmatrix} -22 & -29 & 52 \\ 3 & 4 & -7 \\ 4 & 5 & -9 \end{bmatrix}$.

7. $\begin{bmatrix} 1 & -2 & 1 & 1 \\ 2 & -1 & 1 & -2 \\ 4 & -2 & 3 & -5 \\ -5 & 3 & -3 & 5 \end{bmatrix}$.

9. No, see Chapter 7, Exercise 6, problem 13.

Chapter 8. Exercise 3, page 264

1. (a).   **3.** (b), (e), (f), (g).

9. 4/7; approximately 57 percent of weekends.

$$
\begin{array}{c}
\quad A \quad B \quad C \quad D \\
\begin{array}{c} A \\ B \\ C \\ D \end{array}
\begin{bmatrix}
0 & \frac{1}{3} & \frac{1}{3} & \frac{1}{3} \\
\frac{1}{2} & 0 & \frac{1}{4} & \frac{1}{4} \\
\frac{1}{2} & 0 & 0 & \frac{1}{2} \\
0 & \frac{2}{3} & \frac{1}{3} & 0
\end{bmatrix}
\end{array}
$$

11. (a)   , (b) 7/36,

(c) $\left[\dfrac{42}{171}, \dfrac{44}{171}, \dfrac{40}{171}, \dfrac{45}{171}\right].$

Chapter 8. Exercise 4, page 273

1. (c) 2, (d) 2, (e) 2, (f) 3. In (b) the sum of the elements in the second row is not 1.

3. (c) $N = \begin{bmatrix} \dfrac{4}{3} & 0 \\ \dfrac{4}{3} & 2 \end{bmatrix}, \qquad C = \begin{bmatrix} 0 & 1 \\ 0 & 1 \end{bmatrix},$

(d) $N = \begin{bmatrix} \dfrac{9}{2} & 2 \\ \dfrac{3}{2} & 2 \end{bmatrix}, \qquad C = \dfrac{1}{2}\begin{bmatrix} 1 & 1 \\ 1 & 1 \end{bmatrix},$

(e) $N = \begin{bmatrix} \dfrac{4}{3} & \dfrac{1}{2} & \dfrac{1}{3} \\ \dfrac{8}{9} & \dfrac{4}{3} & \dfrac{2}{9} \\ \dfrac{4}{9} & \dfrac{2}{3} & \dfrac{10}{9} \end{bmatrix}, \qquad C = \begin{bmatrix} \dfrac{2}{3} & \dfrac{1}{3} \\ \dfrac{4}{9} & \dfrac{5}{9} \\ \dfrac{2}{9} & \dfrac{7}{9} \end{bmatrix},$

(f) $N = \begin{bmatrix} \dfrac{5}{3} & \dfrac{2}{3} \\ \dfrac{5}{3} & \dfrac{8}{3} \end{bmatrix}, \qquad C = \dfrac{1}{3}\begin{bmatrix} 1 & 1 & 1 \\ 1 & 1 & 1 \end{bmatrix}.$

5. (a) 1/2, (b) 4.
7. 3/31.
11. (a) 8, (b) 5, (c) 15/7, (d) 2/7.

Chapter 8. Exercise 5, page 279

3. (a) [1/25, 8/25, 16/25],
   (b) [1/100, 18/100, 81/100],
   (c) [1/4, 1/2, 1/4].

7. $P = \begin{bmatrix} 1 & 0 & 0 \\ \dfrac{1}{2} & \dfrac{1}{2} & 0 \\ 0 & 1 & 0 \end{bmatrix}, \qquad P^3 = \begin{bmatrix} 1 & 0 & 0 \\ \dfrac{7}{8} & \dfrac{1}{8} & 0 \\ \dfrac{3}{4} & \dfrac{1}{4} & 0 \end{bmatrix}.$

9.
$$\begin{bmatrix} 1 & 0 & 0 & 0 & 0 & 0 \\ 0 & 1 & 0 & 0 & 0 & 0 \\ \frac{1}{16} & \frac{1}{16} & \frac{1}{4} & \frac{1}{8} & \frac{1}{4} & \frac{1}{4} \\ 0 & 0 & 1 & 0 & 0 & 0 \\ \frac{1}{4} & 0 & \frac{1}{4} & 0 & \frac{1}{2} & 0 \\ 0 & \frac{1}{4} & \frac{1}{4} & 0 & 0 & \frac{1}{2} \end{bmatrix}.$$

11. (a) $3/4$, **(b)** $1/4$, $29/6$.

Chapter 9. Exercise 1, page 289

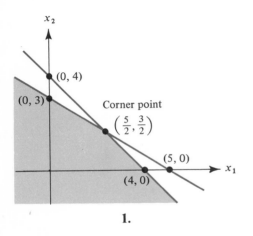

1.

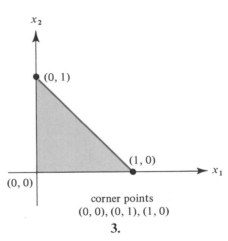

corner points
$(0, 0), (0, 1), (1, 0)$

3.

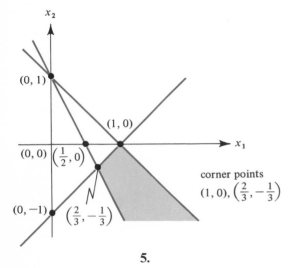

5.

7. No solution.

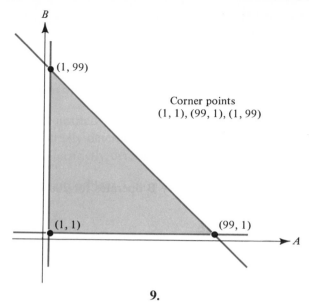

9.

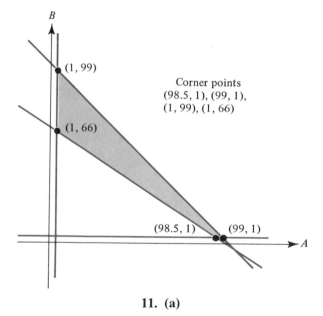

11. (a)

11. (c) No solution.

Chapter 9. Exercise 2, page 295

In problems 1 through 4 answers are in dollars.
1. (a) Minimum 0, maximum 375/4;
   (c) Minimum 75, maximum 625;
   (e) Minimum 0, maximum 1150.

11. $x = 6$; $\quad P\star = \left[\dfrac{3}{5}, \dfrac{2}{5}\right]$, $\quad Q\star = \begin{bmatrix} \dfrac{3}{5} \\ \dfrac{2}{5} \end{bmatrix}$.

Chapter 10. Exercise 5, page 336

1. $v = \dfrac{12}{5}$. $\quad P\star = \left[\dfrac{4}{5}, 0, \dfrac{1}{5}\right]$, $\quad Q\star = \begin{bmatrix} \dfrac{3}{5} \\ \dfrac{2}{5} \end{bmatrix}$,

3. $v = \dfrac{3}{8}$. $\quad P\star = \left[\dfrac{3}{8}, \dfrac{2}{8}, \dfrac{3}{8}\right]$, $\quad Q\star = \begin{bmatrix} \dfrac{1}{2} \\ \dfrac{1}{8} \\ \dfrac{3}{8} \end{bmatrix}$,

5. $v = \dfrac{1}{6}$. $\quad P\star = \left[\dfrac{1}{6}, \dfrac{1}{2}, 0, \dfrac{1}{3}\right]$, $\quad Q\star = \begin{bmatrix} 0 \\ \dfrac{7}{78} \\ \dfrac{47}{78} \\ \dfrac{24}{78} \end{bmatrix}$,

7. $v = 0$. $\quad P\star = [1,0,0,0]$, $\quad Q\star = \begin{bmatrix} 1 \\ 0 \\ 0 \\ 0 \end{bmatrix}$,

9. $33\dfrac{1}{3}$ percent for apartments, $66\dfrac{2}{3}$ percent for houses.

Chapter 11. Exercise 1, page 349

1. Yes. If $S = [s_1, s_2, \ldots]$, then s_k is the total number of errors made on the kth trial. **2.** No.

3. $E(T) = 10/3$, $c = 1/4$.

5. **(a)** $1/21 \approx 0.048$, $(5/7)^k(8/21) \approx 0.272$, $\quad 0.194$, $\quad 0.139$,
  (b) $\qquad\qquad 0.000$, $\qquad\qquad\qquad\quad 0.333$, $\quad 0.167$, $\quad 0.000$.

7. $P[r_{k,1}] = (5/8)^k(1/2)$. For $n \geq 2$, $P[r_{k,n}] = (3/4)^n(5/8)^k(1/9)$.
  (a) 0.312, $\quad 0.195$, $\quad 0.122$; $\qquad\qquad 0.039$, $\quad 0.029$.
  (b) 0.500, $\quad 0.167$, $\quad 0.167$; $\qquad\qquad 0.000$, $\quad 0.167$.

17. $(1 - g)/c$.

19. $c = 1/4$, **(a)** $1/16$, **(b)** $5/48$, **(c)** $5/16$, **(d)** $25/48$.

21. (a) $15/7$, **(b)** 2.

Chapter 11. Exercise 2, page 363

1. 150 A, 335 M, 205 T.

5. $A\star = \begin{bmatrix} 0.5 & 0.02 & 0.1 \\ 1.0 & 0.6 & 1.0 \\ 0.1 & 0.04 & 0.6 \end{bmatrix}$.

7. $X_0 = \begin{bmatrix} 36 \\ 4 \\ 6 \end{bmatrix}$.

9. $N(A) = 2.5$, $(I - A)^{-1} = \begin{bmatrix} 2 & 8 \\ 0 & 2 \end{bmatrix}$, yes.

11. $A^n = \begin{bmatrix} \dfrac{1}{2^n} & \dfrac{n}{2^{n-2}} \\ 0 & \dfrac{1}{2^n} \end{bmatrix}$.

13. No, no.

17. $\dfrac{5}{6} \begin{bmatrix} 0 & -3 & -3 \\ -4 & -2 & -4 \\ -4 & -2 & 2 \end{bmatrix}$.

Chapter 12. Exercise 1, page 372

1. (a) $V = \{A_1, A_2, A_3, A_4, A_5\}$,
$E = \{A_i A_j \mid i < j, i = 1, \ldots, 4, j = 2, \ldots, 5\}$,
(c) V the same as in **(a)**,
$E = \{A_1 A_3, A_2 A_4, A_3 A_5, A_4 A_1, A_5 A_2\}$.

3. (a) $\begin{bmatrix} 0 & 1 & 0 & 1 \\ 1 & 0 & 1 & 1 \\ 0 & 1 & 0 & 0 \\ 1 & 1 & 0 & 0 \end{bmatrix}$, **(c)** $\begin{bmatrix} 0 & 0 & 0 & 1 & 1 & 1 \\ 0 & 0 & 0 & 1 & 1 & 1 \\ 0 & 0 & 0 & 1 & 1 & 1 \\ 1 & 1 & 1 & 0 & 0 & 0 \\ 1 & 1 & 1 & 0 & 0 & 0 \\ 1 & 1 & 1 & 0 & 0 & 0 \end{bmatrix}$.

5. $n(n - 1)/2$.

7. mn,

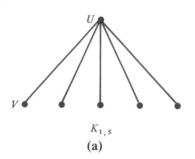

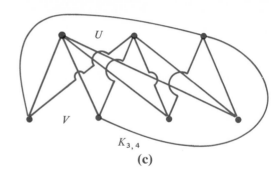

$K_{1,5}$

(a)

$K_{3,4}$

(c)

11. This graph is also $K_{3,3}$ shown in Fig. 4. To go from Fig. 7 to Fig. 4, use the replacements $A_1 \longrightarrow A_1$, $A_2 \longrightarrow B_1$, $A_3 \longrightarrow A_2$, $A_4 \longrightarrow B_2$, $A_5 \longrightarrow A_3$, and $A_6 \longrightarrow B_3$.

13. Each line in G is counted twice.

Chapter 12. Exercise 2, page 378

1. $A^3 = \begin{bmatrix} 3 & 0 & 0 & 1 & 0 \\ 0 & 2 & 0 & 2 & 1 \\ 1 & 0 & 1 & 0 & 2 \\ 1 & 1 & 0 & 1 & 0 \\ 0 & 0 & 1 & 0 & 2 \end{bmatrix}$.

3. $A^3 = \begin{bmatrix} 0 & 0 & 0 & 1 & 3 \\ 0 & 0 & 0 & 0 & 1 \\ 0 & 0 & 0 & 0 & 0 \\ 0 & 0 & 0 & 0 & 0 \\ 0 & 0 & 0 & 0 & 0 \end{bmatrix}$.

4. D.

5. (a) $A_4A_3A_2A_3$, (b) $A_4A_3A_4A_3$, (c) $A_4A_5A_6A_3$, (d) $A_4A_6A_2A_3$, (e) $A_4A_6A_4A_3$, and (f) $A_4A_6A_5A_3$.

11. $A^2 = \begin{bmatrix} 1 & 0 & 1 \\ 0 & 2 & 0 \\ 1 & 0 & 1 \end{bmatrix}$, $A^{2n+1} = 2^n A$.

Chapter 12. Exercise 3, page 383

1. The standard definition is $D(A_i, A_i) = 0$.

3. For the points A_1, A_5 and A_9, we have $3 \leq 2 + 3$.

5. This is a matter of taste, and in this case the authors do not agree.

7. 5.

9. No. $S(C_1) = 40$.

Chapter 12. Exercise 4, page 388

1. $n(n - 1)/2$.

3. No. No. $A^2 = \begin{bmatrix} 0 & 0 & 1 & 2 & 2 & 2 \\ 1 & 0 & 0 & 2 & 2 & 2 \\ 0 & 1 & 0 & 2 & 2 & 2 \\ 0 & 0 & 0 & 0 & 1 & 0 \\ 0 & 0 & 0 & 0 & 0 & 1 \\ 0 & 0 & 0 & 1 & 0 & 0 \end{bmatrix}$.

5. **(a)** 35,   **(b)** 56,   **(c)** 120,   **(d)** 286;   **(a)** 14,   **(b)** 20,   **(c)** 40,   **(d)** 91.

7. If A_1 has the highest score and B won over A_1, then there is an A_j in the contest such that A_1 won over A_j and A_j won over B.

9. **(a)** $A_1 A_2 A_3$,   **(b)** $A_1 A_2 A_3$,   **(c)** $A_1 A_6 A_5 A_4 A_3 A_7 A_2$,
   (d) $A_1 A_2 A_3 A_4 A_6 A_5$.

13. 33 transitive triples and 2 cyclic triples.

Chapter 12. Exercise 5, page 395

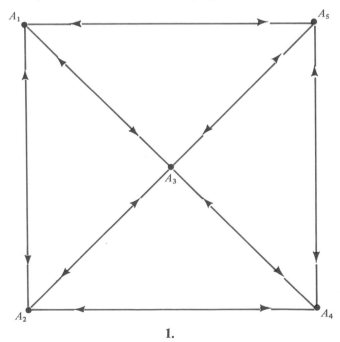

1.

3. $\begin{bmatrix} 0 & 1 & 1 & 1 & 1 \\ 1 & 0 & 1 & 1 & 1 \\ 1 & 1 & 0 & 1 & 1 \\ 1 & 1 & 1 & 0 & 0 \\ 1 & 1 & 1 & 0 & 0 \end{bmatrix}$, $b_{11}^{(3)} = 10$.

5. None.

7. $\{A_1, A_2, A_6\}$, $\{A_2, A_3, A_4\}$, $\{A_4, A_5, A_6\}$, $\{A_2, A_4, A_6, A_7\}$.

Appendix 1. Exercise 1, page 404

1. D, all x. $R = \{y \mid y \geq 0\}$.
2. $D = \{x \mid -6 \leq x \leq 6\}$, $R = \{y \mid 0 \leq y \leq 6\}$.
3. D, all x, $R = \{y \mid y \geq 2\}$.   4. D, all x, $R = \{y \mid 0 < y \leq 2\}$.
5. $D = \{x \mid x \geq -2\}$, $R = \{y \mid y \geq 7\}$.
6. $D = \{x \mid -3 < x < 3\}$, $R = \{y \mid y \geq 12\}$.
7. D is the collection of all finite sets. R is the set of all nonnegative integers.
8. Yes.   9. No; given x, it may be that y is not uniquely determined.
10. Yes.
11. In a monogamous society, yes, if D is the set of married men. Otherwise, no.
12. Yes.
13. No.

Appendix 1. Exercise 2, page 406

1. (a) -2, (b) 2, (c) 14, (d) -4, (e) -26.
2. Same as problem 1.
8. 5, -1.   9. (a) -2, (b) 0, 1, (c) 0, 13.
10. 4, 7, 9, 11, 4, 4.
11. $S = 6x^2$.
12. $S = 6V^{2/3}$.

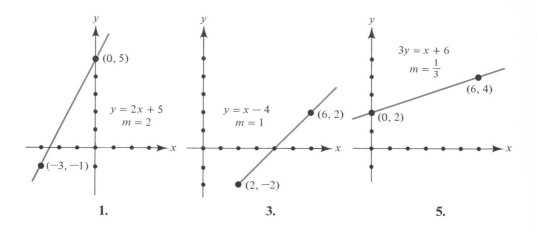

1. **3.** **5.**

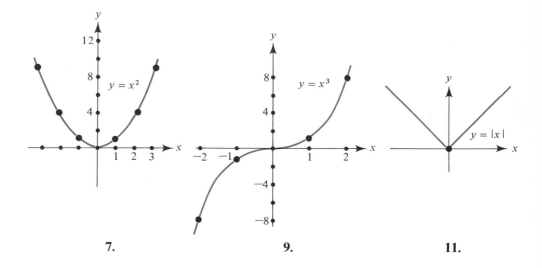

7. **9.** **11.**

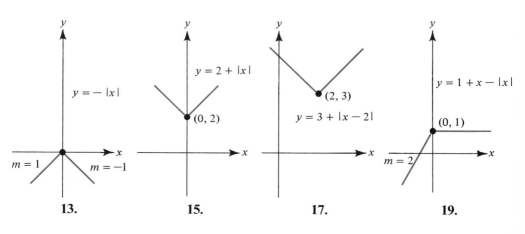

13. **15.** **17.** **19.**

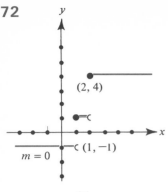

21.

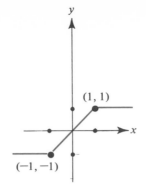

23.

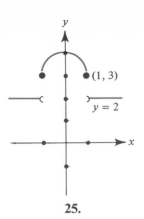

25.

Appendix 2. Exercise 1, page 419

1. $\sqrt{19} + \sqrt{21}$. **2.** $\sqrt{11} - \sqrt{8}$. **3.** $5\sqrt{7}$. **4.** $\sqrt[3]{23}$. **7.** $a = 1$.
8. $2a = 5b$.
9. $c = d$. **10.** $c = d$. **11.** $a = b$. **12.** $a = b$. **13.** $x = 2y$.
14. $x = y = z$.
15. $c = d$. **16.** $a = 3b$. **17.** $c = d$. **18.** $A = D, B = C$.
19. 10, 12, 13, 14, 15.

Appendix 2. Exercise 2, page 423

1. Either $x < 0$ or $x > 1$. **2.** $-1 < x < 8$.
3. $x < -2$. **4.** $1/3 \leqq x$. **5.** $3 \leqq x \leqq 5$.
6. $1/4 < x < 3$. **7.** Either $-4 \leqq x \leqq 0$ or $4 \leqq x$.
8. Either $-5 \leqq x \leqq -1$ or $3 \leqq x$.
9. Either $x \leqq -3$, or $-2 \leqq x \leqq 2$, or $3 \leqq x$.
10. $5 - \sqrt{3} \leqq x \leqq 5 + \sqrt{3}$.
13. Either $x < -1$ or $0 < x < 8$.
14. Either $x < 0$ or $3 \leqq x \leqq 5$.
15. Either $x \leqq -5$, or $-1 \leqq x < 0$, or $3 \leqq x$.
16. Either $x < 0$, or $1 \leqq x < 2$, or $8 \leqq x$.
17. Either $x < 1$, or $3 \leqq x < 5$, or $7 \leqq x$.

Appendix 3. Exercise 1, page 429

15. $n = 1, n \geqq 5$.
16. $n = 1, n \geqq 10$.

index of selected symbols

symbol	name or meaning	page
italic and roman letters		
A^{-1}	inverse of the matrix A	250
A^T	transpose of the matrix A	234
$G\star$	complement of the graph G	371
H_0, H_1	hypotheses	189
I, I_n	identity matrix	231
K_n	complete graph	373
$K_{m,n}$	bipartite graph	373
U	matrix with each $a_{ij} = 1$	373
greek letters		
α	significance level	191
Π	product	98
Σ	sum	96
simple symbols		
$\ldots$	et-cetera	3
$|\ |$	absolute value	5

473

symbol	name or meaning	page
■	Halmos symbol	6
$\wedge$	and	10
$\vee$	or	10
$\sim$	not	10
$\longrightarrow$	if ..., then ...	16
$\longleftrightarrow$	if and only if	17
$\Longrightarrow$	implication	21
$\Longleftrightarrow$	equivalence	21
$\equiv$	equivalence	22
$\therefore$	therefore	29
$\in$	belongs to	40
$\notin$	does not belong to	40
$\mid$	such that	40
$\varnothing$	the empty set	41
$\subset$	is contained in	44
$\supset$	contains	44
$\cup$	union	45
$\cap$	intersection	46
$'$	complement	54
$<$	less than	413
$\leqq$	less than or equal to	414
$>$	greater than	413
$\geqq$	greater than or equal to	414
$>>$	dominates	385
$\mathbf{0}$	zero vector (matrix)	210

compound symbols

symbol	name or meaning	page
$[a_1, a_2, \ldots, a_n]$	vector	210
$\lvert A \rvert$	magnitude of the vector A	217
$[a_{ij}]$	matrix	225
$A \cdot B$	product of vectors	219
$A \times B$	Cartesian product of sets A and B	61
AB	line segment joining A and B	212
$\overline{AB}$	directed distance from A to B	213
$\overrightarrow{AB}$	vector from A to B	214
$A(X)$	area under normal curve	198
$b(k, n, p)$	probability of k successes on n trials	158

symbol	name or meaning	page
$C(n, k)$	number of combinations of n things taken k at a time	81
$d(A_i)$	degree of the vertex A_i	373
$D(A_i, A_j)$	distance in a graph	381
$\lim\limits_{n \to \infty}$	limit as n approaches infinity	162
$n(A)$	number of elements in the set A	60
$N(A)$	norm of the matrix A	357
$n!$	n factorial	4
$\binom{n}{k}$	binomial coefficient	102
$\binom{n}{n_1, n_2, \ldots, n_r}$	multinomial coefficient	108
$P(n, k)$	number of permutations of n things taken k at a time	81
$P[E]$	probability of the event E	124
$P[A \mid B]$	probability of A, given B	143
X_S	characteristic function for the set S	117

index

A

Absolute value, 5–6, 418, 420
Absorbing barriers, 180
Absorbing Markov chains, 266–74
Addition
  of matrices, 225–28
    associative law for, 226
    commutative law for, 226
    sum notation for, 95–98
  of vectors, 209, 215
    associative law for, 211
    commutative law for, 210–11
    parallelogram law for, 215
Algebra
  of matrices, 221–37
  of sets, 39–77
  of statements, 23–25, 36
  of vectors, 207–21
All-or-none-model, 342, 349
Alternative hypothesis, 189, 190
Analysis of the problem, 417
"And," 8, 9, 10
  truth table for, 13 *t.*
A posteriori probability, 129

A priori probability, 128
Areas under unit normal curve,
    438–39 *t.*
Argument, 28–33
  fallacy of, 28
  validity of, 28–33
    truth table for, 29 *t.*
Associative law, 23
  for addition
    of matrices, 226
    of vectors, 211
  for intersection of sets, 49
  for multiplication of matrices, 232
  for union of sets, 49
Augmented matrix, 240–44
Axiom(s), 341–42, 352

B

Barriers, 180
Bayes' theorem, 152–57, 221
Bernoulli trials, 157–62, 168, 188, 200
  defined, 158
  standard deviation of sequence of,
    200, 201–202

479

notes